ENGINEERING
ECONOMY

**McGRAW–HILL
BOOK COMPANY**
New York
St. Louis
San Francisco
Auckland
Düsseldorf
Johannesburg
Kuala Lumpur
London
Mexico
Montreal
New Delhi
Panama
Paris
São Paulo
Singapore
Sydney
Tokyo
Toronto

ANTHONY J. TARQUIN, P.E.
Associate Professor of Environmental Engineering
Department of Civil Engineering
University of Texas at El Paso

LELAND T. BLANK, P.E.
Associate Professor of Industrial Engineering
Department of Mechanical and Industrial Engineering
University of Texas at El Paso

Engineering Economy

A BEHAVIORAL APPROACH

This book was set in Press Roman
by Hemisphere Publishing Corporation.
The editor was B. J. Clark
and the production supervisor was Milton J. Heiberg.
Kingsport Press, Inc., was printer and binder.

Library of Congress Cataloging in Publication Data

Tarquin, Anthony J.
    Engineering economy—a behavioral approach.

    Includes bibliographies and index.
    1. Engineering economy.   I. Blank, Leland T., joint author.   II. Title.
TA177.4.B58   658.1'52'02462   75-16257
ISBN 0-07-062934-X

**ENGINEERING
ECONOMY**

1 2 3 4 5 6 7 8 9 0   K P K P   7 9 8 7 6

To our parents

EDNA
ALVIN
VIRGINIA
ERCOLE

for all their encouragement

CONTENTS

Preface   XV

LEVEL I

1   Basic Concepts   3

    Criteria
    Explanation of Material
1.1 Basic Terminology
1.2 Interest Calculations
1.3 Equivalence
1.4 Alternatives
1.5 Minimum Attractive Rate of Return
1.6 Simple and Compound Interest
    Solved Examples
    Bibliography
    Problems

2   Symbols and Cash-Flow Diagrams   **15**

    Criteria
    Explanation of Material
2.1 Symbols and Their Meaning
2.2 Time Placement of P, F, and A
2.3 Description and Tabulation of Cash Flow
2.4 Cash-Flow Diagrams
    Solved Examples
    Bibliography
    Problems

3   Factors and Their Use   **26**

    Criteria
    Explanation of Material
3.1 Derivation of Single-Payment Formulas
3.2 Derivation of the Uniform-Series Present-Worth Factor and the
    Capital-Recovery Factor
3.3 Derivation of the Uniform-Series Compound-Amount Factor and the
    Sinking-Fund Factor
3.4 Use of Interest Tables
3.5 Interpolation in Interest Tables
3.6 Present-Worth, Future-Worth, and Equivalent Uniform Annual
    Series Calculations
3.7 Calculation of Unknown Interest Rates
3.8 Calculation of Unknown Years
    Solved Examples
    Bibliography
    Problems

4   Nominal and Effective Interest Rates   **48**

    Criteria
    Explanation of Material
4.1 Nominal and Effective Rates
4.2 Effective Interest-Rate Formulation
4.3 Calculation of Effective Interest Rates
4.4 Calculations for Compounding Periods Shorter than Payment Periods
4.5 Calculations for Compounding Periods Longer than Payment Periods
    Solved Examples
    Bibliography
    Problems

5   Use of Multiple Factors   **64**

    Criteria
    Explanation of Material

5.1 Location of Present Worth and Future Worth
5.2 Calculations for a Uniform Series That Begins After Year 1
5.3 Calculations Involving Uniform-Series and Randomly Distributed Amounts
5.4 Equivalent Uniform Annual Series of Both Uniform and Single Payments
Solved Examples
Bibliography
Problems

6 Uniform Gradients   **82**

Criteria
Explanation of Material
6.1 Definition and Cash Flow of Uniform Gradients
6.2 Derivation of Gradient Formulas
6.3 Use of Gradient Tables
6.4 Location of Gradient Present Worth
6.5 Present Worth and Equivalent Uniform Annual Series of Conventional
Gradients
6.6 Present Worth and Equivalent Annual Series of Shifted Gradients
6.7 Decreasing Gradients
Solved Examples
Bibliography
Problems

7 Depreciation and Depletion   **108**

Criteria
Explanation of Material
7.1 Depreciation Terminology
7.2 Additional First-Year Depreciation and Investment Tax Credit
7.3 Sinking-Fund Depreciation
7.4 Straight-Line Depreciation
7.5 Sum-of-Year-Digits Depreciation
7.6 Declining-Balance Depreciation
7.7 Group and Composite Methods of Depreciation
7.8 Income Tax Calculations
7.9 Depletion Methods
Solved Examples
Bibliography
Problems

LEVEL II

8 Present-Worth and Capitalized-Cost Evaluation   **131**

Criteria
Explanation of Material

8.1 Present-Worth Comparison of Equal-Lived Alternatives
8.2 Present-Worth Comparison of Different-Lived Alternatives
8.3 Capitalized-Cost Calculations
8.4 Capitalized-Cost Comparison of Two Alternatives
Solved Examples
Bibliography
Problems

9 **Equivalent Uniform Annual Cost Evaluation** 147
Criteria
Explanation of Material
9.1 Study Period for Alternatives Having Different Lives
9.2 Salvage Sinking-Fund Method
9.3 Salvage Present-Worth Method
9.4 Capital-Recovery-Plus-Interest Method
9.5 Comparing Alternatives by EUAC
9.6 EUAC of a Perpetual Investment
Solved Examples
Bibliography
Problems

10 **Rate-of-Return Evaluation** 162
Criteria
Explanation of Material
10.1 Tabulation of Cash Flow
10.2 Rate-of-Return Calculations by the Present-Worth Method
10.3 Rate-of-Return Calculations by the EUAC Method
10.4 Interpretation of Rate of Return on Extra Investment
10.5 Alternative Evaluation by Incremental-Investment Analysis
10.6 Alternative Evaluation by the EUAC Method
Solved Examples
References
Bibliography
Problems

LEVEL III

11 **Benefit/Cost Ratio and Service-Life Evaluation** 185
Criteria
Explanation of Material
11.1 Classification of Benefits, Costs, and Disbenefits
11.2 Benefits, Disbenefits, and Cost Calculations
11.3 Alternative Comparison by Benefit/Cost Analysis
11.4 Benefit/Cost Analysis for Multiple Alternatives
11.5 Purpose and Formulas of Service-Life Analysis

11.6 Use of Service Life to Determine Required Life
11.7 Comparison of Two Alternatives Using Service-Life Computation
Solved Examples
Reference
Bibliography
Problems

12 Replacement, Retirement, and Breakeven Analysis   **202**

Criteria
Explanation of Material
12.1 The Defender/Challenger Concept
12.2 Replacement Analysis Using a Specified Planning Horizon
12.3 Computation of Replacement Value for a Defender
12.4 Determination of Minimum Cost Life
12.5 Computation of Breakeven Points Between Alternatives
Solved Examples
References
Bibliography
Problems

13 Bonds   **225**

Criteria
Explanation of Material
13.1 Bond Classifications
13.2 Bond Terminology and Interest
13.3 Bond Present-Worth Calculations
13.4 Rate of Return on Bond Investment
Solved Examples
Bibliography
Problems

LEVEL IV

14 Accounting for Engineering Economists   **239**

Criteria
Explanation of Material
14.1 The Balance Sheet
14.2 The Income Statement and the Cost-of-Goods-Sold Statement
14.3 Accounting Ratios
14.4 Allocation of Factory Expense
14.5 Factory-Expense Computation and Variance
Solved Examples
Bibliography
Problems

15 Corporate Tax Structure 255

Criteria
Explanation of Material
15.1 Some Tax Definitions
15.2 Basic Tax Formulas and Computations
15.3 Tax Laws for Capital Gains and Losses
15.4 Tax Laws for Investment Tax Credit
15.5 Tax Laws on Operating Losses
15.6 Tax Effects of Depreciation Models
15.7 Using an Inflated Before-Tax Rate of Return
Solved Examples
References
Bibliography
Problems

16 After-Tax Economic Analysis 273

Criteria
Explanation of Material
16.1 Tabulation of Cash Flow After Taxes
16.2 After-Tax Analysis Using Present-Worth or EUAC Analysis
16.3 Computation of After-Tax Rate of Return
16.4 After-Tax Replacement Analysis
16.5 Tax Effect of Depletion Laws
Solved Examples
Reference
Bibliography
Problems

17 Evaluation of Multiple Alternatives 291

Criteria
Explanation of Material
17.1 Selection from Mutually Exclusive Alternatives
17.2 Selection Using Incremental Rate of Return
17.3 Selection Using Incremental Benefit/Cost Ratio
17.4 The Capital-Budgeting Problem
17.5 Rate-of-Return Solution to Capital-Budgeting Problems
17.6 Present-Worth Solution to Capital-Budgeting Problems
Solved Examples
References
Bibliography
Problems

18 Establishing the Minimum Attractive Rate of Return 314

Criteria
Explanation of Material

18.1 Types of Capital Financing
18.2 The Cost of Capital
18.3 Variations in MARR
18.4 Cost of Capital for Debt Financing
18.5 Cost of Capital for Equity Financing
18.6 Computation of an Average Cost of Capital
18.7 Effect of the Debt/Equity Ratio on the Cost of Capital
    Solved Examples
    References
    Bibliography
    Problems

19 Sensitivity and Risk Analysis   **332**

    Criteria
    Explanation of Material
19.1 The Approach of Sensitivity Analysis
19.2 Determination of the Sensitivity of Estimates
19.3 Sensitivity of Alternatives Using Three Estimates of Factors
19.4 Economic Uncertainty and the Expected Value
19.5 Expected Value of Economy Alternatives
    Solved Examples
    Bibliography
    Problems

APPENDIXES

A Compound Interest Factors   **353**

B Continuous Compounding   **385**

    Criteria
    Explanation of Material
B-1 Effective Interest for Continuous Compounding
B-2 Interest Factors for Continuous Compounding
B-3 Computations Using Continuous Compounding
B-4 Uniform Flow of Funds
    Bibliography

C Answers to Problems   **394**

Bibliography   **423**

Index   **425**

PREFACE

When the seemingly infinite phases of economic decision analyses are considered, it becomes obvious that condensing hundreds of years' worth of knowledge into the pages of one textbook would be impossible. There is probably no single engineering course that contributes more to the overall education of an undergraduate student than does engineering economy. The tremendous scope of economic analysis dictates that considerable judgment be used in selecting the most appropriate material for inclusion in an undergraduate engineering-economy text. In this regard, this book strongly emphasizes the fundamentals of economic analysis and covers those topics that are relevant to the work of an average engineer.

Although this book has been developed from material that formed the backbone of a self-paced course, the format that has been adopted allows its use in a lecture course equally well. Due to its organization, this text may be used quite easily as a study or reference guide by the practicing engineer or manager who wishes to learn or review certain aspects of engineering economy.

The book is divided into 19 chapters, with each chapter containing an overall objective, the criteria which explain what the reader should be able to do after completion of the chapter, and the study material. The chapters have been combined into four levels, as shown in the flow chart below. The material of Level I emphasizes

Chapter Completion Order

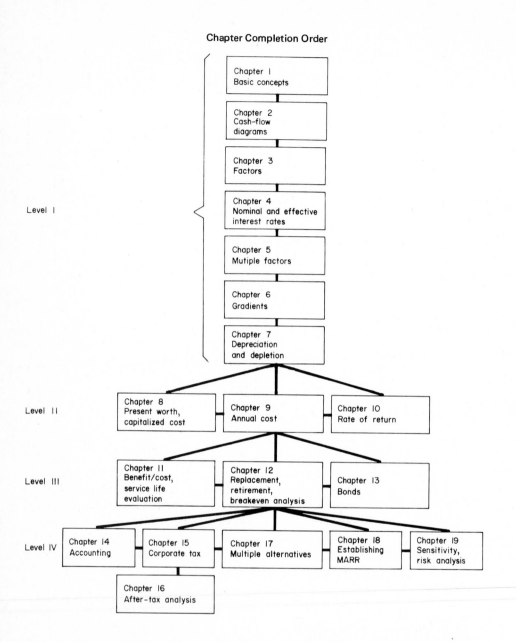

basic computational skills, necessitating completion of the chapters in actual numerical order. Level II discusses applications of the techniques learned in Level I for evaluating alternatives by three of the most common methods of analysis. Level III utilizes the concepts learned in the first two levels to extend the analytical capabilities of the reader to other methods of analysis. Level IV briefly discusses some areas that may be considered peripheral to basic engineering-economy analysis.

The number of each section heading within each chapter corresponds to the number of the criterion that specifies what the reader is expected to perform upon completion of the section. Thus, the material that satisfies criterion 1 in Chap. 3 is included in Sec. 3.1. Even though most sections include worked example problems with commentary about the solution, there are several solved examples, with comments where appropriate, at the end of each chapter. The solved examples that relate to a particular section are listed in bold type at the section end. Numerous unsolved problems for which answers are provided in Appendix C are also included at the end of each chapter. Both the solved examples and the unsolved problems are cross-referenced in bold type to the section containing the appropriate material. This type of organization allows the student either to read all the chapter material before studying the solved examples and working the problems or to study the solved examples and work the problems after reading each section.

To the maximum extent possible, the chapters have been organized so that each chapter builds on preceding chapters. Thus, in order to work the problems the student must have mastered the behaviors taught in previous chapters; this reinforces desirable behavior through repetition.

A bibliography of pertinent texts is included at the end of the volume; appropriate pages are referenced in the brief bibliographies that follow each chapter.

The authors wish to acknowledge the helpful comments and useful suggestions of many students and teachers. Particular thanks must be extended to Ms. Karen Vinson for the overtime hours she spent in designing and typing the manuscript and to Dr. John Levosky for his understanding and timely assistance. Finally, we wish to thank our wives, Brynn and Sandy, for their encouragement and continued moral support.

ANTHONY J. TARQUIN

LELAND T. BLANK

LEVEL I

In the first seven chapters of this book, you learn how to account for the fact that money has different values over different time intervals. This involves the derivation and use of engineering-economy factors, which greatly simplify computations that would otherwise be rather laborious.

Chapter	Subject
1	Basic concepts
2	Cash-flow diagrams
3	Economy factors
4	Nominal and effective interest rates
5	Multiple factors
6	Gradients
7	Depreciation and depletion

1
BASIC CONCEPTS

The purpose of this chapter is to provide you with the basic terminology of engineering economy and the fundamental concepts that form the basis for economic analysis.

CRITERIA

In order to complete this chapter you must be able to do the following:

1 Define the terms *time value of money, interest,* and *principal.*
2 Define *interest rate* and *interest period* and calculate the interest that has been accrued in one interest period, given the principal and the interest rate or total accrued amount.
3 Define *equivalence* and (*a*) calculate the amount of money equivalent to a present sum, given the present sum, the future or past date of equivalence, and the interest rate; and (*b*) calculate the interest rate per year at which different sums separated by one year are equivalent.
4 Define *alternative, evaluation criterion,* and *intangible factors.*

5 State the relations between the minimum attractive rate of return, the rate of return obtainable from a safe investment, and the expected rate of return on a new proposal.
6 Define *simple interest* and *compound interest* and calculate the total amount of money accrued after one or more years using simple and compound interest methods, given the annual interest rate and the original principal.

EXPLANATION OF MATERIAL

1.1 Basic Terminology

Before we begin to develop the terminology and fundamental concepts upon which engineering economy are based, it would be appropriate to define what is meant by *engineering economy*. In the simplest terms, engineering economy is a collection of mathematical techniques which simplify economic comparisons. With these techniques, a rational, meaningful approach to evaluating the economic aspects of different methods of accomplishing a given objective can be developed. Engineering economy is, therefore, a decision assistance tool by which one method will be chosen as the most economical one.

In order for you to be able to apply the techniques, however, it is necessary for you to understand the basic terminology and fundamental concepts that form the foundation for engineering-economy studies. Some of these terms and concepts are described below.

It is often said that "money makes money." The statement is indeed true, for if a person elects to invest his money today (for example, in a bank or savings and loan association), by tomorrow he will have accumulated more money than he had originally invested. This accumulation of money, called the *time value of money*, is the most important concept in engineering economy. You should also realize that if a person or company finds it necessary to borrow money today, by tomorrow more money than the original loan will be owed. This fact is also explained by the time value of money.

The manifestation of the time value of money is termed *interest*, which is a measure of the increase between the original sum borrowed or invested and the final amount owed or accrued. Thus, if you had invested money at some time in the past, the interest would be

$$\text{Interest} = \text{total amount accumulated} - \text{original investment} \qquad (1.1)$$

On the other hand, if you had borrowed money at some time in the past, the interest would be

$$\text{Interest} = \text{present amount owed} - \text{original loan} \qquad (1.2)$$

In either case, there is an increase in the amount of money that was originally invested or borrowed, and the increase over the original amount in the interest. The original investment or loan is referred to as *principal.*

Problems P1.1, P1.2

1.2 Interest Calculations

When interest is expressed as a percentage of the original amount per unit time, the result is an *interest rate*. This rate is calculated as follows:

$$\text{Percent interest rate} = \frac{\text{interest accrued per unit time}}{\text{original amount}} \times 100\% \qquad (1.3)$$

By far the most common time period used in expressing interest as interest rate is one year. However, since interest rates are often expressed over periods of time shorter than one year (i.e., 1% per month), the time unit used in expressing an interest rate is termed an *interest period*. The following two examples illustrate the computation of interest and interest rate.

Example 1.1 The Get-Rich-Quick (GRQ) Company invested $100,000 on May 1 and withdrew a total of $106,000 exactly one year later. Compute (*a*) the interest gained from the original investment and (*b*) the interest rate from the investment.

SOLUTION

(*a*) Using Eq. (1.1)

$$\text{Interest} = 106{,}000 - 100{,}000 = \$6{,}000$$

(*b*) Equation (1.3) is used to obtain

$$\text{Percent interest rate} = \frac{6{,}000/\text{year}}{100{,}000} \times 100\% = 6\% \text{ per year}$$

COMMENT For borrowed money, computations are similar to those shown above, except that interest is computed by Eq. (1.2). For example, if GRQ borrowed $100,000 now and repaid $110,000 in one year, using Eq. (1.2), interest is $10,000 and the interest rate from Eq. (1.3) is 10% per year.   ////

Example 1.2 Joe Bilder plans to borrow $20,000 for one year at 5% interest. Compute (*a*) the interest and (*b*) the total amount due after one year.

SOLUTION

(*a*) Equation (1.3) may be solved for the interest accrued to obtain

$$\text{Interest} = 20{,}000(0.05) = \$1{,}000$$

(b) Total amount due is the sum of principal and interest, or

$$\text{Total due} = 20{,}000 + 1{,}000 = \$21{,}000$$

COMMENT Note that in (b) above, the total amount due may also be computed as

$$\text{Total due} = \text{principal}(1 + \text{interest rate}) = 20{,}000(1.05) = \$21{,}000 \qquad ////$$

In all the preceding examples the interest period was one year and the interest was calculated as of the end of one period. When more than one yearly interest period is involved (for example, if we want to know the amount of interest GRQ would owe on the above loan after three years), it becomes necessary to determine whether the interest is payable on a *simple* or *compound* basis. The concepts of simple and compound interest are discussed in Sec. 1.6.

Examples 1.5, 1.6
Problems P1.3–P1.5

1.3 Equivalence

The time value of money and interest rate utilized together generate the concept of *equivalence*, which means that different sums of money at different times can be equal in economic value. For example, if the interest rate is 6% per year, $100 today (i.e., at present) would be equivalent to $106 one year from today since

$$\text{Amount accrued} = 100 + 100(0.06) = 100(1 + 0.06) = 100(1.06) = \$106$$

Thus, if someone offered you a gift of $100 today or $106 one year from today, it would make no difference which offer you accepted, since in either case you would have $106 one year from today! The two sums of money, therefore, are equivalent to each other when the interest rate is 6% per year. At either a higher or a lower interest rate, however, $100 today is not equivalent to $106 one year from today. In addition to considering future equivalence, one can apply the same concepts for determining equivalence in previous years. Thus, $100 now would be equivalent to $100/1.06 = \$94.34$ one year ago if the interest rate is 6% per year. From these examples, it should be clear that $94.34 last year, $100 now, and $106 one year from now are equivalent when the interest rate is 6% per year. The fact that these sums are equivalent can be established by computing the interest rate as follows:

$$\frac{106}{100} = 1.06 \text{ or } 6\% \text{ per year}$$

and

$$\frac{100}{94.34} = 1.06 \text{ or } 6\% \text{ per year}$$

Examples 1.7, 1.8
Problems P1.6, P1.7

1.4 Alternatives

An *alternative* is a stand-alone solution for a given situation. We are faced with alternatives in virtually everything we do, from selecting the method of transportation we use to get to work every day to deciding between buying a house or renting one. Similarly, in engineering practice, there are always several ways of accomplishing a given task, and it is necessary to be able to compare them in a rational manner so that the most economical alternative can be selected. The alternatives in engineering considerations usually involve such items as asset purchase cost (first cost), the anticipated life of the asset, the yearly costs of maintaining the asset (annual operating cost), the anticipated asset resale value (salvage value), and the interest rate. After the facts and all the relevant estimates have been collected, an engineering-economy analysis will be conducted to determine which alternative is best from an economic point of view. It should be clear that if there were not alternative ways of accomplishing a particular task, there would be no need for an engineering-economy analysis. The procedures developed in this book will enable you to make accurate economic decisions when one or more alternatives are being considered. As in most engineering and scientific work, the accuracy of the solution is directly proportional to the accuracy of the data and estimates that were used in arriving at the solution. Just as in the cases of computer solutions, "garbage in means garbage out."

In order to be able to compare different methods of accomplishing a given objective, however, it is necessary to have an *evaluation criterion* that could be used as a basis for judging the alternatives. In engineering economy, *dollars* are used as the basis for comparison. Thus, when there are several ways of accomplishing a given objective, the method that has the lowest overall cost is *usually* selected.

In most cases, the alternatives involve *intangible factors*, which cannot be expressed in terms of dollars, such as the effect of a process change on employee morale. When the alternatives available have approximately the same equivalent cost, the nonquantifiable, or intangible, factors may be used as the basis for selecting the best alternative.

Problem P1.8

1.5 Minimum Attractive Rate of Return

In order for a proposed investment to seem "profitable" to investors, they must expect to receive more money than they invest. In other words, they expect to receive a fair *rate of return* on investment. When the interest period is equal to or less than a year, the rate of return (RR) in percent for the interest period is

$$RR = \frac{\text{total amount of money received} - \text{original investment}}{\text{original investment}} \times 100\% \qquad (1.4)$$

$$RR = \frac{\text{profit}}{\text{original investment}} \times 100\%$$

Note that the calculation for rate of return is exactly the same as the calculation for interest rate, Eq. (1.3). The two terms can be used interchangeably, but generally *rate of return* is used when determining the profitability of a proposed or past investment, while *interest rate* is used when borrowing capital or when a fixed rate has been established, such as that on a bond. This will become very clear in subsequent chapters. As stated earlier, investors must expect to make a "reasonable" profit or rate of return on investment before making a capital commitment. The "reasonable" rate of return must, therefore, be greater than some stated rate of return. This stated rate is usually that which could be received from a bank or other safe investment; thus, the "reasonable" rate is usually higher than the bank rate, since most other investments would involve some risks or uncertainties. This "reasonable" rate is called the minimum required rate of return or *minimum attractive rate of return* (MARR).

For most commercial and industrial organizations, the amount of investment capital available is the limiting resource; that is, there are many investment opportunities that would yield a rate of return greater than even the MARR. Since only limited investment funds are available, however, the projects that are undertaken usually have a projected rate of return considerably higher than the MARR. In addition, several projects that cannot be funded immediately because of limited capital also have projected rates of return greater than the MARR. Therefore, new projects that are under consideration are not to be undertaken unless their expected rate of return is at least as great as the rate of return *on the least attractive proposal that has not yet been funded*. Figure 1.1 is a presentation of the relation between the different rate-of-return values.

The important concept that must be understood is that a MARR must be established; then, proposed projects that are not expected to yield rates of return that would at least equal or exceed the MARR can be regarded as economically unacceptable.

Problems P1.9, P1.10

1.6 Simple and Compound Interest

The concepts of interest and interest rate were introduced in Secs. 1.1 and 1.2 and used in Sec. 1.3 to calculate, for one interest period, past and future sums of money equivalent to a present sum (principal). When more than one interest period is involved, the terms *simple* and *compound* interest must be considered.

Simple interest is calculated using the principal only, ignoring any interest that had been accrued in preceding interest periods. The total interest can be computed using the relation

$$\text{Interest} = (\text{principal})(\text{number of periods})(\text{interest rate}) = Pni \qquad (1.5)$$

Example 1.3 If you borrow \$1,000 for three years at 6% per year simple interest, how much money will you owe at the end of three years?

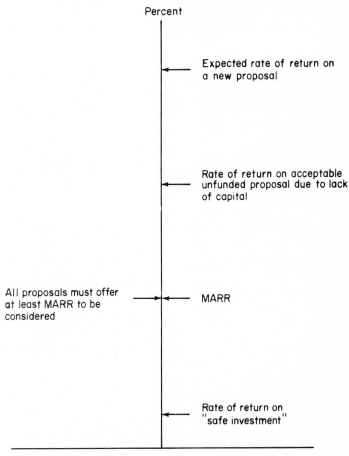

FIG. 1. Relation between different rate-of-return values.

SOLUTION   The interest for each of the three years is

$$\text{Interest per year} = 1{,}000(0.06) = \$60$$

Total interest for three years from Eq. (1.5) is

$$\text{Total interest} = 1{,}000(3)(0.06) = \$180$$

Finally, amount due after three years is

$$1{,}000 + 180 = \$1{,}180$$

COMMENT   The $60 interest that accrued in the first year and the $60 accrued in the second year did not earn interest. The interest due was calculated on the principal only. The results of this loan are tabulated in Table 1.1. The end-of-year figure of zero represents the present, that is, when the money is borrowed. Note that

Table 1.1 SIMPLE–INTEREST COMPUTATION

End of year (1)	Amount borrowed (2)	Interest (3)	Amount owed (4) = (2) + (3)	Amount paid (5)
0	$1,000	–	–	–
1	–	$60	$1,060	$ 0
2	–	60	1,120	0
3	–	60	1,180	1,180

no payment is made by the borrower until the end of year 3. Thus, the amount owed each year increases uniformly by $60, since interest is figured only on the principal of $1,000.   ////

In calculations of *compound interest*, the interest for an interest period is calculated on the principal *plus the total amount of interest accumulated in previous periods*. Thus compound interest means "interest on top of interest" (i.e., it reflects the effect of the time value of money).

Example 1.4 If, as in the preceding example, you borrow $1,000 at 6% per year *compound* interest, compute the total amount due after a three-year period.

SOLUTION The interest and total amount due for each year is computed as follows:

Interest, year 1 = 1,000(0.06) = $60
Total amount due after year 1 = 1,000 + 60 = $1,060

Interest, year 2 = 1,060(0.06) = $63.60
Total amount due after year 2 = 1,060 + 63.60 = $1,123.60

Interest, year 3 = 1,123.60(0.06) = $67.42
Total amount due after year 3 = 1,123.60 + 67.42 = $1,191.02

COMMENT The details are shown in Table 1.2. The repayment scheme is the same as for the simple-interest example; that is, no amount is repaid until the principal, plus all interest, is due at the end of year 3. This time value of money is recognized in compound interest. Thus, with compound interest, the original $1,000 would accumulate an extra $1,191.02 − 1,180.00 = $11.02 compared to simple interest in the three-year period.   ////

In Chap. 3, formulas are developed which simplify compound-interest calculations. The same concepts are involved when the interest period is less than a year. A discussion of this case, however, is deferred until Chap. 4. Since "real-world" calculations almost always involve compound interest, the interest rates specified hereafter in this book refer to compound interest unless specified otherwise.

Example 1.9
Problems P1.11–P1.21

Table 1.2   COMPOUND–INTEREST COMPUTATION

End of year (1)	Amount borrowed (2)	Interest (3)	Amount owed (4) = (2) + (3)	Amount paid (5)
0	$1,000	–	–	–
1	–	$60.00	$1,060.00	$ 0
2	–	63.60	1,123.60	0
3	–	67.42	1,191.02	1,191.02

SOLVED EXAMPLES

You should refer to the section noted in boldface type after each example if you don't understand the solution.

Example 1.5   Calculate the interest and total amount accrued after one year if $2,000 is invested at an interest rate of 5% per year.

SOLUTION

$$\text{Interest earned} = 2,000(0.05) = \$100$$

$$\text{Total amount accrued} = 2,000 + 2,000(0.05) = 2,000(1 + 0.05)$$

$$= \$2,100 \qquad ////$$

Sec. 1.2

Example 1.6   (*a*) Calculate the amount of money that had to have been deposited one year ago for you to have $1,000 now at an interest rate of 5% per year. (*b*) Calculate the interest that was earned in the same time period.

SOLUTION

(*a*)   Total amount accrued = original deposit + (original deposit)(interest rate)
Let X = original deposit

$$1,000 = X + X(0.05) = X(1 + 0.05)$$

$$1,000 = 1.05X$$

$$X = \frac{1,000}{1.05} = 952.38$$

Original deposit = $952.38
(*b*)   By using Eq. (1.1), we have

$$\text{Interest} = \$1,000 - \$952.38 = \$47.62 \qquad ////$$

Sec. 1.2

Example 1.7 Calculate the amount of money that had to have been deposited one year ago for the investment to earn $100 in interest in the one year, if the interest rate is 6% per year.

SOLUTION

$$\text{Let } a = \text{total amount accrued}$$
$$b = \text{original deposit}$$
$$\text{Interest} = a - b$$
$$a = b + b(\text{interest rate})$$
$$\text{Interest} = b + b(\text{interest rate}) - b$$
$$\text{Interest} = b(\text{interest rate})$$
$$\$100 = b(0.06)$$
$$b = \frac{100}{0.06} = \$1,666.67 \qquad ////$$

Secs. 1.2, 1.3

Example 1.8 Make the calculations necessary to show which of the statements below are true and which are false, if the interest rate is 5% per year:

(*a*) $98 now is equivalent to $105.60 one year from now.
(*b*) $200 one year past is equivalent to $205 now.
(*c*) $3,000 now is equivalent to $3,150 one year from now.
(*d*) $3,000 now is equivalent to $2,887.14 one year ago.
(*e*) Interest accumulated in one year on an investment of $2,000 is $100.

SOLUTION

(*a*) Total amount accrued $= 98(1.05) = \$102.90 \neq \$105.60$; therefore false. Another way to solve this is

$$\text{Required investment} = \frac{105.60}{1.05} = \$100.57 \neq \$98$$

therefore false.

Sec. 1.3

(*b*) Required investment $= \dfrac{205.00}{1.05} = \$195.24 \neq \$200$; therefore false.

Secs. 1.2, 1.3

(*c*) Total amount accrued $= 3,000(1.05) = \$3,150$; therefore true.

Secs. 1.2, 1.3

(*d*) Total amount accrued $= 2,887.14(1.05) = \$3,031.50 \neq \$3,000$; therefore false.

Secs. 1.2, 1.3

(*e*) Interest $= 2,000(0.05) = \$100$; therefore true.   ////

Secs. 1.2, 1.3

Example 1.9 Calculate the total amount due after two years if \$2,500 is borrowed now and the compound interest rate is 8% per year.

SOLUTION The results may be presented as follows:

End of year	Amount borrowed	Interest	Amount owed	Amount paid
0	\$2,500	–	–	–
1	–	\$200	\$2,700	\$ 0
2	–	216	2,916	2,916

Total amount due = \$2,916.   ////

BIBLIOGRAPHY

Full citations are found in the Bibliography at the back of the book.

Barish, pp. 49–50.
DeGarmo and Canada, pp. 111–116.
Emerson and Taylor, pp. 1-6 – 1-13, 2-5 – 2-6, 3-10 – 3-11.
Fabrycky and Thuesen, pp. 41–47, 56–58, 83–91.
Grant and Ireson, pp. 25–32.
Ostwald, p. 319.
Park, pp. 17–19.
Reisman, pp. 9–11.
Riggs, pp. 158–161, 163–165.
Smith, pp. 35–37, 90.
Taylor, pp. 18–24, 28, 37.
Thuesen, Fabrycky, and Thuesen, pp. 51–56, 80–82.

PROBLEMS

P1.1   What is meant by *the time value of money*?
P1.2   What is the difference between the terms *principal* and *present amount owed*?
P1.3   If a bank advertises 4% per year interest compounded semiannually, what is the interest period?

P1.4   Find the interest due after one year on a loan of $5,000, if interest is 12% per year.

P1.5   Find the original amount of a loan if interest is $1\frac{1}{2}\%$ per month payable monthly and the borrower just made the first monthly payment of $25 in interest.

P1.6   At what interest rate are $450 a year ago and $550 one year from now equivalent?

P1.7   How do you explain the fact that two different amounts of money can be equivalent to each other?

P1.8   List three items that might be regarded as intangible factors.

P1.9   If the Get-Rich-Quik Company invested $50,000 in a new process one year ago and has just now realized a profit of $7,500, what was its rate of return based on this investment?

P1.10 Why is the minimum attractive rate of return of a business organization greater than the interest rate obtainable from a bank or savings and loan association?

P1.11 Assume that you have been offered an investment opportunity in which you may invest $1,000 at 7% per year simple interest for three years or you may invest the same $1,000 at 6% per year compound interest for three years. Which investment offer would you accept?

P1.12 (*a*) How much interest would you pay if you borrowed $600 at $1\frac{1}{2}\%$ per month compounded monthly for three months?
(*b*) What percent of the principal is this interest amount?

P1.13 Work the two parts of problem P1.12 for $1\frac{1}{2}\%$ per month simple interest.

P1.14 How much money would your friend owe after four years if she borrows $1,000 now at 7% per year simple interest?

P1.15 How much money would be owed after two years if a person borrows $500 at 1% per month simple interest?

P1.16 How much money can you borrow now if you will repay the lender $850 in two years and the interest rate is 6% per year compounded yearly?

P1.17 If you borrow $1,500 now and must repay $1,800 two years from now, what is the interest rate on your loan? Assume interest is compounded yearly.

P1.18 If you invest $10,000 now in a business venture that promises to return $14,641, how soon must you receive the $14,641 in order to make at least 10% per year compounded yearly on your investment?

P1.19 Your friend tells you he has just repaid a loan he got three years ago at 10% per year simple interest. If, upon questioning, you learn his payment was $195, how much did he borrow?

P1.20 If you invest $3,500 now in return for a guaranteed $5,000 income at a later date, when must you receive your money in order to earn at least 8% per year simple interest?

P1.21 Finally, here is a somewhat harder problem. You learned in Sec. 1.6 that $1,000 at 6% per year simple interest is equivalent to $1,180 in three years. Now, find the compound interest rate per year for which this equivalence is also correct.

2

SYMBOLS AND CASH–FLOW DIAGRAMS

This chapter will teach you the meaning of the symbols used in engineering economy and how to construct a cash-flow diagram. The material you learn here will be used through the remainder of this book. In particular, you will find the cash-flow diagram exceptionally useful in simplifying complicated descriptive problems.

CRITERIA

In order to complete this chapter you must be able to do the following:

1 Define and recognize in a problem statement the economy symbols P, F, A, n, and i.
2 State what is meant by *end-of-year convention*.
3 Define *cash flow* and tabulate cash flows, given a statement of the cash-flow sequence.
4 Construct a cash-flow diagram, given a statement describing the amount and the times at which the cash flows take place.

EXPLANATION OF MATERIAL

2.1 Symbols and Their Meaning

The mathematical relations used in engineering economy employ the following symbols:

P = value or sum of money at a time denoted as the present
F = value or sum of money at some future time
A = a series of periodic, equal amounts of money
n = number of interest periods
i = interest rate per interest period

The symbols P and F represent single-time occurrence values; A occurs each interest period for a specified number of periods with the same dollar value. The units of the symbols aid in clarifying their meaning. The present sum P and future sum F are expressed in dollars, while A is referred to in dollars per interest period. It is important to note here that in order for a series to be represented by the symbol A, it must be uniform (i.e., the dollar value must be the same for each period) and the uniform dollar amounts must extend through *consecutive* periods. Both conditions must exist before the dollar values can be represented by an A. Since n is commonly expressed in years, A is usually expressed in units of dollars per year. The compound interest rate i is expressed in percent per interest period, for example, 5% per year. Except where noted otherwise, this rate applies throughout the entire n years or n interest periods. The most common engineering-economy problems involve the use of n and i and at least two of the three terms P, F, and A. The following four examples illustrate the use of the symbols.

Example 2.1 If you borrow $2,000 now and must repay the loan plus interest at a rate of 7% per year in five years, what is the total amount you must pay? List the values of P, F, A, n, and i.

SOLUTION In this situation only P and F are involved and not A, since all transactions are single payments. The values are as follows:

$P = \$2,000$
$F = ?$
$i = 7\%$ per year
$n = 5$ years   ////

Example 2.2 If you borrow $2,000 now at 7% per year for five years and must repay the loan in equal yearly payments, what will you be required to pay? Determine the value of the symbols involved.

SOLUTION

$$P = 2,000$$
$$A = ? \text{ per year for 5 years}$$
$$i = 7\% \text{ per year}$$
$$n = 5 \text{ years}$$
$$\text{There is no } F \text{ value involved.} \qquad ////$$

In both examples above, the *P* value of $2,000 is a receipt and *F* or *A* is a disbursement. It is equally correct to use these symbols in reverse roles, as in the examples below.

Example 2.3 If you deposit $500 in a savings account on May 1, 1975, which pays 7% per year, what annual amount can you withdraw for the following ten years? List the symbol values.

SOLUTION

$$P = \$500$$
$$A = ? \text{ per year for 10 years}$$
$$i = 7\% \text{ per year}$$
$$n = 10 \text{ years}$$

COMMENT The values for the $500 disbursement *P* and receipt *A* are given the same symbol names as before, but considered in a different context. Thus, a *P* value may be a receipt (Examples 2.1 and 2.2) or a disbursement (this example).   ////

Example 2.4 If you deposit $100 in a savings account each year for seven years at an interest rate of 6% per year, what single amount will you be able to withdraw after the seven years? Define the symbols and their roles.

SOLUTION In this example, the equal annual deposits are in a series *A* and the withdrawal is a future sum, or *F* value. There is no *P* value here. Thus,

$$A = \$100 \text{ per year for 7 years}$$
$$F = ?$$
$$i = 6\% \text{ per year}$$
$$n = 7 \text{ years} \qquad ////$$

Example 2.10
Problems P2.1–P2.3

2.2 Time Placement of *P, F,* and *A*

The dollar amounts of *P, F,* or *A* are always considered to occur at the *end of the interest period*. This is merely a simplifying convention, useful for formula derivation and problem solution. However, it should be understood that this does not mean that the end of the year is December 31. In the situation of Example 2.3, since investment took place on May 1, 1975, the withdrawals will take place on May 1, 1976, and each succeeding May 1 for ten years (the last withdrawal will be on May 1, 1985, *not* 1986). Thus, *end of the year* means one year from the date of the transaction (whether it be receipt or disbursement).

Because the most common interest period is a year, the symbol A will indicate an end-of-year amount (disbursement or receipt) continuing for n consecutive years. With this time frame, it is helpful to define our symbols more accurately, with common units in parentheses:

> n = number of interest periods (usually years)
> i = interest rate per interest period (% per year)
> P = a single sum of money at the present ($)
> F = a single sum of money at the end of n interest periods ($)
> A = one of a series of consecutive, equal, end-of-period amounts
>   ($ per year)

In the next chapter you learn how to determine the equivalent relations between $P, F,$ and A values at different times.

Problem P2.4

2.3 Description and Tabulation of Cash Flow

Every person or company has cash receipts (income) and cash disbursements (costs). The result of income and costs is conveniently called *cash flow* and may be defined as the net receipt or net disbursement resulting from receipts and disbursements occurring in the same interest period. Algebraically,

$$\text{Cash flow} = \text{receipts} - \text{disbursements} \qquad (2.1)$$

Thus, a positive cash flow indicates a net receipt in a particular interest period or year, while a negative cash flow indicates a net disbursement in that period.

Example 2.5 Suppose you borrowed $1,000 on May 1, 1974, and agreed to repay the loan in one lump sum of $1,402.60 at the end of four years at 7% per year. Tabulate your yearly cash flows.

SOLUTION

Date	Receipt	Disbursement	Cash flow
May 1, 1974	$1,000.00	$ 0	$ 1,000.00
May 1, 1975	0	0	0
May 1, 1976	0	0	0
May 1, 1977	0	0	0
May 1, 1978	0	1,402.60	−1,402.60

////

Example 2.6 If you buy a new television in 1975 for $300, maintain it for three years at a cost of $20 per year, and then sell it for $50, what are your cash flows?

SOLUTION

Year	Receipt	Disbursement	Cash flow
1975	$ 0	$300	$−300
1976	0	20	−20
1977	0	20	−20
1978	50	20	+30

COMMENT It is important for you to remember that all receipts and disbursements and thus, cash-flow values, are assumed to be end-of-period amounts. Therefore, 1975 is the present (now) and 1978 is the end of year 3.　　////

Example 2.11
Problems P2.5–P2.7

2.4 Cash-Flow Diagrams

A *cash-flow diagram* is simply a graphical representation of cash flows drawn on a time scale. The diagram should represent the statement of the problem and should include what is given and what is to be found. That is, after the cash-flow diagram has been drawn, an outside observer should be able to work the problem by looking at the cash-flow diagram. Time zero is considered to be the present and time 1 the end of time period 1. (We will assume the periods are in years.) The time scale of Fig. 2.1 is set up for five years. Since it is assumed that cash flows occur only at the end of the year, we will be concerned only with the times marked 0, 1, 2, . . . , 5.

The direction of the arrows on the cash-flow diagram is important to problem solution. Therefore, in this text, we use a vertical arrow pointing up to indicate a positive cash flow. Conversely, an arrow pointing down indicates a negative cash flow. The cash-flow diagram in Fig. 2.2 illustrates a receipt (income) at the end of year 1 and a disbursement at the end of year 2.

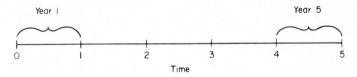

FIG. 2.1 A typical cash-flow time scale.

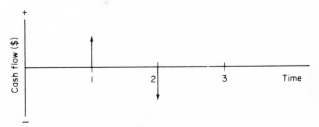

FIG. 2.2 Example of positive and negative cash flows.

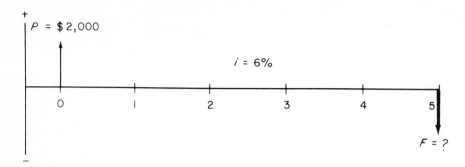

FIG. 2.3 Cash-flow diagram for Example 2.7.

It is important that you thoroughly understand the meaning and construction of a cash-flow diagram, since it is a valuable tool in problem solution. The three examples below illustrate the construction of cash-flow diagrams.

Example 2.7 Consider the situation presented in Example 2.1 where $2,000 (*P*) is borrowed and *F* is to be found after five years. Construct the cash-flow diagram for this case, assuming an interest rate of 6% per year.

SOLUTION Figure 2.3 presents the cash-flow diagram.

COMMENT While it is not necessary to use an exact scale on the cash-flow axes, you will probably avoid errors later on if you make a neat diagram. Notice also that the present sum *P* is a *receipt* at year zero and the future sum *F* is a *disbursement* at the end of year 5.   ////

Example 2.8 If you start now and make five deposits of $1,000 per year (*A*) in a 7%-per-year account, how much money will be accumulated immediately after you have made the last deposit? Construct the cash-flow diagram.

SOLUTION The cash flows are shown in Fig. 2.4.

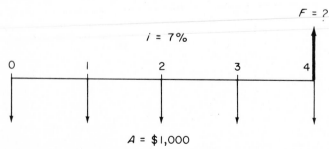

FIG. 2.4 Cash-flow diagram for Example 2.8.

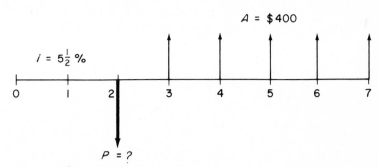

FIG. 2.5   Cash-flow diagram for Example 2.9.

COMMENT   Since you have decided to start now, the first deposit is at year zero and the fifth deposit and withdrawal occur at the end of year 4. Note that in this example, the amount accumulated after the fifth deposit is to be computed; thus, the future amount is represented by a question mark (i.e., $F = ?$).   ////

Example 2.9   Assume that you want to deposit an amount P into an account two years from now in order to be able to withdraw $400 per year for five years starting three years from now. Assume the interest rate is $5\frac{1}{2}\%$ per year. Construct the cash-flow diagram.

SOLUTION   Figure 2.5 presents the cash flows, where P is to be found.

COMMENT   Note that the diagram shows what was given and what is to be found. The diagram, therefore, illustrates very clearly which calculations must be made. Such cash-flow diagrams are extremely important in later chapters, where lengthy descriptive problems are presented.   ////

Examples 2.12, 2.13
Problems P2.8–P2.20

SOLVED EXAMPLES

Example 2.10   Assume you plan to make a lump-sum deposit of $5,000 now into an account that pays 6% per year, and you plan to withdraw an equal end-of-year amount of $1,000 for five years starting next year. At the end of the sixth year, you plan to close your account by withdrawing the remaining money. Define the engineering-economy symbols involved.

SOLUTION

$$P = \$5,000$$
$$A = \$1,000 \text{ per year for 5 years}$$
$$F = ? \text{ at end of year 6}$$
$$i = 6\% \text{ per year}$$
$$n = 5 \text{ years for } A$$

////

Sec. 2.1

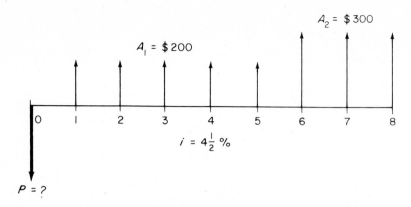

$A_2 = \$300$

$A_1 = \$200$

$i = 4\frac{1}{2}\%$

$P = ?$

FIG. 2.6   Cash-flow diagram for two different A values, Example 2.12.

Example 2.11   The GRQ Company invested \$2,500 in a new air compressor seven years ago. Annual income from the compressor was \$750. During the first year \$100 was spent on maintenance, and this cost increased each year by \$25. The company plans to sell the compressor for salvage at the end of next year (year 8) for \$150. Tabulate the yearly cash flows for the piece of equipment.

SOLUTION

End of year	Income	Cost	Cash flow
0	\$ 0	\$2,500	\$–2,500
1	750	100	650
2	750	125	625
3	750	150	600
4	750	175	575
5	750	200	550
6	750	225	525
7	750	250	500
8	750 + 150	275	625

////

Sec. 2.3

Example 2.12   Suppose that you want to make a deposit into your account now such that you can withdraw an equal annual amount of $A_1 = \$200$ per year for the first five years starting one year after your deposit and a different annual amount of $A_2 = \$300$ per year for the following three years. How would the cash-flow diagram appear if i is $4\frac{1}{2}\%$ per year?

SOLUTION   The cash flows would appear as shown in Fig. 2.6.

COMMENT   The first withdrawal (positive cash flow) occurs at the end of year 1, exactly one year after P is deposited.   ////

Sec. 2.4

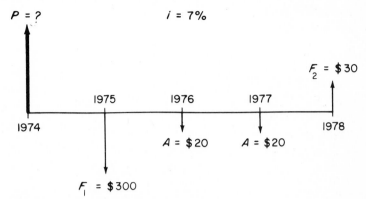

FIG. 2.7   A cash-flow diagram for Example 2.13.

Example 2.13   Consider the cash flows of Example 2.6. Diagram the cash flows and label each arrow as P, F, or A with its respective dollar value if you want to find the single amount in 1974 that would be equivalent to all the cash flows shown. Assume an interest rate of 7%.

SOLUTION   Figure 2.7 presents the cash-flow diagram.

COMMENT   The two $20 negative cash flows form a series of two equal end-of-year values. As long as the dollar values are equal and in two or more consecutive periods, they can be represented by A, regardless of where they begin or end. However, the $30 positive cash flow in 1978 is a single-occurrence value in the future and, therefore, is labeled an F value. It is possible, however, to view all the individual cash flows as F values. The diagram could be drawn as shown in Fig. 2.8. In general, however, if two or more equal end-of-period amounts occur consecutively, by the definition in Sec. 2.2, they should be labeled A values because, as is described in Chap. 3, the use of A values when possible simplifies calculations considerably. Thus, the interpretation pictured by the diagram of Fig. 2.8 is discouraged and will not be used further in this text.   ////

Sec. 2.4

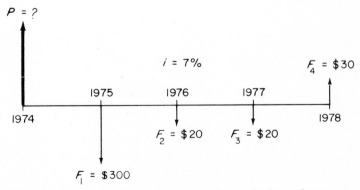

FIG. 2.8   A cash flow for Example 2.13 considering all values as future sums.

BIBLIOGRAPHY

DeGarmo and **Canada**, pp. 116–117.
Emerson and **Taylor**, pp. 2-3 – 2-4, 2-7.
Fabrycky and **Thuesen**, pp. 47–48.
Grant and **Ireson**, pp. 33–34.
Riggs, pp. 165–166.
Smith, pp. 37–41.
Taylor, pp. 26, 27.
Thuesen, Fabrycky, and **Thuesen**, pp. 56–67.

PROBLEMS

P2.1 Five equal deposits of $1,000 will be made every two years starting next year at 10% per year and the total accrued amount withdrawn when the last deposit is made. List the economy symbols and values involved in this problem.

P2.2 The GRQ Company plans to deposit $709.90 now at 6% per year and withdraw $100 per year for the next five years and $200 per year for the following two years. What are the economy symbols and their respective values?

P2.3 How many years would it take for $1,400 to triple in value at an interest rate of 6% per year? Define the economy symbols.

P2.4 What is meant by *end-of-year convention*?

P2.5 Tabulate the yearly net cash flows for Example 2.1.

P2.6 Tabulate the yearly net cash flows for Example 2.3.

P2.7 Assume that you have developed an investment plan that is carried out as follows: Invest $500 now and every other year for ten years and withdraw $300 every year starting five years from now and continuing for eight years. Tabulate the yearly net cash flows.

P2.8 Construct the cash-flow diagram for Example 2.11. Label each amount as *P*, *F*, or *A*.

P2.9 If you plan to make a deposit now such that you will have $3,000 in your account five years from now, how much must you deposit if the interest rate is 8%? Draw the cash-flow diagram.

P2.10 Your uncle has agreed to deposit $700 per year in a savings account for the next five years starting now. You have in turn agreed not to withdraw any money until the end of year 9, at which time you plan to remove $3,000 from the account. Further, you plan to withdraw the remaining amount in three equal year-end installments after the initial withdrawal. Diagram the cash flows for your uncle and yourself.

P2.11 The president of GRQ wants to make two equal lump-sum deposits, one two years and the second four years from now, so he can withdraw $100 per year for five years starting when the second deposit is made. Further, he plans to withdraw an additional $500 the year after the withdrawal series. Draw his cash-flow diagram.

P2.12 You want to invest money at 8% per year so that six years from now you can withdraw an amount F in a lump sum. The investment consultant at the bank has developed the following two plans for you: (*1*) Deposit $351.80 now and $351.80 three years from now. (*2*) Deposit $136.32 per year starting next year and ending in year 6. Draw the cash-flow diagram for each plan if F is to be found.

P2.13 How much could you spend now in order to avoid spending $580 eight years from now if the interest rate is 6%? Draw the cash-flow diagram.

P2.14 If you deposit $100 per year for five years starting one year from now, how much will you have in your account 15 years from now, if the interest rate is 6% per year? Draw the cash-flow diagram.

P2.15 What is the present worth of an expenditure of $1,200 five years from now and $2,200 eight years from now if the interest rate is 10% per year? Tabulate the net cash flows and construct the cash-flow diagram.

P2.16 Calculate the present worth of an expenditure of $85 per year for six years that starts three years from now if the interest rate is $4\frac{1}{2}$%. Construct the cash-flow diagram.

P2.17 If you invest $10,000 now in a real estate venture, how much must you sell your property for ten years from now if you want to make a 12% rate of return on your investment? Define the economy symbols and draw the cash-flow diagram.

P2.18 If you invest $4,100 now and receive $7,500 five years from now, what is the rate of return on your investment? Define the economy symbols and construct the cash-flow diagram.

P2.19 How much money would be accumulated in six years if a person deposits $500 now and amounts increasing by $50 per year for the next six years? Assume i is 6% per year and draw the cash-flow diagram.

P2.20 What uniform payment for eight years beginning one year from now would be equivalent to spending $4,500 now, $3,300 three years from now, and $6,800 five years from now if the interest rate is 8% per year? Define the economy symbols and draw the cash-flow diagram.

3
FACTORS AND THEIR USE

The objective of this chapter is to teach you the derivation of the engineering-economy factors and the use of these basic factors in economy computations. This chapter is probably the most important in the book, since the concepts presented here will be used throughout the remainder of the text.

CRITERIA

To complete this chapter you must be able to define and derive the formulas for the following:

1 Single-payment compound-amount factor and single-payment present-worth factor
2 Uniform-series present-worth factor and capital-recovery factor using the single-payment present-worth factor
3 Uniform-series compound-amount factor and sinking-fund factor using the single-payment compound-amount formula and the capital-recovery factor

You must also be able to do the following:

4 Find the correct numerical value of a factor in a table, given the standard factor notation.

5   Linearly interpolate to find a correct factor value, given an interest rate and/or year value not listed in the tables.

6   Calculate the present worth P, future worth F, or equivalent uniform annual series A of an investment, given the interest rate i, the number of years n, and the monetary value of one of the above symbols.

7   Calculate the interest rate (rate of return) on a sequence of cash flows, given the number of years n and two of the following:
(a)   present worth P
(b)   future worth F, or
(c)   uniform series A starting at the end of year 1 and ending at the end of year n.

8   Determine the number of years n for a sequence of cash flows, given the interest rate i and two of the following:
(a)   present worth P,
(b)   future worth F, or
(c)   uniform series A starting at the end of year 1 and ending at the end of year n.

EXPLANATION OF MATERIAL

3.1   Derivation of Single-Payment Formulas

In this section, a formula is developed which allows determination of the amount of money that is accumulated (F) after n years from a *single* investment (P) when interest is compounded one time per year (or period).

You will recall from Chap. 1 that compound interest refers to interest paid on top of interest. Therefore, if an amount of money P is invested at some time $t = 0$, the amount of money that will be accumulated one year hence (F) would be

$$F_1 = P + Pi$$
$$F_1 = P(1 + i)$$

where F_1 = total amount accumulated after one year. At the end of the second year, the amount of money accumulated (F_2) would be equal to the amount that had accumulated after year 1 plus the interest from the end of year 1 to the end of year 2. Thus,

$$F_2 = F_1 + F_1 i$$
$$= P(1 + i) + P(1 + i)i \qquad (3.1)$$

or

$$F_2 = P(1 + i + i + i^2)$$
$$= P(1 + 2i + i^2)$$
$$= P(1 + i)^2$$

Similarly, the amount of money accumulated at the end of year 3, using Eq. (3.1) would be

$$F_3 = F_2 + F_2 i$$
$$= [P(1 + i) + P(1 + i)i] + [P(1 + i) + P(1 + i)i]i$$
$$= P(1 + i) + 2P(1 + i)i + P(1 + i)i^2$$

Factoring out $P(1 + i)$, we have

$$F_3 = P(1 + i)(1 + 2i + i^2)$$
$$= P(1 + i)(1 + i)^2$$
$$= P(1 + i)^3$$

From the preceding values, it is evident by mathematical induction that the formula can be generalized for n years as

$$F = P(1 + i)^n \qquad (3.2)$$

The expression $(1 + i)^n$ is called the *single-payment compound-amount factor* (SPCAF) and will yield the future amount F of an initial investment P after n years at interest rate i.

Expressing P from Eq. (3.2) in terms of F results in the expression

$$P = F \left[\frac{1}{(1 + i)^n} \right] \qquad (3.3)$$

The expression in brackets is known as the *single-payment present-worth factor* (SPPWF). This expression will allow determination of the present worth P of a given future amount F after n years at interest rate i. The cash-flow diagram for this formula is shown in Fig. 3.1. Conversely, if you were to use the SPCAF for the diagram in Fig. 3.1, you could find F, given P.

It is important to note that the two formulas derived here are *single-payment* formulas; that is, they are used to find the present or future amount when only one

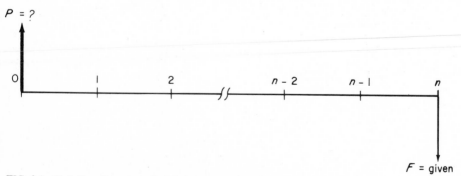

FIG. 3.1 Cash-flow diagram to find P given F using the SPPWF.

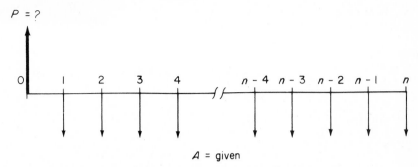

FIG. 3.2   Diagram used to determine the present worth of a uniform series.

payment or receipt is involved. In the following two sections, formulas are developed for calculating the present or future worth when several payments or receipts must be considered.

3.2   Derivation of the Uniform-Series Present-Worth Factor and the Capital-Recovery Factor

The present worth of the uniform series shown in Fig. 3.2 can be determined by considering each A value as a future worth F in the single-payment present-worth factor and then summing the present-worth values. The general formula is

$$P = A \left[\frac{1}{(1 + i)^1} \right] + A \left[\frac{1}{(1 + i)^2} \right] + A \left[\frac{1}{(1 + i)^3} \right]$$

$$+ \cdots + A \left[\frac{1}{(1 + i)^{n-1}} \right] + A \left[\frac{1}{(1 + i)^n} \right]$$

where the terms in brackets represent the SPPWF for years 1 through n, respectively. Factoring out A,

$$P = A \left[\frac{1}{(1 + i)^1} + \frac{1}{(1 + i)^2} + \frac{1}{(1 + i)^3} + \cdots + \frac{1}{(1 + i)^{n-1}} + \frac{1}{(1 + i)^n} \right] \qquad (3.4)$$

Equation (3.4) may be simplified by multiplying both sides by $1/(1 + i)$ to yield

$$\frac{P}{1 + i} = A \left[\frac{1}{(1 + i)^2} + \frac{1}{(1 + i)^3} + \frac{1}{(1 + i)^4} \right.$$

$$\left. + \cdots + \frac{1}{(1 + i)^n} + \frac{1}{(1 + i)^{n+1}} \right] \qquad (3.5)$$

Subtracting Eq. (3.4) from Eq. (3.5) yields

$$\frac{P}{1 + i} - P = A \left[-\frac{1}{(1 + i)^1} + \frac{1}{(1 + i)^{n+1}} \right]$$

Factoring out P and rearranging, we have

$$P\left(\frac{1}{1 + i}\right) - 1 = A \left[\frac{1}{(1 + i)^{n+1}} - \frac{1}{1 + i} \right]$$

Simplifying both sides of the equation yields

$$P\left(\frac{-i}{1 + i}\right) = A \frac{1}{1 + i} \left[\frac{1}{(1 + i)^n} - 1 \right]$$

Dividing by $-i/(1 + i)$ yields

$$P = A\left(\frac{1}{1 + i}\right) \frac{\{1/[(1 + i)^n]\} - 1}{-i/(1 + i)}$$

$$= A\left(\frac{1}{-i}\right) \left[\frac{1 - (1 + i)^n}{(1 + i)^n} \right]$$

$$= A \left[\frac{(1 + i)^n - 1}{i(1 + i)^n} \right] \tag{3.6}$$

The term in brackets is called the *uniform-series present-worth factor* (USPWF). This factor will give the present worth P of an equivalent uniform annual series A which begins *at the end of year 1* and extends for n years at an interest rate i.

By rearranging Eq. (3.6), A can be expressed in terms of P:

$$A = P \left[\frac{i(1 + i)^n}{(1 + i)^n - 1} \right] \tag{3.7}$$

The term in brackets is called the *capital-recovery factor* (CRF) and yields the equivalent uniform annual cost A over n years of a given investment P when the interest rate is i.

It is very important to commit to memory the fact that these formulas were derived with the present worth P and the first uniform annual-cost value A *one year apart*. That is, the present sum P *must always* be located one period *prior* to the first A. The correct use of these factors is illustrated in Sec. 3.6.

3.3 Derivation of the Uniform-Series Compound-Amount Factor and the Sinking-Fund Factor

While the *sinking-fund factor* (SFF) and the *uniform-series compound-amount factor* (USCAF) could be derived using the SPCAF, the simplest way to derive the formula is

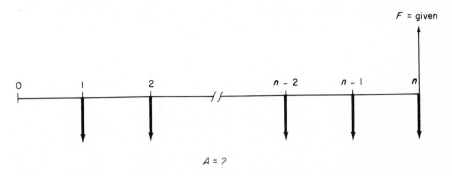

FIG. 3.3 Transformation of a given F value into an equivalent A series.

by substituting into those already developed. Thus, if P from Eq. (3.3), which uses the SPPWF, is substituted into Eq. (3.7), the following formula results:

$$A = F \left[\frac{1}{(1 + i)^n} \right] \left[\frac{i(1 + i)^n}{(1 + i)^n - 1} \right] = F \left[\frac{i}{(1 + i)^n - 1} \right] \qquad (3.8)$$

The expression in brackets in Eq. (3.8) is the SFF. Equation (3.8) is used to determine the uniform annual series that would be equivalent to a given future worth F. This is shown graphically in Fig. 3.3. Note that the uniform series A begins at the end of year 1 and continues through the year of the given F. This is unlike the uniform-series present-worth formulas in the preceding section, where the P and the first A were always one year apart.

Equation (3.8) can be rearranged to express F in terms of A:

$$F = A \left[\frac{(1 + i)^n - 1}{i} \right] \qquad (3.9)$$

The term in brackets is called the *uniform-series compound-amount factor* (USCAF) and when multiplied by the given uniform annual amount A yields the future worth of the uniform series. The cash-flow diagram for this case would be just the opposite of that shown in Fig. 3.3. Again, it is important to remember that the future amount F occurs in the same year as the last A.

Problem P3.1

3.4 Use of Interest Tables

To avoid the cumbersome task of writing out the formulas each time one of the factors is used, a standard notation has been adopted which represents the various factors. This standard notation, which also includes the interest rate and the number of periods, will always be of the general form $(X/Y,i\%,n)$. The first letter inside the

Table 3.1 STANDARD FACTOR NOTATIONS

Factor name	Standard notation
Single-payment present-worth (SPPWF)	$(P/F,i\%,n)$
Single-payment compound-amount (SPCAF)	$(F/P,i\%,n)$
Uniform-series present-worth (USPWF)	$(P/A,i\%,n)$
Capital-recovery (CRF)	$(A/P,i\%,n)$
Sinking-fund (SFF)	$(A/F,i\%,n)$
Uniform-series compound-amount (USCAF)	$(F/A,i\%,n)$

parentheses X represents what you "want to find," while the second letter (Y) represents what is "given." For example, F/P means "find F when given P." The i is the interest rate in percent and the n represents the number of periods involved. Thus, "$(F/P,6\%,20)$" means obtain the factor which when multiplied by a given P allows you to find the future amount of money F that will be accumulated in 20 periods if the interest rate is 6%.

The standard notation is simpler than factor names for identifying factors and will be used exclusively hereafter. Table 3.1 shows the standard notation for the formulas that were derived in the first three sections of this chapter.

For ready reference the formulas used in computations are collected in Table 3.2.

In order to simplify the routine engineering-economy calculations involving the above factors, tables of factor values have been prepared for interest rates from $\frac{1}{4}\%$ to 50% and time periods from 1 to 100 years. These tables, found in Appendix A and identified as Tables A-1 through A-24, are arranged with various factors across the top and the number of years n down the left and right column. Thus, for a given factor, interest rate, and time, the correct factor value would be found in the respective interest rate table at the intersection of the given factor and n. For example, the value of the factor $(P/A,5\%,10)$ is found in the P/A column of Table A-9 at year 10 as 7.7216. The value 7.7216 could, of course, have been computed using the mathematical expression for the USPWF as follows:

Table 3.2 COMPUTATIONS USING STANDARD NOTATION

To find	Given	Factor	Formula
P	F	$(P/F,i\%,n)$	$P = F(P/F,i\%,n)$
F	P	$(F/P,i\%,n)$	$F = P(F/P,i\%,n)$
P	A	$(P/A,i\%,n)$	$P = A(P/A,i\%,n)$
A	P	$(A/P,i\%,n)$	$A = P(A/P,i\%,n)$
A	F	$(A/F,i\%,n)$	$A = F(A/F,i\%,n)$
F	A	$(F/A,i\%,n)$	$F = A(F/A,i\%,n)$

Table 3.3 USE OF INTEREST TABLES

Standard notation	i	Years, n	Table	Factor value
$(F/A,10\%,3)$	10%	3	A-14	3.310
$(A/P,7\%,20)$	7%	20	A-11	0.09439
$(P/F,25\%,35)$	25%	35	A-19	0.0004

$$(P/A,5\%,10) = \frac{i(1 + i)^n}{(1 + i)^n - 1} = \frac{0.05(1 + 0.05)^{10}}{(1 + 0.05)^{10} - 1} = 7.7216$$

Table 3.3 lists several examples of the use of the interest tables.

Throughout this chapter we will assume the n value is given in years. In Chap. 4 you will learn how to use the tables in Appendixes A and B for interest periods other than a year.

Problem P3.2

3.5 Interpolation in Interest Tables

Sometimes it is necessary to locate a factor value for an interest rate i or year n that is not in the interest tables. When this occurs, it is necessary to interpolate between the tabulated values on both sides of the unlisted, desired value. *Linear* interpolation is considered sufficient as long as the values of i or n are not too far distant from each other. The first step in linear interpolation is to set up the known and the unknown values as shown in Table 3.4. A ratio equation is then set up and solved for c, as follows (refer to Table 3.4):

$$\frac{a}{b} = \frac{c}{d} \qquad c = \frac{a}{b}d \qquad (3.10)$$

where a, b, c, and d represent the differences between the numbers shown in the interest tables. The value of c from Eq. (3.10) is added to or subtracted from value no. 1, depending on whether the factor is increasing or decreasing in value, respectively. The following examples illustrate the procedure just described.

Table 3.4 LINEAR INTERPOLATION SET-UP

		i or n	Factor	
		tabulated	value no. 1	
b	a	desired	unlisted	c d
		tabulated	value no. 2	

Example 3.1   Determine the value of the A/P factor for an interest rate of 7.3% and n of 10 years, that is, $(A/P,7.3\%,10)$.

SOLUTION   The values of the A/P factor for interest rates of 7% and 8% are listed in Tables A-11 and A-12, respectively. Thus we have the following situation:

$$
b\left[\,{}^{a}\left[\begin{array}{l}\text{7\%}\\ \text{7.3\%}\\ \text{8\%}\end{array}\right.\right.
\qquad
\begin{array}{l}0.14238\\ X\\ 0.14903\end{array}\left.{}^{c}\right]\,d
$$

The unknown X is the desired factor value. From Eq. (3.10)

$$
c = \frac{a}{b}\,d
$$

$$
= \frac{7.3 - 7}{8 - 7}\,(0.14903 - 0.14238)
$$

$$
= \frac{0.3}{1}\,0.00665
$$

$$
= 0.00199
$$

Since the factor is increasing in value as the interest rate increases from 7% to 8%, the value of c must be *added* to the value of the 7% factor. Thus,

$$
X = 0.14238 + 0.00199 = 0.14437
$$

COMMENT   It is good practice to check the "reasonableness" of your final answer by verifying that X lies between the values of the known factors used in the interpolation in approximately the correct proportions. In this case, since 0.14437 is less than 0.5 of the distance between 0.14238 and 0.14903, the answer seems reasonable. A simpler procedure in some cases may be to use the formula to compute the factor value directly, rather than interpolating.   ////

Example 3.2   Find the value of the $(P/F,4\%,48)$ factor.

SOLUTION   From Table A-8 for 4% interest, the values of the P/F factor for 45 and 50 years can be found as follows:

$$
b\left[\,{}^{a}\left[\begin{array}{l}\text{45}\\ \text{48}\\ \text{50}\end{array}\right.\right.
\qquad
\begin{array}{l}0.1712\\ X\\ 0.1407\end{array}\left.{}^{c}\right]\,d
$$

Again, from Eq. (3.10)

$$
c = \frac{a}{b}\,d = \frac{48 - 45}{50 - 45}\,(0.1712 - 0.1407) = 0.0183
$$

Since the value of the factor decreases as n increases, the value of c must be *subtracted* from the value for $n = 45$. Thus,

$$X = 0.1712 - 0.0183 = 0.1529$$

COMMENT The calculated value lies approximately midway between the two given values and, therefore, seems reasonable. ////

Example 3.11
Problems P3.3–P3.5

3.6 Present-Worth, Future-Worth, and Equivalent Uniform Annual Series Calculations

The first and probably most important step in solving engineering-economy problems is construction of a cash-flow diagram. In addition to more clearly illustrating "the problem," the cash-flow diagram immediately shows which formulas should be used and whether the conditions of cash flow presented allow straightforward application of the formulas as derived in the preceding sections. Obviously, the formulas can be used only when the cash flow of the problem conforms exactly to the cash-flow diagram for the formulas listed in Table 3.2. For example, the uniform-series factors could not be used if payments or receipts occurred *every other year* instead of every year. It is very important, therefore, to remember the conditions for which the formulas apply. The correct use of the formulas for finding P, F, or A is illustrated in the examples that follow. All equations used are taken from Table 3.2. See the Solved Examples for cases in which some of these formulas cannot be applied.

Example 3.3 If a woman deposits $600 now, $300 two years from now, and $400 five years from now, how much will she have in her account ten years from now if the interest rate is 5%?

SOLUTION The first step is to draw the cash-flow diagram (Fig. 3.4), which indicates that an F value is to be computed. Since each value is different and they do

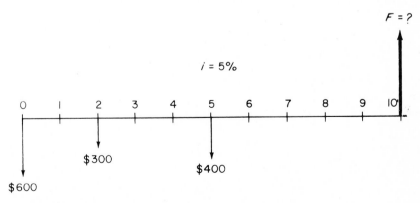

FIG. 3.4 Diagram for a future value, Example 3.3.

not take place each year, the future worth F is equal to the sum of the individual single payments at year 10. Thus,

$$F = 600(F/P,5\%,10) + 300(F/P,5\%,8) + 400(F/P,5\%,5)$$
$$= 600(1.6289) + 300(1.4774) + 400(1.2763)$$
$$= \$1,931.08$$

COMMENT The problem could also be solved by finding the present worth in year zero of the $300 and $400 deposits using the P/F factors and then finding the future worth of the total.

$$P = 600 + 300(P/F,5\%,2) + 400(P/F,5\%,5)$$
$$= 600 + 300(0.9070) + 400(0.7835)$$
$$= \$1,185.50$$

Then,

$$F = 1,185.50(F/P,5\%,10)$$
$$= 1,185.50(1.6289)$$
$$= \$1,931.06$$

It should be obvious that there are a number of ways the problem could be worked, since any year could be used to find the equivalent total of the deposits before finding the value at year 10. As an exercise, you should work the problem using year 2 and year 5 for finding the equivalent total before determining the final amount in year 10. All answers should be the same, except for round-off error.　////

Example 3.4 How much money would a man have in his account after eight years if he deposited $100 per year for eight years at 4% starting one year from now?

SOLUTION The cash-flow diagram is shown in Fig. 3.5. Since the payments start at the end of year 1 and end in the year the future worth is desired, the F/A formula can be used. Thus,

$$F = 100(F/A,4\%,8) = 100(9.214) = \$921.40$$

COMMENT While the problem could be solved using F/P factors as in Example 3.3, it is much simpler to use the uniform-series formulas when possible.　////

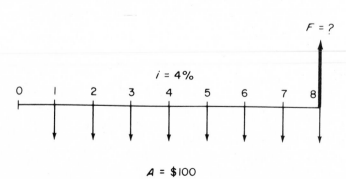

FIG. 3.5 Diagram to find F for a uniform series, Example 3.4.

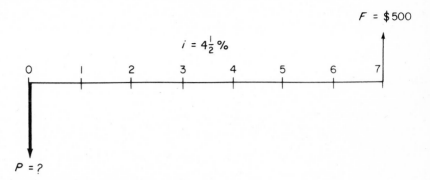

FIG. 3.6 Diagram for Example 3.5.

Example 3.5 How much money would you be willing to spend now in order to avoid spending $500 seven years from now if the interest rate is $4\frac{1}{2}\%$?

SOLUTION The cash-flow diagram appears in Fig. 3.6. The problem might be easier if it were stated in another manner, such as, what is the present worth of $500 seven years from now if the interest rate is $4\frac{1}{2}\%$; or, what present amount would be equivalent to $500 seven years hence if the interest is $4\frac{1}{2}\%$; or, what initial investment is equivalent to spending $500 seven years from now at an interest rate of $4\frac{1}{2}\%$? In all cases F is given and P is to be computed.

$$P = 500(P/F, 4.5\%, 7) = 500(0.7353) = \$367.65$$

COMMENT Although there are several ways to state the same problem, the cash-flow diagram remains the same in each case. ////

Example 3.6 How much money would you be willing to pay now for a note that will yield $600 per year for nine years starting next year if the interest rate is 7%?

SOLUTION The cash-flow diagram is shown in Fig. 3.7. Since the cash-flow diagram fits the P/A uniform-series formula, the problem can be solved directly.

$$P = 600(P/A, 7\%, 9) = 600(6.5152) = \$3,909.12$$

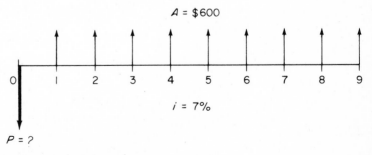

FIG. 3.7 Diagram for Example 3.6.

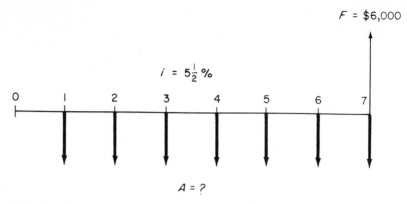

FIG. 3.8   Diagram for Example 3.7.

COMMENT   Again you should recognize that P/F factors could be used for each of the nine years and the resulting present worths added to get the correct answer. Another way would be to find the future worth F of the $600 payments and then find the present worth of the F value. As pointed out in previous examples, there are many ways to solve an engineering-economy problem. Only the most direct method will be presented here, but you should work the problems in at least one other way to become more familiar with the use of the formulas.   ////

Example 3.7   How much money must a person deposit every year starting one year from now at $5\frac{1}{2}\%$ per year in order to accumulate $6,000 seven years from now?

SOLUTION   The cash-flow diagram is shown in Fig. 3.8. The cash-flow diagram fits the A/F formula as derived. Thus,

$$A = 6,000(A/F,5.5\%,7) = 6,000(0.12098) = \$725.88 \text{ per year}$$

COMMENT   Work the problem another way by first finding the present worth of the $6,000 and then using the A/P factor.   ////

Example 3.12
Problems P3.6–P3.20

3.7   Calculation of Unknown Interest Rates

In some cases, the amount of money invested and the amount of money received after a specified number of years are known, and it is desired to determine the interest rate or rate of return. When only a single payment and a single receipt, or a uniform series of payments or receipts are involved, the unknown interest rate can be determined by direct solution of the economy equation. When nonuniform payments or several factors are involved, however, the problem must be solved by the trial-and-error method. In this section, only single-payment or uniform-series cash-flow problems are considered. The more complicated trial-and-error problems are deferred until Chap. 10, which deals with rate-of-return analysis.

Although the single-payment and uniform-series formulas can be rearranged and expressed in terms of i, it is generally simpler to *solve for the value of the factor* and then look up the interest rate in the interest tables. This method is illustrated in the examples that follow.

Example 3.8

(*a*) If a person can make a business investment requiring an expenditure of $3,000 now in order to receive $5,000 five years from now, what would be the rate of return on the investment?

(*b*) If the same person can receive 7% interest from certificates of deposit, which investment should be made?

SOLUTION

(*a*) The cash-flow diagram is shown in Fig. 3.9. The interest rate can be found by setting up the P/F or F/P equations and solving directly for the factor value. Using P/F,

$$P = F(P/F,i\%,n)$$
$$3,000 = 5,000(P/F,i\%,5)$$
$$(P/F,i\%,5) = \frac{3,000}{5,000} = 0.6000$$

From the interest tables, a P/F factor of 0.6000 for n equal to 5 lies between 10% and 12%. Interpolating between these two values using Eq. (3.10), we have

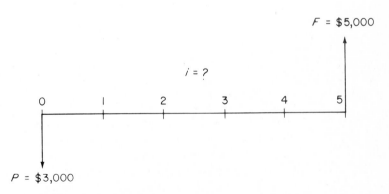

FIG. 3.9 Diagram used to determine the rate of return, Example 3.8.

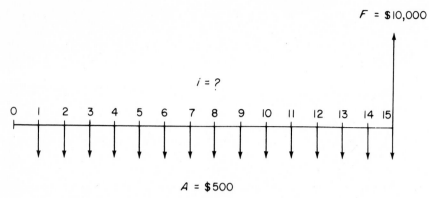

FIG. 3.10   Diagram to determine the rate of return, Example 3.9.

$$c = \left(\frac{0.6209 - 0.6000}{0.6209 - 0.5674}\right)(12 - 10) = \frac{0.0209}{0.0535}(2) = 0.7813$$

Therefore,

$$i = 10 + 0.78 = 10.78\%$$

(*b*) Since 10.78% is greater than the 7% available in certificates of deposit, the person should make the business investment.

COMMENT Since the higher rate of return would be received on the business investment, the investor would probably select this option instead of the certificates of deposit. However, the degree of risk associated with the business investment was not specified. Obviously, the amount of risk associated with a particular investment is an important parameter and oftentimes causes selection of the lower-rate-of-return investment. Unless specified to the contrary, the problems in this text will assume equal risks for all alternatives.   ////

Example 3.9 A couple wishing to save money for their child's education purchased an insurance policy that will yield $10,000 fifteen years from now. The parents must pay $500 per year for the 15 years starting one year from now. What will be the rate of return on their investments?

SOLUTION The cash-flow diagram is shown in Fig. 3.10. Either the A/F or F/A factor could be used. Using A/F,

$$A = F(A/F,i\%,n)$$
$$500 = 10,000(A/F,i\%,15)$$
$$(A/F,i\%,15) = 0.0500$$

From the interest tables under the A/F column for 15 years, the value 0.0500 is found to lie between 3% and 4%. By interpolation,

$$i = 3.98\%$$

COMMENT You should rework this problem using the F/A factor to satisfy yourself that it is immaterial which ratio you use. ////

Problems P3.21–P3.27

3.8 Calculation of Unknown Years

In breakeven economic analysis, it is sometimes necessary to determine the number of years required before an investment "pays off." Other times it is desirable to be able to determine when given amounts of money will be available from a proposed investment. In these cases, the unknown value is n, and techniques similar to those of the preceding section on unknown interest rates can be used to find n.

Though these problems can be solved directly for n by proper manipulation of the single-payment and uniform-series formulas, it is generally easier to solve for the factor value and interpolate in the interest tables, as illustrated below.

Example 3.10 How long would it take for $1,000 to double if the interest rate is 5%?

SOLUTION The cash-flow diagram is shown in Fig. 3.11. The problem can be solved using either the F/P or P/F factor. Using the P/F factor,

$$P = F(P/F,i\%,n)$$
$$1,000 = 2,000\,(P/F,5\%,n)$$
$$(P/F,5\%,n) = 0.500$$

From the 5% interest table, the value 0.500 under the P/F column lies between 14 and 15 years. By interpolation, $n = 14.2$ years.

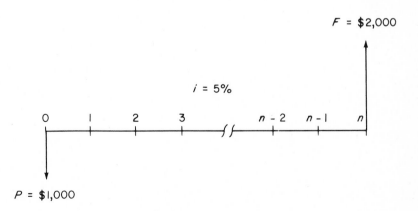

FIG. 3.11 Diagram to determine an n value, Example 3.10.

COMMENT Problems of this type become more complicated when two or more nonuniform payments are involved. See the Solved Examples for an illustration using trial and error. ////

Example 3.13
Problems P3.28–P3.32

SOLVED EXAMPLES

Example 3.11 A new building has been purchased by Waldorf Concession Stands, Inc. The present worth of future maintenance costs is to be calculated with a P/A factor. If $i = 13\%$ per year and the life is expected to be 42 years, what factor value is correct?

SOLUTION This P/A factor requires two-way interpolation, for i and n. First, we find the P/A factor for $i = 13\%$ at $n = 40$ and $n = 45$.

i	$n = 40$		$n = 45$	
12%	8.2438		8.2825	
13%	X_{40}	c_{40} 1.6020	X_{45}	c_{45} 1.6282
15%	6.6418		6.6543	

The subscripts correspond to the n value for which the factor is computed.

$$c_{40} = 1/3(1.6020) = 0.5340 \qquad X_{40} = 8.2438 - 0.5340 = 7.7098$$

$$c_{45} = 1/3(1.6282) = 0.5427 \qquad X_{45} = 8.2825 - 0.5427 = 7.7398$$

Now compute the P/A factor for $n = 42$.

P/A	n
7.7098	40
X_{42}	42
7.7398	45

$$X_{42} = 7.7098 + 2/5(0.0300) = 7.7218$$

Thus, we have

$$(P/A, 13\%, 42) = 7.7218$$

(Often an "eyeball" factor is good enough.) ////

Sec. 3.5

Example 3.12 Explain why the uniform-series factors *cannot* be used to compute P or F *directly* for the cash flows of Fig. 3.12.

SOLUTION

(*a*) The P/A factor cannot be used to compute P since the $100-per-year receipt does not occur each year from year 1 through year 5.

(*b*) Since there is no $A = \$550$ in year 5, the F/A factor cannot be used. The relation $F = 550(F/A,i\%,4)$ would furnish the future worth in year 4, not year 5 as desired.

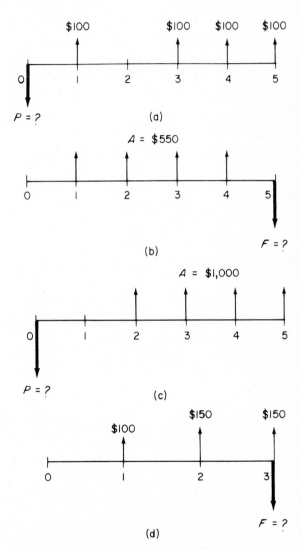

FIG. 3.12 Cash-flow diagrams, Example 3.12.

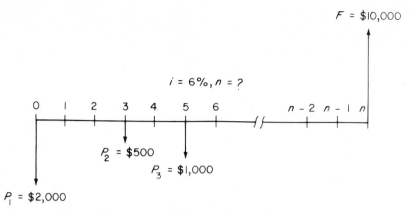

FIG. 3.13   Diagram to determine n for a nonuniform series, Example 3.13.

(c) The first $A = \$1,000$ value occurs in year 2. Use of the relation $P = 1,000(P/A,i\%,4)$ will compute P in year 1, not year zero.

(d) The receipt values are unequal; thus the relation $F = A(F/A,i\%,3)$ cannot be used to compute F.

COMMENT   Naturally, there are ways to compute P or F without resorting to only P/F and F/P factors; these methods are discussed in Chap. 5.   ////

Sec. 3.6

Example 3.13   If an investor deposits $2,000 now, $500 three years from now, and $1,000 five years from now, how many years will it take from now for his total investment to amount to $10,000 if the interest rate is 6%?

SOLUTION   The cash-flow diagram (Fig. 3.13) requires that the following equation be satisfied:

$$F = P_1(F/P,i\%,n) + P_2(F/P,i\%,n - 3) + P_3(F/P,i\%,n - 5)$$
$$10,000 = 2,000(F/P,6\%,n) + 500(F/P,6\%,n - 3)$$
$$+ 1,000(F/P,6\%,n - 5)$$

This relation must be solved by selecting various values of n and solving until the equation is satisfied. Interpolation for n will be necessary to obtain an exact equality.

Table 3.5   TRIAL-AND-ERROR SOLUTION FOR n

n	$2,000(F/P,6\%,n)$	$500(F/P,6\%,n\text{-}3)$	$1,000(F/P,6\%,n\text{-}5)$	F	Comment
5	$2,676.40	$ 561.80	$1,000.00	$ 4,238.20	Too small
15	4,793.00	1,006.10	1,790.80	7,589.90	Too small
20	6,414.20	1,346.35	2,396.50	10,157.05	Too large

The procedure is shown in Table 3.5, which indicates that 20 years is too long and 15 years is too short. Therefore, we interpolate between 15 and 20 years, as below.

F	Years
7,589.90	15
10,000.00	n
10,157.05	20

$$c = \frac{10,000 - 7,589.90}{10,157.05 - 7,589.90} (20 - 15) = 4.69$$

$$n = 15 + c$$
$$= 19.69 \text{ years}$$

COMMENT The final answer is close to 20 years, as it should be since the F value calculated for $n = 20$ in Table 3.5 is close to the desired value of $10,000.

////

Sec. 3.8

BIBLIOGRAPHY

Barish, pp. 51–56.
DeGarmo and Canada, pp. 117–124.
Emerson and Taylor, pp. 2-7 – 2-13, 3-5 – 3-10.
Fabrycky and Thuesen, pp. 76–79, 98–105.
Grant and Ireson, pp. 34–38, 46–54.
Ostwald, pp. 319–324.
Park, pp. 19–25.
Reisman, pp. 11–15, 20–22.
Riggs, pp. 166–171.
Smith, pp. 43–47, 64–70.
Taylor, pp. 28–36, 52–56.
Thuesen, Fabrycky, and Thuesen, pp. 57–60, 63–65, 82–90.

PROBLEMS

P3.1 Construct the cash-flow diagrams and derive the SPPWF, USPWF, and USCAF formulas for beginning-of-year amounts, rather than the end-of-year convention. The P value should take place at the same time as for the end-of-year convention.

P3.2 Find the correct numerical value for the following factors from the interest tables:

(a) $(A/F, 6\%, 20)$

(b) $(F/P, 8\%, 5)$

(c) $(P/A, 20\%, 8)$

(d) $(A/P, 10\%, 25)$

P3.3 Linearly interpolate to find the value of the following factors:
  (a) $(P/A,8.5\%,13)$
  (b) $(F/A,37\%,24)$
  (c) $(P/F,7.7\%,9)$
  (d) $(A/F,49\%,28)$

P3.4 Linearly interpolate in order to find the value of the following factors:
  (a) $(F/P,3\%,39)$
  (b) $(A/P,10\%,9.8)$
  (c) $(A/F,6\%,52)$
  (d) $(P/F,18\%,37)$

P3.5 Linearly interpolate to find the value of the following factors:
  (a) $(P/F,3.8\%,7.7)$
  (b) $(P/A,9.7\%,68)$
  (c) $(F/A,23\%,11.6)$
  (d) $(A/F,17\%,23)$

P3.6 How much money would Mr. Jones have in his bank account in 12 years if he deposits $3,500 now at an interest rate of 7%?

P3.7 If Ms. James wants to have $8,000 in her account eight years from now to buy a new sports car, how much money would she have to deposit every year starting one year from now if the interest rate is 6%?

P3.8 What is the present worth of $700 now, $1,500 four years from now, and $900 six years from now at an interest rate of $5\frac{1}{2}\%$?

P3.9 How much money would be accumulated in 14 years if $1,290 were deposited each year starting one year from now at an interest rate of 5%?

P3.10 If Mr. Savum borrowed $4,500 with a promise to make ten equal annual payments starting one year from now, how much would his payments be if the interest rate were 6%?

P3.11 How much money must be deposited in a lump sum four years from now in order to accumulate $20,000 eighteen years from now if the interest rate is 8%?

P3.12 Ms. Lendup would like to know the present worth of a 35-year, $600-per-year annuity at an interest rate of $6\frac{1}{2}\%$.

P3.13 How much money can you borrow now if you promise to pay $600 per year for seven years at an interest rate of 7%?

P3.14 How much money now would be equivalent to $5,000 six years from now at an interest rate of 7%?

P3.15 What uniform annual amount would you have to deposit for five years to have an equivalent present-investment sum of $9,000 at an interest rate of 10%?

P3.16 How much money would be accumulated in 25 years if $800 were deposited one year from now, $2,400 six years from now, and $3,300 eight years from now, all at an interest rate of 8%?

P3.17 What is the future worth of a uniform annual series of $1,000 for ten years at an interest rate of $8\frac{3}{4}\%$?

P3.18 What is the present worth of $600 per year for 52 years at an interest rate of 10%?

P3.19 How much money would be accumulated in 43 years from an annual deposit of $1,200 per year starting one year from now if the interest rate is $9\frac{1}{4}\%$?

P3.20 I plan to buy some property which my uncle has generously offered to me. The payment scheme is $700 every other year for eight years starting two years from now. What is the present worth of this generous offering if the interest rate is 7%?

P3.21 If a person invests $3,000 now and will receive $5,000 twelve years from now, what is the rate of return?

P3.22 The Playmore Company, a group of investors, is considering the attractiveness of purchasing a piece of property for $18,000. The group anticipates that the value of the property will increase to $21,500 in five years. What is the rate of return on this investment?

P3.23 The Juicer Utility Company has a retirement program in which employees invest $1,200 every year for 25 years, starting one year after their initial employment. If the company guarantees at least $50,000 at the time of retirement, what is the rate of return on the investment?

P3.24 If a person borrows $6,000 for a new car with an agreement to pay the loan company $2,500 per year for three years, what is the interest rate on the loan?

P3.25 At what interest rate would $1,500 accumulate to $2,500 in five years?

P3.26 If a person purchased stock for $8,000 12 years ago and received dividends of $1,000 per year, what is the rate of return on the investment?

P3.27 At what rate of return would $900 per year for nine years accumulate to (*a*) $8,100? (*b*) $10,000?

P3.28 How many years would it take for $1,750 to triple in value if the interest rate is 12%?

P3.29 If a person deposits $5,000 now at 8% interest and plans to withdraw $500 per year every year starting one year from now, how long can the full withdrawals be made?

P3.30 What is the minimum number of years a person must deposit $400 per year in order to have at least $10,000 on the date of the last deposit? Use an interest rate of 8% and round off to the higher integer year.

P3.31 What is the minimum number of year-end deposits that have to be made before the total value of the deposits is at least ten times greater than the value of a single year-end deposit if the interest rate is $12\frac{1}{2}\%$?

P3.32 How many years would it take for a $800 deposit now and a $1,600 deposit three years from now to accumulate to $3,500 at an interest rate of 8%?

4

NOMINAL AND EFFECTIVE INTEREST RATES

This chapter teaches you how to make engineering-economy computations using interest periods other than one year. The material of this chapter is often helpful for handling personal financial matters.

CRITERIA

To complete this chapter you must be able to:

1 Define *compounding period*, *payment period*, *nominal interest rate*, and *effective interest rate*.
2 Write the formula for computing the effective interest rate and define each term.
3 Compute the effective interest rate and find the numerical value of any specific engineering-economy factor for that rate, given the nominal interest rate and number of compounding periods.
4 Calculate the present worth or future worth of a specified cash flow when the compounding period is *shorter* than the payment period, given the amount and times of the payments, the compounding period, and the nominal interest rate.

5 Calculate the present worth or future worth of a specified cash flow when the compounding period is *longer* than the payment, given the amount and times of the payments, the compounding period, and the nominal interest rate.

EXPLANATION OF MATERIAL

4.1 Nominal and Effective Rates

In Chap. 1, the concepts of simple and compound interest rates were introduced; the basic difference between the two is that compound interest includes interest on the interest earned in the previous year. In essence, nominal and effective interest rates have the same relationship as simple and compound interest; the difference is that nominal and effective interest rates are used when the *compounding period* (or interest period) is less than one year. Thus, when an interest rate is expressed over a period of time shorter than a year, such as 1% per month, the terms *nominal* and *effective* interest rates must be considered. Specifically, the *nominal interest rate* is defined as the period interest rate multiplied by the number of periods per year. Thus a period interest rate listed as 1% per month could also be expressed as a *nominal* 12% per year (i.e., 1% per month × 12 months per year). Obviously, the calculation for the nominal interest rate ignores the time value of money, similar to the calculation of simple interest. When the time value of money is taken into consideration in calculating annual interest rates from period interest rates, the annual rate is called the *effective interest rate*.

In addition to considering the interest or compounding period, it is also necessary to consider the frequency of the payments or receipts within the one-year interval. For simplicity, the frequency of the payments or receipts is known as the *payment period*. It is important to distinguish between the compounding period and the payment period because in many instances the two do not coincide. For example, if a person deposits money each month into a savings account that pays a nominal interest rate of 6% per year compounded semiannually, the payment period would be one month while the compounding period would be six months. Similarly, if a person deposits money each year into a savings account which compounds interest quarterly, the payment period is one year, while the compounding period is three months. If the payment and compounding periods are the same, the rate is expressed as in preceding chapters (e.g., 1% per month, where the compounding period is a month and payments are to be made at the end of each month).

Problems P4.1, P4.2

4.2 Effective Interest-Rate Formulation

To illustrate the difference between nominal and effective interest rates, the future worth of $100 after one year is determined using both rates. If a bank pays 8% interest

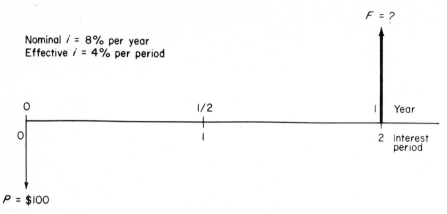

FIG. 4.1 Cash-flow diagram for semiannual compounding.

compounded semiannually, the future worth of $100 using a nominal interest rate of 8% per year is

$$F = P(1 + i)^n = 100(1.08)^1 = \$108.00 \qquad (4.1)$$

On the other hand, if the interest is compounded semiannually, the future worth must include the *interest on the interest earned in the first period*. An interest rate of 8% per year compounded semiannually means that the bank will pay 4% interest two times per year (i.e., every six months). Figure 4.1 is the cash-flow diagram for semiannual compounding for a nominal interest rate of 8% per year. Equation (4.1) obviously ignores the interest earned in the first period. Taking this into consideration, therefore, the future worth of the $100 would actually be

$$F = 100(1 + 0.04)^2 = 100(1.04)^2$$
$$= 100(1.0816) \qquad (4.2)$$
or $$F = \$108.16$$

where 4% is the *effective semiannual interest rate* found by taking 8%/2 = 4%, since there are two compounding periods per payment period. The effective annual interest rate, therefore, would be 8.16%, instead of 8%, since $8.16 interest would be earned. The equation for obtaining the effective interest rate from the nominal interest may be obtained by recognizing that the first term in parentheses in Eq. (4.2) is equal to (1 + i) when the interest period is one year. Setting this term equal to the generalized form of the term in parentheses in Eq. (4.1) yields

$$(1 + i) = \left(1 + \frac{r}{t}\right)^t$$

$$i = \left(1 + \frac{r}{t}\right)^t - 1 \qquad (4.3)$$

where $i =$ effective period interest rate
$\quad r =$ nominal interest rate
$\quad t =$ number of compounding periods

Equation (4.3) is referred to as the effective-interest-rate equation. As the number of compounding periods increases, t approaches infinity, in which case the equation represents the interest rate for *continuous compounding*. See Appendix B for a discussion of this subject.

Problem P4.3

4.3   Calculation of Effective Interest Rates

For an annual interest rate that is compounded over a specified number of interest periods, the effective interest rate can be calculated by substituting into Eq. (4.3), developed in the preceding section. After the effective rate has been determined, the factors in the interest tables of Appendix A can be used in the normal manner (Chap. 3). As is shown in the next section, this is only one of two methods that could be used when the interest period is shorter than the payment period. Since the annual effective interest rate is usually found to be a rate for which no tables are prepared, it is necessary to interpolate between interest rates to determine the factor value.

Example 4.1   A university credit union advertises that its interest rate on loans is 1% per month. Calculate the annual effective interest rate and find the corresponding P/F factor for $n = 8$.

SOLUTION   Substituting into Eq. (4.3) yields

$$i = (1 + 0.01)^{12} - 1$$
$$= 1.1268 - 1$$
$$= 0.1268 \quad (12.68\%)$$

In order to find the P/F factor, it is necessary to interpolate between $i = 12\%$ and $i = 15\%$. Thus,

$$c = \frac{0.68}{3}(0.0770) = 0.0175$$

Then,

$$(P/F, 12.68\%, 8) = 0.4039 - 0.0175 = 0.3864$$

The term r/t is equal to 0.01 in the effective-interest-rate equation because the interest rate stated in the problem is already expressed as a period interest rate, rather than a nominal rate. You might find it easier to think of the interest rate of 1% per month as being the same as an annual nominal rate of 12% per year compounded monthly, in which case Eq. (4.3) would be

$$i = \left(1 + \frac{0.12}{12}\right)^{12} - 1 = 0.1268 \quad (12.68\%)$$

COMMENT   It is important for you to remember that 1% per month is already an effective interest rate, if you consider the interest period as one month. In this case, the P/F factor desired above would be represented as $(P/F,1\%,96)$. This is discussed more thoroughly in the next section. Appendix Table B-1 is a tabulation of Eq. (4.3) for various nominal interest rates (r) and compounding frequencies from semiannual to continuous.   ////

Example 4.6
Problems P4.4–P4.9

4.4   Calculations for Compounding Periods Shorter than Payment Periods

When the compounding period of an investment (or loan) does not coincide with the payment period, it becomes necessary to manipulate the interest rate and/or payment period in order to determine the correct amount of money accumulated or paid at various times. Remember that if the payment and compound periods do not agree, the interest tables cannot be used until appropriate corrections are made. In this section, we consider the situation where the payment periods are equal to or less frequent than the compounding periods. The two conditions that occur, as discussed below, are

1   The cash flows require the use of the single-payment factors $(P/F,F/P)$.
2   The cash flows require the use of the uniform-series factors.

4.4.1   **Single-payment factors**   There are two correct procedures that can be used when only single factors are involved. The first is the procedure that was illustrated in Example 4.1, where the effective interest rate coinciding with the payment period or one-year period (whichever comes first) is found from Eq. (4.3) or Table B-1, and the interest tables are used for the appropriate number of periods or years. The second procedure is to *divide* the nominal interest rate r by the number of compounding periods per year t and *multiply* the number of years by the number of compounding periods per year. Using standard notation, the single-payment equations would appear as

$$P = F\left(P/F, \frac{r}{t} \%, tn\right) \qquad (4.4)$$

$$F = P\left(F/P, \frac{r}{t} \%, tn\right) \qquad (4.5)$$

where n is the number of years and r and t are as defined in Sec. (4.2). Example 4.2 demonstrates both of these procedures.

Example 4.2 If a man deposits $1,000 now, $3,000 four years from now, and $1,500 six years from now at an interest rate of 6% per year compounded semiannually, how much money will he have in his account ten years from now?

SOLUTION The cash-flow diagram is shown in Fig. 4.2. According to the first procedure specified above, the effective interest per year should be calculated and then used to find F in year 10. Thus, from Table B-1, with $n = 6\%$ and semiannual compounding, effective $i = 6.09\%$; or by Eq. (4.3),

$$i = \left(1 + \frac{0.06}{2}\right)^2 - 1 = (1.03)^2 - 1 = 0.0609 \quad (6.09\%)$$

Then,

$$F = 1,000(F/P,6.09\%,10) + 3,000(F/P,6.09\%,6)$$
$$+ 1,500(F/P,6.09\%,4)$$
$$= \$7,983.70$$

By the second procedure, Eq. (4.5) yields

$$F = 1,000\left[F/P, \frac{6}{2}\%, 2(10)\right] + 3,000\left[F/P, \frac{6}{2}\%, 2(6)\right]$$

$$+ 1,500\left[F/P, \frac{6}{2}\%, 2(4)\right]$$

$$= 1,000(F/P,3\%,20) + 3,000(F/P,3\%,12) + 1,500(F/P,3\%,8)$$

$$= \$7,983.70$$

COMMENT The second procedure is the easier of the two since it usually does not require interpolation. ////

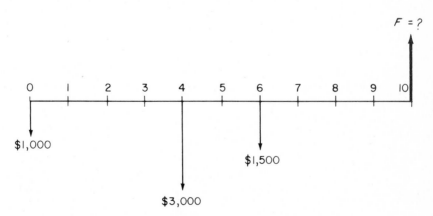

FIG. 4.2 Cash-flow diagram, Example 4.2.

Table 4.1 FINDING i AND n FOR COMPOUNDING PERIOD EQUAL TO PAYMENT PERIOD

Payment scheme A	Nominal interest rate r	$A(P/A,i\%,n)$
$500 semiannually for 5 years	6% per year compounded semiannually	$500(P/A,3\%,10)$
$ 70 monthly for 3 years	12% per year compounded monthly	$75(P/A,1\%,36)$
$180 quarterly for 15 years	8% per year compounded quarterly	$180(P/A,2\%,60)$

4.4.2 Uniform-series factors When the cash flow of the problem dictates the use of one or more of the uniform-series factors, the only procedure that can be used is to *express the effective interest rate over the same time period as the payments*. When a nominal interest rate is given and the *compounding period is equal to the payment period* (for example, *semiannual* payments at an interest rate of 8% per year compounded *semiannually*), then, as for the single-payment factor, simply divide the interest rate by the number of compounding periods per year and set n equal to the total number of payments. For example, the P/A factor would appear as $(P/A,r/t\%,tn)$. The procedure is illustrated in Table 4.1, where the standard notation for the P/A factor is shown for three payment schemes and interest rates.

When the compounding period is more frequent than the payment period (such as *semiannual* payments at an interest rate of 8% per year compounded *quarterly*), the procedure is a little more difficult, since the effective interest rate per Eq. (4.3) must be used. This procedure is illustrated in Example 4.3.

Example 4.3 If a woman deposits $500 every six months for seven years, how much money will she have in her account after she makes her last deposit if the interest rate is 8% per year compounded quarterly?

SOLUTION The cash-flow diagram is shown in Fig. 4.3. Since interest is compounded quarterly, the effective interest rate per payment period must be determined (i.e., the effective semiannual rate). The effective semiannual rate can be

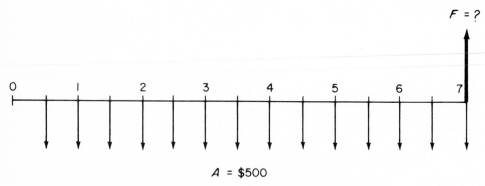

FIG. 4.3 Diagram of semiannual deposits used to determine F, Example 4.3.

obtained by finding the quarterly interest rate and "plugging" it into the effective interest rate Eq. (4.3). Here $r/t = 0.08/4 = 0.02$, and since there are two quarters in one semiannual period,

$$i = (1 + 0.02)^2 - 1 = 0.0404 \quad (4.04\%)$$

Alternatively, the effective interest rate could have been obtained from Table B-1 by using the r value of 4% with "semiannual" compounding to get $i = 4.04\%$.

The value $i = 4.04\%$ seems reasonable since we would expect the effective rate to be slightly higher than the nominal rate of 4% per six-month period. The effective rate can now be used in the F/A formula to find the future worth of the semiannual deposits, where $n = 2(7) = 14$ periods. Thus,

$$
\begin{aligned}
F &= A(F/A,4.04\%,14) \\
&= 500(18.344) \\
&= \$9,172.12 \quad\quad\quad (4.6)
\end{aligned}
$$

COMMENT   It is important to note that the effective interest *per payment period* (six months) was used for i and that the *number of payment periods* was used for n.

////

The procedure to be followed when the compounding period is equal to or more frequent (i.e., shorter) than the payment period can be summarized as follows:

1 Determine whether the single-payment formulas or the uniform-series formulas must be used.
2 If the single-payment formulas are required, use one of the methods specified in Sec. 4.4.1 above for single-payment factors.
3 If uniform-series formulas are required, use one of the methods specified in Sec. 4.4.2 for uniform payments, depending on whether the compounding period is equal to or shorter than the payment period.
4 Use the factor obtained in step *2* or *3* to find the desired P, F, or A.

Example 4.7, 4.8
Problems P4.10–P4.23

4.5 Calculations for Compounding Periods Longer than Payment Periods

When the compounding period is less frequent than the payment period, there are three possible ways to calculate the future amount or present worth, depending on the conditions specified (or assumed) regarding the interperiod compounding. *Interperiod compounding*, as used here, refers to the handling of the payments made *between* compounding periods. The three cases are as follows:

1 There is no interest paid on the money deposited (or withdrawn) between compounding periods.

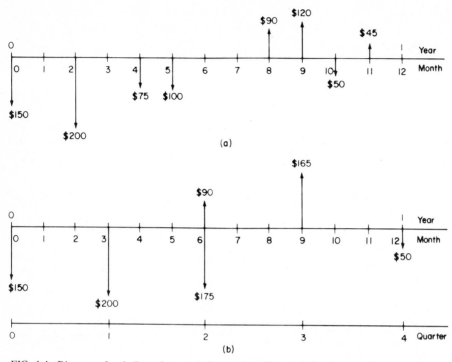

FIG. 4.4   Diagram of cash flows for quarterly compounding periods using case 1.

2 The money deposited (or withdrawn) between compounding periods earns simple interest. That is, interest is not paid on the interest earned in the preceding interperiod.

3 The money deposited (withdrawn) between compounding periods earns interest at an interperiod rate equivalent to the compounding period's effective rate.

4.5.1 No interperiod interest   In the first case, money deposited or withdrawn between compounding periods is regarded as having been *deposited* at the *beginning of the next compounding period* or *withdrawn* at the *end of previous compounding period*. This is the usual mode of operation of banks and other lending institutions. Thus, if the compounding period were a *quarter*, the actual transactions shown in Fig. 4.4(*a*) would be treated as shown in Fig. 4.4(*b*).

In order to find the present worth of the cash flow represented by Fig. 4.4(*b*), the nominal yearly interest rate is divided by four (since interest is compounded quarterly) and the appropriate *P/F* or *F/P* factor is used.

4.5.2 Simple interperiod interest   If the money deposited between compounding periods earns simple interest, then in order to obtain the interest earned in the interperiod *each interperiod deposit* must be *multiplied* by

$$\left(\frac{m}{N}\right)i \qquad (4.7)$$

where N = number of periods in a compounding period
$\quad\quad m$ = number of periods prior to the end of a compounding period
$\quad\quad\; i$ = interest rate per compounding period

Note that the value obtained by Eq. (4.7) yields only the interest that has accumulated and is not the total end-of-period amount. Example 4.4 illustrates the calculations described here.

Example 4.4 Calculate the amount of money that would be in a person's savings account if deposits were made as shown in Fig. 4.5. Assume the bank pays 6% per year compounded semiannually and pays simple interest on the interperiod deposits.

SOLUTION The first step is to find the amount of money that will be accumulated at each compounding period (i.e., every six months) using the effective rate of 3% semiannually. The future or present worth of those deposits can then be calculated with the regular interest formulas. Thus, for the deposits made within the first compounding period, the total value (F_6) after six months is

$$F_6 = \left[100 + 100\left(\frac{5}{6}\right)0.03\right] + \left[90 + 90\left(\frac{3}{6}\right)0.03\right] + 80$$

$$= [100 + 2.50] + [90 + 1.35] + 80$$

$$= \$273.85$$

Similarly, the amount accumulated in the second compounding period (F_{12}) is

$$F_{12} = \left[75 + 75\left(\frac{5}{6}\right)0.03\right] + \left[85 + 85\left(\frac{4}{6}\right)0.03\right] + \left[70 + 70\left(\frac{1}{6}\right)0.03\right]$$

$$= [75 + 1.88] + [85 + 1.70] + [70 + 0.35]$$

$$= \$233.93$$

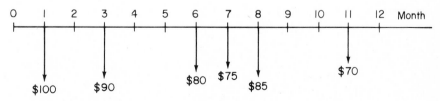

FIG. 4.5 Actual deposits made with simple interperiod interest paid using case 2.

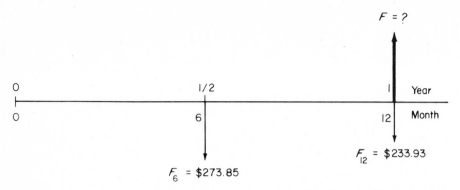

FIG. 4.6 Equivalent deposits after simple interest computed for interperiods, Example 4.4.

The cash-flow diagram has now been reduced to that shown in Fig. 4.6. Thus, the future worth F at the end of the year is

$$F = 273.85(F/P,3\%,1) + 233.93 = \$516$$

A lot of trouble, right?

COMMENT In calculating the amounts of money accumulated after six months (F_6) and 12 months (F_{12}), note that the amount of the deposit was added to each interest term, since Eq. (4.7) represents only the *interest accumulated*, not the total amount. ////

Determination of the compound-period amount (i.e., end-of-period amount) is simplified when the interperiod payments are regular and uniform. That is, if the same amount of money is deposited in *each* interperiod, the equivalent compound-period amount can be calculated from the following equation.

$$A = ND + \frac{N-1}{2}(iD) \qquad (4.8)$$

where A = equivalent uniform compound-period payment
  D = value of the deposit made each interperiod

See the Solved Examples for an illustration.

4.5.3 Compound interperiod interest The third case assumes that the money deposited between compounding periods earns interest which is compounded each payment period. Therefore, an interperiod interest rate must be calculated which is equivalent to the compound-period effective interest rate. This is just the reverse of the situation in which a nominal rate is specified and the effective rate is desired.

Therefore, the effective interest-rate formula, Eq. (4.3), can be used to find the nominal interest rate r. The next example illustrates these calculations.

Example 4.5 What is the present worth of $100 deposited every three months for eight years if the interest rate is 8% per year compounded semiannually and interperiod interest is compounded?

SOLUTION The cash-flow diagram is left to the student. The first step in solving the problem is to find the interperiod rate (i.e., the three-month rate) which will be equivalent to an effective semiannual rate of 4%. This can be done with the effective interest-rate formula as follows:

$$i = \left(1 + \frac{r}{t}\right)^t - 1$$

$$0.04 = \left(1 + \frac{r}{2}\right)^2 - 1$$

$$1.04 = \left(1 + \frac{r}{2}\right)^2$$

$$1.0198 = \left(1 + \frac{r}{2}\right)$$

$$r = 0.0396 \quad (3.96\% \text{ per six months})$$

This value seems reasonable, since the nominal rate should be less than the effective rate. Therefore, a quarterly interest rate of 1.98% compounded quarterly is equivalent to 4% compounded semiannually. The quarterly interest rate can now be used with the factors in the interest tables:

$$P = 100(P/A,1.98\%,32) = 100(23.5398) = \$2,354$$

COMMENT Note that n is equal to 32 in the P/A formula since the interest rate used is the quarterly rate for eight years; $n = 4(8) = 32$ deposits. You should also realize that in using the effective interest formula you cannot go directly to a quarterly rate (i.e., use $t = 4$), because the compounding is semiannual, not quarterly. ////

Example 4.9
Problems P4.24–P4.33

SOLVED EXAMPLES

Example 4.6 Mr. Jones plans to deposit money in a certificate of deposit that pays 8% per year compounded daily. What effective rate will he receive (*a*) yearly and (*b*) semiannually?

SOLUTION

(a) Using Eq. (4.3), with $t = 365$,

$$i = \left(1 + \frac{0.08}{365}\right)^{365} - 1$$

$$= 0.08328 \quad (8.328\%)$$

That is, Mr. Jones will get an effective 8.328% per year on his deposit.

(b) Here $r = 0.04$ per six months and $t = 182$ days:

$$i = \left(1 + \frac{0.04}{182}\right)^{182} - 1$$

$$= 0.04081 \quad (4.081\%) \qquad ////$$

Sec. 4.3

Example 4.7 Ms. Warren wants to purchase a new compact car for $4,500. She plans to borrow the money from her credit union and to repay it monthly over a period of three years. If the nominal interest rate is 12% per year compounded monthly, what will be her monthly installments?

SOLUTION Since the compounding period equals the payment period, the effective monthly rate is

$$i = \frac{12\%}{12} = 1\% \text{ per month}$$

$$\text{for } n = 12(3) = 36 \text{ payments}$$

Thus, the monthly payments are

$$A = 4,500(A/P,1\%,36) = \$149.49$$

COMMENT If the compounding period had been anything other than a month, we would have had to compute an effective monthly interest rate to find A. ////

Sec. 4.4

Example 4.8 Compute the monthly deposit required to accumulate $5,000 in five years at a nominal 6% per year compounded daily.

SOLUTION The payment period is a month and the compounding period is a day. Since a uniform-series factor is involved and since compounding is more frequent than payment, the effective interest rate per month must be determined from Eq. (4.3) or Table B-1. Assuming 30 days per month, an interest rate of $6\%/12 = \frac{1}{2}\%$ per month compounded daily is the effective interest rate:

$$i = \left(1 + \frac{0.005}{30}\right)^{30} - 1 = 0.00501 \quad (0.501\%)$$

A total of $12(5) = 60$ deposits will be made in the five years. Using the A/F factor yields

$$A = F(A/F, 0.501\%, 60) = 5{,}000(0.01434) = \$71.70 \text{ per month} \qquad ////$$

Sec. 4.4

Example 4.9 Calculate the equivalent compound-period amount if $200 is deposited every month at an interest rate of 6% per year compounded semiannually with simple interest paid on interperiod deposits.

SOLUTION   From Eq. (4.8) and a semiannual interest rate of 3%,

$$A = 6(200) + \frac{5}{2}(0.03)(200) = \$1{,}215$$

Thus, $200 per month is equivalent to $1,215 every six months since the compounding period is semiannual.

COMMENT   The present worth or future worth is calculated using the P/A or F/A factors for $i = 3\%$ and $n =$ number of compounding periods (for example, three years is six compounding periods). ////

Sec. 4.5

BIBLIOGRAPHY

Barish, pp. 58–60.
DeGarmo and Canada, pp. 132–137.
Emerson and Taylor, p. 2-16.
Fabrycky and Thuesen, pp. 68–76.
Grant and Ireson, pp. 38–40.
Park, p. 28.
Riggs, pp. 162–163.
Smith, pp. 53–55, 74–75.
Taylor, pp. 36, 37.
Thuesen, Fabrycky, and Thuesen, pp. 65–67.

PROBLEMS

P4.1   What is the nominal interest rate per year if interest is (*a*) 0.50% every two weeks and (*b*) 2% every semiannual period?

P4.2 What time period is usually used in expressing a nominal interest rate?

P4.3 What is the dimensional unit of the term r/t in the effective interest-rate equation?

P4.4 Calculate the nominal and effective interest rates for a finance charge of $1\frac{1}{2}\%$ per month.

P4.5 What are the nominal and effective interest rates for an interest charge of 4% every six months?

P4.6 What effective interest rate is equivalent to a nominal rate of 12% per year compounded semiannually?

P4.7 What effective interest rate is equivalent to a nominal rate of 16% per year compounded quarterly?

P4.8 Calculate the nominal and effective interest rates for a finance charge of $1\frac{3}{4}\%$ per month and find the value of the P/A factor for ten years for each rate.

P4.9 What quarterly interest rate is equivalent to an effective annual rate of 6%?

P4.10 How much money would be accumulated in eight years if an investor deposits $2,500 now at a nominal interest rate of 8% compounded semiannually?

P4.11 If a person buys a car for $5,500 and must make monthly payments of $200 for 36 months, what are the nominal and effective interest rates for this transaction?

P4.12 How much should you be willing to pay for an annuity that will provide $300 every three months for six years starting three months from now if you want to make a nominal 12% per year?

P4.13 If a person deposits $75 into a savings account every month, how much money will be accumulated after ten years if the interest rate is a nominal 12% per year compounded monthly?

P4.14 If a person borrows $3,000 and must repay the loan in two years with equal monthly installments, how much is the monthly payment if the interest rate is 1% per month?

P4.15 If a person made a lump-sum deposit 12 years ago which has accumulated to $9,500 now, how much was the original deposit if the interest rate received was a nominal 4% per year compounded semiannually?

P4.16 What is the present worth of $10,000 now, $6,000 eight years from now, and $9,000 twelve years from now if the interest rate is a nominal 6% per year compounded quarterly?

P4.17 What would be the value of the deposits in problem P4.16 25 years from the date of the original deposit?

P4.18 A man has been presented with the opportunity to buy a second mortgage note valued at $1,500 for $1,300. The note is due four months from now. If he purchases the note, what nominal and effective rate of return will he make?

P4.19 What monthly interest rate is equivalent to an effective semiannual rate of 4%?

P4.20 If a pants manufacturing company spends $14,000 in order to improve the efficiency of a sewing operation, how much must it save each month in reduced manpower costs in order to recover its investment in two and one-half years if the effective interest rate is 12.68% per year?

P4.21 A woman who has just won $45,000 in a lottery wants to deposit enough of her winnings into a savings account so that she will have $10,000 for her son's college education. Assume her son just turned three years old and will begin college when he is 18 years of age. How much must the woman deposit if she can earn 7% interest compounded quarterly on her investment?

P4.22 How much money can be withdrawn semiannually for 20 years from a retirement fund which earns 5% interest compounded semiannually and has a present amount of $36,000 in it?

P4.23 What year-end payment is equivalent to the monthly payments that would be paid on a $2,000 loan that must be repaid in one year at an interest rate of 1% per month?

P4.24 Draw two cash-flow diagrams illustrating the timing of cash flow if banks paid interest on interperiod deposits but not on withdrawals. Assume interest is payable quarterly and represent deposits by X's and withdrawals by Y's, with at least one deposit and one withdrawal per interest period. Draw one diagram for actual deposits and withdrawals and another illustrating deposits and withdrawals from the bank's point of view.

P4.25 How much money would be in a savings account in which a person had deposited $100 every month for five years at an interest rate of 5% per year compounded quarterly? Assume interperiod interest is compounded.

P4.26 How much money would the person in problem P4.25 have if interest were compounded (*a*) monthly? (*b*) daily?

P4.27 Calculate the future worth of the transactions shown in Fig. 4.4(*a*) if interest is 6% per year compounded semiannually and interperiod interest is not paid.

P4.28 Rework problem P4.27, except assume simple interest is paid on interperiod deposits but not withdrawals.

P4.29 A tool-and-die company expects to have to replace one of its lathes in five years at a cost of $18,000. How much would the company have to deposit every month in order to accumulate $18,000 in five years if the interest rate is 6% per year compounded semiannually? Assume simple interperiod interest.

P4.30 How much money would be in a savings account in which a person had deposited $300 every three months for seven years at an interest rate of 6% compounded semiannually? Assume interperiod compounding.

P4.31 What monthly deposit would be equivalent to a deposit of $600 every three months for two years if the interest rate is 6% compounded semiannually? Assume interperiod compounding.

P4.32 How much money would be in the savings account of a person who had deposited $150 every month and withdrawn $300 every six months for two years? Use an interest rate of 4% per year compounded semiannually and assume interperiod interest is not paid.

P4.33 How many monthly deposits of $75 would a person have to make in order to accumulate $15,000 if the interest rate is 6% per year compounded semiannually? Assume simple interperiod interest is paid.

5
USE OF MULTIPLE FACTORS

The purpose of this chapter is to teach you how to solve problems that may involve the multiplication of several engineering economy factors to find a P, F, or A value.

CRITERIA

To complete this chapter you must be able to:

1 Determine the year in which the present worth or future worth is located, given a statement describing a randomly placed uniform series.
2 Determine the present worth, equivalent uniform annual series, or future worth of a uniform series of disbursements or receipts which begin at a time other than after year 1, given the amount and time of each payment and the interest rate.
3 Calculate the present worth or future worth of randomly distributed single amounts and uniform-series amounts, given the times and amounts of the payments and the interest rate.
4 Calculate the equivalent uniform annual series of randomly distributed single amounts and uniform-series amounts, given the times, amounts, and the interest rate.

EXPLANATION OF MATERIAL

5.1 Location of Present Worth and Future Worth

When a uniform series of payments begins at a time other than at the end of year 1, several methods can be used to find the present worth. For example, the present worth of the uniform series of disbursements shown in Fig. 5.1 could be determined by any of the following methods:

1 Use the single-payment present-worth factor $(P/F,i\%,n)$ to find the present worth of each disbursement at year zero and add them.

2 Use the single-payment compound-amount factor $(F/P,i\%,n)$ to find the future worth of each disbursement in year 13, add them, and then find the present worth of the total using $P = F(P/F,i\%,13)$.

3 Use the uniform-series compound-amount factor $(F/A,i\%,n)$ to find the future amount by $F = A(F/A,i\%,10)$ and then find the present worth using $P = F(P/F,i\%,13)$.

4 Use the uniform-series present-worth factor $(P/A,i\%,n)$ to compute the "present worth" (which will *not* be located in year zero!) and then find the present worth in year zero by using the $(P/F,i\%,n)$ factor. (Present worth is enclosed in quotation marks to represent the present worth as determined by the uniform-series present-worth factor and to differentiate it from the present worth in year zero.)

This and the next section illustrate the fourth method for calculating the present worth of a uniform series that does not begin at the end of year 1.

For the cash-flow diagram shown in Fig. 5.1, the "present worth" that would be obtained by using the $(P/A,i\%,n)$ factor would be located in *year 3*, not year 4. This is shown in Fig. 5.2, with the present-worth arrow pointing upward to represent an equivalent amount in year 3 for the uniform series. Note that P is located *one year prior* to the beginning of the first annual disbursement. Why? Because the P/A factor was derived with the P in year zero and the A beginning at the end of year 1; that is, the P must always be *one year ahead* of the first A value (Sec. 3.2). The most common

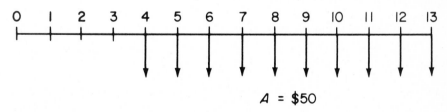

$A = \$50$

FIG. 5.1   A randomly placed uniform series.

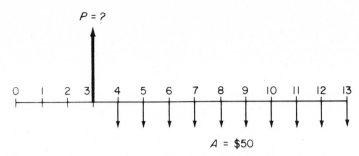

FIG. 5.2 Location of P for the randomly placed uniform series in Fig. 5.1.

mistake made in working problems of this type is improper placement of P in year 4 instead of in year 3. Therefore, it is extremely important that you remember the following rule. *The present worth is always located ONE YEAR PRIOR to the first annual payment when using the uniform-series present-worth factor,* $(P/A, i\%, n)$.

On the other hand, the uniform-series compound-amount factor, $(F/A, i\%, n)$, was derived (Sec. 3.3) with the future worth F located in the *same year* as the last payment. Figure 5.3 shows the location of the future worth when F/A is used for the cash flow shown in Fig. 5.1. Thus, *the future worth is always located in the SAME YEAR as the last annual payment when using the uniform-series compound-amount factor* $(F/A, i\%, n)$.

It is always important to remember that the number of years n that should be used with the P/A or F/A factors is equal to the number of payments. It is generally helpful to *renumber* the cash-flow diagram to avoid counting errors. Figure 5.4 shows the cash-flow diagram of Fig. 5.1 renumbered for determination of n. Note that in this example n equals ten years. Detailed calculations using the concepts learned in this section are shown in Sec. 5.2.

Example 5.8
Problem P5.1

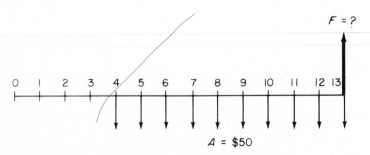

FIG. 5.3 Placement of F for the uniform series of Fig. 5.1.

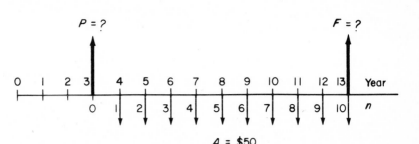

FIG. 5.4   Renumbering of payments in Fig. 5.1 to show that $n = 10$ in the P/A or F/A factors.

5.2   Calculations for a Uniform Series That Begins After Year 1

As stated in Sec. 5.1, there are many methods that can be used to solve problems having a uniform series that begins at a time other than the end of year 1. However, it is generally much more convenient to use the uniform-series formulas than it is to use the single-payment formulas.

There are specific steps which should be followed in solving problems of this type in order to avoid unnecessary errors:

1 Draw a cash-flow diagram of the receipts and disbursements of the problem.
2 Locate the present worth or future worth on the cash-flow diagram.
3 Determine *n* by renumbering the cash-flow diagram.
4 Draw the cash-flow diagram representing the desired equivalent cash flow.
5 Set up and solve the equations.

These steps are illustrated in the following two examples.

Example 5.1   A person buys a piece of property for $5,000 down and deferred annual payments of $500 a year for six years starting three years from now. What is the present worth of the investment if the interest rate is 8%?

SOLUTION   The cash-flow diagram is shown in Fig. 5.5. The nomenclature P_A is used throughout this book to represent the present worth of a uniform annual series and P'_A represents the present worth at a time other than year zero. Similarly, P_T represents the total present worth at time zero. The correct placement of P'_A and diagram renumbering to obtain *n* are also indicated in Fig. 5.5. Note that P'_A is located in year 2, not year 3, and $n = 6$, not 8. The solution to this problem would be to first find the value of P'_A:

$$P'_A = \$500(P/A,8\%,6)$$

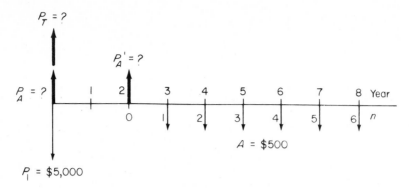

FIG. 5.5 Placement of present-worth values, Example 5.1.

Since P'_A is located in year 2, it is necessary to take it back to year zero:

$$P_A = P'_A(P/F,8\%,2)$$

The total present worth can now be determined by adding P_A and P_1 (the initial investment is always labeled P_1):

$$P_T = P_1 + P_A = 5,000 + 500(P/A,8\%,6)(P/F,8\%,2)$$
$$= 5,000 + 500(4.6229)(0.8573)$$
$$= \$6,981.60$$

COMMENT The most common error in problems of this type is improper location of P'_A. If this part of the problem is properly handled, the solution to the problem is rather routine.   ////

Example 5.2 Calculate the eight-year equivalent uniform annual series at 6% interest for the uniform disbursements shown in Fig. 5.6.

SOLUTION Figure 5.7 shows the original cash-flow diagram and the desired equivalent diagram. In order to convert uniform cash flows that begin sometime after year 1 into an equivalent uniform annual cost over *all* the years, the first step is to convert the cash flow into a present worth or future worth. Then either the conventional capital-recovery factor $(A/P,i\%,n)$ or the sinking-fund factor $(A/F,i\%,n)$

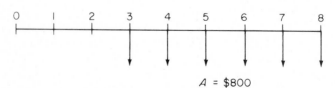

FIG. 5.6 Series of uniform disbursements, Example 5.2.

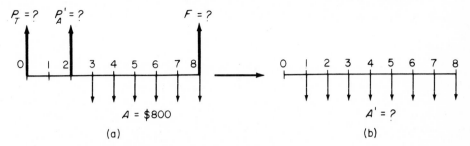

FIG. 5.7   Desired equivalent diagram for the series of Fig. 5.6.

can be used to determine the equivalent uniform annual cost. Both of these methods are illustrated below.

1 Present-worth method (refer to Fig. 5.7):

$$P'_A = 800(P/A,6\%,6)$$
$$P_T = P'_A(P/F,6\%,2) = 800(P/A,6\%,6)(P/F,6\%,2)$$
$$= \$3,501.12$$

where P_T is the total present worth of the cash flow. The equivalent series (A') can now be determined with the A/P factor

$$A' = P_T(A/P,6\%,8) = \$563.82$$

for eight years as shown in Fig. 5.7(*b*).

2 Future-worth method (Fig. 5.7): The first step is to calculate the future worth *F*:

$$F = 800(F/A,6\%,6) = \$5,580$$

The sinking-fund factor $(A/F,i\%,n)$ can now be used to obtain A':

$$A' = F(A/F,6\%,8) = \$563.80$$

COMMENT   In the present-worth method, note that P'_A was located in year 2, not year 3. After the present worth was determined, the equivalent series was calculated using $n = 8$. In the future-worth method, $n = 6$ was used to find F, and $n = 8$ for finding the equivalent series, since the cost must be spread uniformly over *all* the years.   ////

Example 5.9
Problems P5.2–P5.14

5.3 Calculations Involving Uniform-Series and Randomly Distributed Amounts

When a uniform series of payments is included in a cash flow that also contains randomly distributed single amounts, the procedures learned in Sec. 5.2 should be applied to the uniform-series amounts and the single-payment formulas applied to the single-payment amounts. This type of problem, illustrated below, is merely a combination of previous types.

Example 5.3 A couple owning 50 acres of valuable land decided to sell the mineral rights on their property to a mining company. Their primary objective was to obtain long-term investment income and sufficient money to finance the college education of their two children. Since the children were 12 years and two years of age at the time the couple were negotiating the contract, they knew that the children would be in college six years and 16 years from the present. They therefore made a proposal to the company that it pay them $20,000 per year for 20 years beginning one year hence plus $10,000 six years from now and $15,000 sixteen years from now. If the company wanted to pay off its lease immediately, how much would it have to pay now, if the interest rate is 6%?

SOLUTION The cash-flow diagram for this problem is shown in Fig. 5.8. This problem is solved by finding the present worth of the uniform series and adding it to the present worths of the two individual payments. Thus,

$$P = 20{,}000(P/A,6\%,20) + 10{,}000(P/F,6\%,6) + 15{,}000(P/F,6\%,16)$$
$$= \$242{,}352$$

COMMENT In this example, note that the uniform series started at the end of year 1 so that the present worth obtained with the P/A factor represented the present worth at year zero. It was not necessary to use the P/F factor on the uniform series.

////

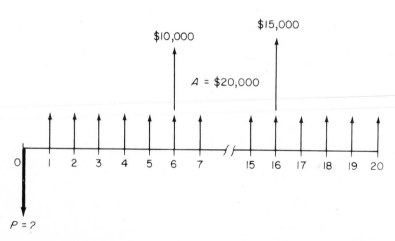

FIG. 5.8 Diagram including a uniform series and single amounts, Example 5.3.

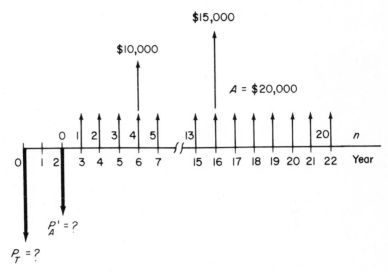

FIG. 5.9 Diagram from Fig. 5.8 with the A series shifted three years.

Example 5.4 If the uniform payments described in Example 5.3 did not begin until three years from the time to contract was signed, what would be the present worth of the receipts?

SOLUTION The cash-flow diagram is shown in Fig. 5.9, with the uniform-series time sequence shown above the time axis. The number of years for the uniform series is still 20.

$$P'_A = 20{,}000(P/A,6\%,20)$$
$$P_T = P'_A(P/F,6\%,2) + 10{,}000(P/F,6\%,6) + 15{,}000(P/F,6\%,16)$$
$$= 20{,}000(P/A,6\%,20)(P/F,6\%,2) + 10{,}000(P/F,6\%,6)$$
$$\quad + 15{,}000(P/F,6\%,16)$$
$$= \$217{,}118$$

COMMENT Displacement of the annual series by three years has decreased the present worth of the cash flows by $25,234.   ////

Example 5.5 Calculate the future worth of the receipts shown in Fig. 5.9 using $i = 6\%$.

SOLUTION The future worth of the uniform series and the single, random receipts can be calculated as follows:

$$F = 20{,}000(F/A,6\%,20) + 10{,}000(F/P,6\%,16) + 15{,}000(F/P,6\%,6)$$
$$= \$782{,}381$$

COMMENT   Although the determination of n is straightforward, make sure that you fully understand how the values 20, 16, and 6 were obtained for calculating the future worth.   ////

Example 5.10
Problems P5.15–P5.19

5.4   Equivalent Uniform Annual Series of Both Uniform and Single Payments

Whenever it is desired to calculate the equivalent uniform annual series of randomly distributed single payments and/or uniform amounts, the most important fact to remember is that the payments *must first be converted to a present worth or future worth*. Then the equivalent uniform annual series can be obtained with the appropriate A/P or A/F factor.

Example 5.6   Calculate the equivalent uniform annual series for the receipts described in Example 5.3 (Fig. 5.8).

SOLUTION   The desired equivalent cash-flow diagram is shown in Fig. 5.10. From the cash-flow diagram shown in Fig. 5.8, it is evident that the uniform-series receipts (i.e., $20,000) are already distributed through all 20 years of the diagram. Therefore, it is necessary only to convert the single amounts to an equivalent uniform annual series and add the value obtained to the $20,000. This can be done by either the present-worth or future-worth method.

(*a*)   Present-worth method:

$$A = 20{,}000 + 10{,}000(P/F,6\%,6)(A/P,6\%,20)$$
$$+ 15{,}000(P/F,6\%,16)(A/P,6\%,20)$$
$$= 20{,}000 + [10{,}000(P/F,6\%,6)$$
$$+ 15{,}000(P/F,6\%,16)](A/P,6\%,20)$$
$$= \$21{,}129 \text{ per year}$$

(*b*)   Future-worth method:

$$A = 20{,}000 + 10{,}000(F/P,6\%,14)(A/F,6\%,20)$$
$$+ 15{,}000(F/P,6\%,4)(A/F,6\%,20)$$
$$= 20{,}000 + [10{,}000(F/P,6\%,14)$$
$$+ 15{,}000(F/P,6\%,4)](A/F,6\%,20)$$
$$= \$21{,}129 \text{ per year}$$

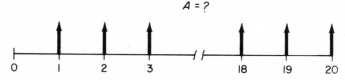

FIG. 5.10   Desired equivalent series, Example 5.6.

COMMENT Note that it was necessary to take the single payments to either end of the time scale before annualizing. Failure to do so would result in unequal receipts in some years. ////

Example 5.7 Convert the cash flow shown in Fig. 5.9 to an equivalent uniform annual series. Use $i = 6\%$.

SOLUTION Since the uniform-series receipts are not distributed through all 20 years of the time scale, it is first necessary to find the present worth or future worth of the series. This was done in Examples 5.4 and 5.5, respectively. The equivalent uniform annual series can now be obtained by multiplying the values previously obtained by the $(A/P,6\%,22)$ factor or the $(A/F,6\%,22)$ factor as follows:

$$A = P_T(A/P,6\%,22) = 217,118(A/P,6\%,22) = \$18,032$$

or

$$A = F(A/F,6\%,22) = 782,381(A/F,6\%,22) = \$18,034$$

COMMENT When a uniform series begins at a time other than at the end of year 1, or when intermediate single amounts are involved, it is most important to remember that the equivalent present or future worth of the uniform series must be determined before the equivalent uniform annual series can be obtained. ////

Example 5.11
Problems P5.20–P5.25

SOLVED EXAMPLES

Example 5.8 A family decides to buy a new refrigerator on credit. The payment scheme calls for a $100 down payment now (the month is March) and $55 a month from June to November with interest at $1\frac{1}{2}\%$ per month compounded monthly. Construct the cash-flow diagram and mark the month in which you can compute an equivalent value using one P/A and one F/P factor. Give the n values for all computations. Try this problem yourself before you read the solution.

SOLUTION Since the payment period equals the compounding period, the interest tables of Appendix A can be used. Figure 5.11 solves the problem by

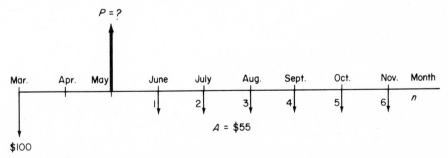

FIG. 5.11 Placement of an equivalent amount using only P/A and F/P factors, Example 5.8.

placement of P in May. The relation using only the two factors, with the correct n values, is

$$P = 100(F/P,1.5\%,2) + 55(P/A,1.5\%,6)$$

COMMENT The control for placement of P must be retained by a uniform series since the P/A factor is inflexible in the computational procedure for P.   ////

Sec. 5.1

Example 5.9 Consider the two uniform series shown in Fig. 5.12. Compute the present worth at 15% using at least three different methods.

SOLUTION There are numerous ways to find the present worth. The two simplest are probably the future-worth and present-worth methods. For a third method, the use of the *intermediate-year method* at year 7 is demonstrated.

(*a*) Present-worth method—see Fig. 5.13(*a*):
The use of P/A factors for the uniform series and P/F factors to obtain the actual present worths allows us to find P_T.

$$P_T = P_{A1} + P_{A2}$$

where $P_{A1} = P'_{A1}(P/F,15\%,2) = A_1(P/A,15\%,3)(P/F,15\%,2)$
$= 1,000(2.2832)(0.7561)$
$= \$1,726$
$P_{A2} = P'_{A2}(P/F,15\%,8) = A_2(P/A,15\%,5)(P/F,15\%,8)$
$= 1,500(3.3522)(0.3269)$
$= \$1,644$

$$P_T = 1,726 + 1,644 = \$3,370$$

(*b*) Future-worth method—Fig. 5.13(*b*):
Using the F/A, F/P, and P/F factors, we have

$$P_T = (F_{A1} + F_{A2})(P/F,15\%,13)$$

where $F_{A1} = F'_{A1}(F/P,15\%,8) = A_1(F/A,15\%,3)(F/P,15\%,8)$
$= 1,000(3.472)(3.0590)$
$= \$10,621$

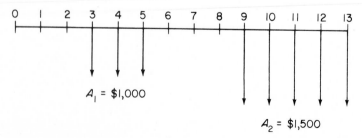

FIG. 5.12 Uniform series used to compute a present worth by several methods, Example 5.9.

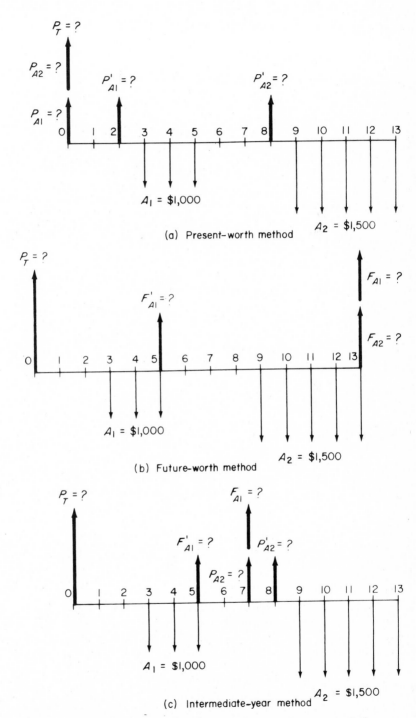

FIG. 5.13 Computation of the present worth of Fig. 5.12 by three methods.

$$F_{A2} = A_2(F/A,15\%,5) = 1,500(6.742)$$
$$= \$10,113$$

$$P_T = 20,734(0.1625)$$
$$= \$3,369$$

(c) Intermediate-year method—Fig. 5.13(c):
If we find the present worth of both series at year 7 and use the P/F factor, we have

$$P_T = (F_{A1} + P_{A2})(P/F,15\%,7)$$

The P_{A2} value is computed as a present worth, but to find the value at year zero, it must be treated as an F value. Thus,

$$F_{A1} = F'_{A1}(F/P,15\%,2) = A_1(F/A,15\%,3)(F/P,15\%,2)$$
$$= 1,000(3.472)(1.3225)$$
$$= \$4,592$$
$$P_{A2} = P'_{A2}(P/F,15\%,1) = A_2(P/A,15\%,5)(P/F,15\%,1)$$
$$= 1,500(3.3522)(0.8696)$$
$$= \$4,373$$
$$P_T = (P_{A1} + P_{A2})(P/F,15\%,7)$$
$$= 8,965(0.3759)$$
$$= \$3,370 \qquad\qquad ////$$

Sec. 5.2

Example 5.10 Calculate the present worth of the following series of cash flows if $i = 8\%$.

Year	Cash flow
0	$ +460
1	+460
2	+460
3	+460
4	+460
5	+460
6	+460
7	−5,000

SOLUTION The cash-flow diagram is shown in Fig. 5.14. Since the disbursement in year zero is equal to the disbursements of the A series, the P/A factor can be used for either six or seven years. The problem is worked below both ways.

(a) Using P/A and $n = 6$:
For this case, the disbursement in year zero (P_1) is added to the present worth of the remaining payments, since the P/A factor for $n = 6$ will place P_A in year zero. Thus,

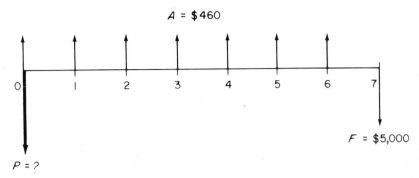

FIG. 5.14 Cash-flow diagram, Example 5.10.

$$P = P_1 + P_A - P_F$$
$$= 460 + 460(P/A,8\%,6) - 5,000(P/F,8\%,7)$$
$$= \$-331$$

Note that the present worth of the $-5,000 cash flow (P_F) is negative, since it is a negative cash flow.

(b) Using P/A and $n = 7$:

By using the P/A factor for $n = 7$, the "present worth" is located in year -1, not year zero, because the P must always be one year ahead of the first A when using the P/A factor. It is, therefore, necessary to move the P_A value one year forward with the F/P factor. Thus,

$$P = 460(P/A,8\%,7)(F/P,8\%,1) - 5,000(P/F,8\%,7)$$
$$= \$-331$$

COMMENT Rework the problem by first finding the future worth of the series and then solving for P. Obviously, you should obtain the same answer as above, if you work it correctly.   ////

Sec. 5.3

Example 5.11 For the cash flows in Fig. 5.12, if, in addition to the A_1 and A_2 series, there is a one-time payment of $3,000 in year 8, find the 15% equivalent uniform annual series for years 4-13 only.

SOLUTION The easiest way to approach this situation is to compute F in year 13 and then use the $(A/F,15\%,10)$ factor to find the equivalent series. In Example 5.9 the future worth of the A_1 and A_2 series–Fig. 5.13(b)–was found to be $F_{A1} = \$10,621$ and $F_{A2} = \$10,113$. For the $3,000 amount,

$$F = 3,000(F/P,15\%,5)$$
$$= \$6,034$$

Now the total future worth (F_T) may be found

$$F_T = F_{A1} + F_{A2} + F$$
$$= \$26,768$$

and

$$A = F_T(A/F,15\%,10)$$
$$= \$1,318$$

for years 4–13 only.

COMMENT   Note the use of $n = 10$ in the A/F factor since not all 13 years are included in the series. You could find A using the present-worth method, but the value at year 3 must be found before the A/P factor could be used. Try it! See if you get $A = \$1,318$ per year. Note that while this example sought the equivalent uniform annual series in years 4–13 only, most problems involving the equivalent uniform annual series require determining the equivalent series through all of the years.   ////

Sec. 5.4

BIBLIOGRAPHY

Barish, pp. 60–62.
DeGarmo and Canada, pp. 125–128.
Fabrycky and Thuesen, pp. 65–68.
Grant and Ireson, pp. 55–62.
Riggs, pp. 172–176.
Smith, pp. 70–71, 73–74.
Thuesen, Fabrycky, and Thuesen, pp. 90–94.

PROBLEMS

P5.1   Construct a cash-flow diagram for each of the following and locate the present worth and future worth of each uniform series *separately*.

(*a*) A $600 deposit for seven years starting four years from now.

(*b*) A $2,000 per month deposit for six months starting next month and a $1,500 per month withdrawal for four months starting three months from now.

(*c*) You had the following transactions in your Christmas Club account for the past 12 months:

Month	Deposit	Withdrawal	Month	Deposit	Withdrawal
January	$20	$ 0	July	$75	$ 25
February	20	0	August	75	25
March	20	10	September	25	10
April	20	10	October	25	10
May	75	10	November	25	10
June	75	25	December	0	250

P5.2 Determine the amount of money a person must deposit now in order to be able to withdraw $3,600 per year for ten years starting 20 years from now if the interest rate is $4\frac{1}{2}\%$.

P5.3 What is the equivalent uniform annual cost of $250 per year for 12 years starting five years from now if the interest rate is 12%?

P5.4 A couple purchased an insurance policy which they plan to use to finance their child's college education. If the policy will provide $10,000 twelve years from now, how much can be withdrawn every year for five years if the child starts college 15 years from now? Assume $i = 6\%$.

P5.5 A woman deposited $700 per year for eight years. Starting in the ninth year, she increased her deposits to $1,200 per year for five more years. How much money did she have in her account immediately after she made her last deposit if the interest rate was 5%?

P5.6 How much money would the woman in problem P5.5 have in her account 30 years from the present time if she made no deposits after the one in year 13?

P5.7 What is the present worth one year prior to the first deposit for the investment specified in problem P5.5?

P5.8 A man plans to begin saving for his retirement such that he will be able to withdraw money every year for 30 years starting 25 years from now. He estimates that he will be able to start saving money one year from now and plans to deposit $500 per year. What uniform annual amount will he be able to withdraw when he retires, if the interest rate is 6%?

P5.9 A businessman purchased a used building and found that the ceiling was poorly insulated. He estimated that with 6 inches of foam insulation, he could cut the heating bill by $15 per month and the air conditioning cost by $10 per month. Assuming that the winter season is the first six months of the year and the summer season is the next six months, how much can he afford to spend on insulation if he expects to keep the building for only two years? Assume $i = 1\frac{1}{2}\%$ per month.

P5.10 A couple plan to make an investment now in order to finance their child's college education. If the child should be able to withdraw $1,000 per year for six years starting 15 years from now such that the last withdrawal will close the account, how much must be invested now if interest is computed at 9%?

P5.11 If the couple in problem P5.10 wanted to make a uniform deposit for 14 years instead of the lump-sum investment, how much would they have to deposit every year starting one year from now?

P5.12 How much money would the heirs of the man in problem P5.8 receive if he died four years after his retirement?

P5.13 How much money would the woman in problem P5.5 have if the interest rate increased from 5% to 6% after five years?

P5.14 Giovanni's Pizza Palace has a ten-year lease on 2,000 square feet of space in an enclosed shopping center. The rent is paid yearly at a rate of $6 per square foot. At the end of the fourth year of the lease, the owner of the pizza shop decides to purchase a building and relocate the business. How much must the owner of the shopping center be paid for the remaining portion of the lease if an interest rate of 9% is used?

P5.15 If a couple open a savings account by depositing $500 now and then deposit $500 every year for 14 years, how much would they have in their account after the last deposit? Use an interest rate of 8%.

P5.16 A company purchases a machine that costs $12,000 with a $2,000 salvage value. Operating expenses for the machine are $800 per year. In addition, a major overhaul is required every five years at a cost of $2,800. What is the equivalent uniform annual cost of the machine if it has a life of 20 years and the interest rate is 12%?

P5.17 Two investors purchased a rundown house for $17,000. In the first month that they owned the house, they spent $3,000 on repairs and remodeling. Immediately after the house was remodeled, they were offered $24,000 to sell the house. After some consideration, they decided to keep the house and rented it for $200 per month starting two months after the purchase. They collected rent for 15 months and then sold the house for $22,000. If the interest rate was 1% per month, how much extra money did they make or lose by not selling the house immediately after it was remodeled?

P5.18 A petroleum investor wants to sell her share of a number of oil wells. Her wells are expected to produce 60 barrels of oil per day 365 days per year for 11 more years. If the investor currently receives $5.25 per barrel, how much would you be willing to pay for the wells if you expect the price of the oil to increase by $3.00 per barrel every three years starting two years from now? Assume the interest rate is 8% for the first three years and 10% thereafter.

P5.19 Calculate the (a) present worth and (b) future worth of the following series of disbursements:

Year	Disbursement	Year	Disbursement
0	$3,500		
1	3,500	6	$ 5,000
2	3,500	7	5,000
3	3,500	8	5,000
4	5,000	9	5,000
5	5,000	10	15,000

Assume $i = 6\%$ per year compounded semiannually.

P5.20 A building contractor purchased a dirt scraper for $35,000. He maintained the scraper at a cost of $2,500 per year. He overhauled the machine four years after the purchase at a cost of $4,000. Two years after the overhaul, he sold the scraper for $18,000. What was his equivalent uniform annual cost if the interest rate was 10%?

P5.21 How much money would you have to deposit for six consecutive years starting one year from now if you want to be able to withdraw $45,000 eleven years from now? Assume the interest rate is 5%.

P5.22 What is the equivalent uniform annual cost of the disbursements shown in problem P5.19?

P5.23 A large manufacturing company purchased a semiautomatic machine which cost $13,000. Its annual maintenance and operation cost was $1,700. Five years after the initial purchase, the company decided to purchase an additional unit for the machine which would make it fully automatic. The additional unit had a first cost of $7,100. The cost for operating the machine in the fully automatic condition was $900 per year. If the company used the machine for a total of 16 years and then sold the automatic addition for $1,800, what was the equivalent uniform annual cost of the machine at an interest rate of 9%?

P5.24 Was the purchase of the additional unit in problem P5.23 justified by the savings realized by automatic operation? Show your calculations.

P5.25 Find the equivalent uniform monthly deposit for the months May through September to be equivalent to the transactions prior to the December withdrawal detailed in problem P5.1(*c*) if interest is computed at 1% per month.

6

UNIFORM GRADIENTS

The objective of this chapter is to introduce you to uniform-gradient cash flow and teach you how to solve problems involving uniform gradients.

CRITERIA

In order to complete this chapter you must be able to do the following:

1 Define and construct the cash-flow diagram for a uniform-gradient series.
2 Derive the present-worth and annual-series gradient formulas using the single-payment present-worth factor.
3 Find the correct gradient present-worth or annual-series factor from the tables, given the interest rate and number of years.
4 On a cash-flow diagram locate the present worth of a gradient series and determine the n value for the gradient factors when the gradient begins at a time other than year 2, given the gradient cash-flow amounts and times.

5 Calculate the present-worth and equivalent uniform annual series of alternatives involving a conventional uniform gradient, given the interest rate and statement of the problem.

6 Calculate the present-worth and equivalent uniform annual series of cash flows involving shifted gradients, given the interest rate and statement of the problem.

7 Calculate the present-worth or equivalent uniform annual series of cash flows which include a decreasing gradient, given the interest rate and statement of the problem.

EXPLANATION OF MATERIAL

6.1 Definition and Cash Flow of Uniform Gradients

A *uniform gradient* is a cash-flow *series* which either increases or decreases *uniformly*. That is, the cash flow, whether income or disbursements, changes by the same amount each year. The *amount* of the increase or decrease is the *gradient*. For example, if a clothing manufacturer predicts that the cost of maintaining a cutting machine will increase by $500 per year until the machine is retired, a gradient series is involved and the amount of the gradient is $500. Similarly, if the company expects income to decrease by $3,000 per year for the next five years, the decreasing income represents a gradient in the amount of $3,000 per year.

The formulas previously developed for uniform-series cash flows were generated on the basis of year-end payments of equal value. In the case of a gradient, each year-end cash flow is different, so a new formula must be derived. In developing a formula which can be used for uniform gradients, it is convenient to assume that the payment that occurs at the end of year 1 does not involve a gradient but is rather a base payment. Thus, the first year-end amount will be called the *base amount* since, in actual applications, the first payment is usually larger or smaller than the gradient increase or decrease. For example, if you purchase a new car with a 12,000-mile complete guarantee, you might reasonably expect to have to pay for only the gasoline during the first year of operation. Let us assume that this cost is $400; that is, $400 is the base amount. After the first year, however, you would have to absorb the cost of repair or replacement yourself, and these costs could reasonably be expected to increase each year that you own the car. So, if you estimate your operation costs to increase by $25 each year, the amount you would pay after the second year is $425, after the third $450, and so on to year n, when the total cost will be $400 + (n-1)25$. The cash-flow diagram for this is shown in Fig. 6.1. Note that the gradient is first

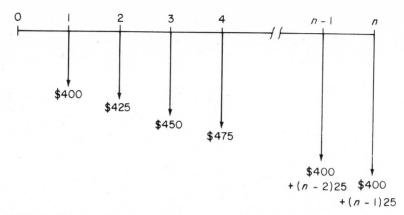

FIG. 6.1   Diagram of uniform-gradient series with a gradient of $25.

observed between year 1 and year 2, and the first or base payment ($400) is not equal to the gradient ($25). We will now define a new symbol for gradients:

$$G = \text{annual arithmetic change in the}$$
$$\text{magnitude of receipts or disbursements}$$

The value of G may be positive or negative. If we were to ignore the base payment, we could construct a generalized uniform increasing-gradient cash-flow diagram as shown in Fig. 6.2. Note that the gradient begins in year 2. This is called a *conventional gradient*. The generalized gradient formula will be derived from the cash-flow diagram of Fig. 6.2.

Example 6.1 The Free Spirit Company expects to realize a revenue of $100,000 next year from the sale of a new product. However, sales are expected to decrease uniformly with new competition to a level of $47,500 in eight years. Determine the gradient and construct the cash-flow diagram.

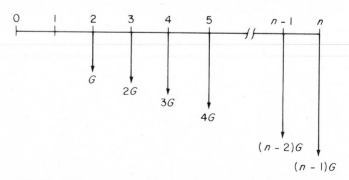

FIG. 6.2   Uniform-gradient series ignoring the base amount.

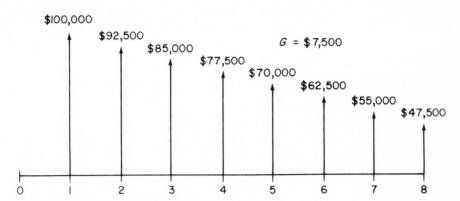

FIG. 6.3 Diagram for gradient series, Example 6.1.

SOLUTION

Base amount   = $100,000
Revenue loss by year 8 = 100,000 − 47,500 = $52,500
Gradient   = loss/$(n - 1)$
                       = 52,500/(8 − 1) = $7,500 per year

The cash-flow diagram is shown in Fig. 6.3.   ////

Problems P6.1, P6.2

6.2 Derivation of Gradient Formulas

There are several ways by which the uniform-gradient formula can be derived. We will derive the formula using the single-payment present-worth factor $(P/F,i\%,n)$ but the same result could be obtained using single-payment compound-amount factors, uniform-series compound-amount factors, or uniform-series present-worth factors.

Referring to Fig. 6.2, the present worth at year zero of the gradient payments would be equal to the sum of the present worths of the individual payments. Thus,

$$P = G(P/F,i\%,2) + 2G(P/F,i\%,3) + 3G(P/F,i\%,4)$$
$$+ \cdots + [(n-2)G]\,(P/F,i\%,n-1) + [(n-1)G]\,(P/F,i\%,n)$$

Factoring out G yields

$$P = G[(P/F,i\%,2) + 2(P/F,i\%,3) + 3(P/F,i\%,4)$$
$$+ \cdots + (n-2)(P/F,i\%,n-1) + (n-1)(P/F,i\%,n)]$$

Replacing the symbols with the appropriate single-payment present-worth factor expression, (Eq. 3.3), yields

$$P = G\left[\frac{1}{(1+i)^2} + \frac{2}{(1+i)^3} + \frac{3}{(1+i)^4} + \cdots + \frac{n-2}{(1+i)^{n-1}} + \frac{n-1}{(1+i)^n}\right] \qquad (6.1)$$

Multiplying both sides of Eq. (6.1) by $1/(1 + i)^{-1}$ to simplify yields

$$\frac{P}{(1 + i)^{-1}} = G\left[\frac{1}{(1 + i)^1} + \frac{2}{(1 + i)^2} + \frac{3}{(1 + i)^3}\right.$$
$$\left. + \cdots + \frac{n - 2}{(1 + i)^{n-2}} + \frac{n - 1}{(1 + i)^{n-1}}\right] \qquad (6.2)$$

Subtracting Eq. (6.1) from Eq. (6.2), noting that the first term of (6.2) and the last term of (6.1) have no matching terms, yields

$$\frac{P}{(1 + i)^{-1}} - P = G\left[\frac{1}{(1 + i)^1} + \frac{(2 - 1)}{(1 + i)^2} + \frac{(3 - 2)}{(1 + i)^3}\right.$$
$$\left. + \cdots + \frac{(n - 1) - (n - 2)}{(1 + i)^{n-1}} - \frac{n - 1}{(1 + i)^n}\right]$$

Rearranging the left side of the equation and the last term on the right side yields

$$P(1 + i)^1 - P = G\left[\frac{1}{(1 + i)^1} + \frac{1}{(1 + i)^2} + \frac{1}{(1 + i)^3}\right.$$
$$\left. + \cdots + \frac{1}{(1 + i)^{n-1}} + \frac{1 - n}{(1 + i)^n}\right]$$

If we write the left side of this equation as $P + Pi - P$ and factor out the n in the last term, we have

$$Pi = G\left[\frac{1}{(1 + i)^1} + \frac{1}{(1 + i)^2} + \frac{1}{(1 + i)^3}\right.$$
$$\left. + \cdots + \frac{1}{(1 + i)^{n-1}} + \frac{1}{(1 + i)^n}\right] - \frac{Gn}{(1 + i)^n}$$

Dividing by i yields

$$P = \frac{G}{i}\left[\frac{1}{(1 + i)^1} + \frac{1}{(1 + i)^2} + \frac{1}{(1 + i)^3}\right.$$
$$\left. + \cdots + \frac{1}{(1 + i)^{n-1}} + \frac{1}{(1 + i)^n}\right] - \frac{Gn}{i(1 + i)^n}$$

The expression in the brackets is the present worth of a uniform series of 1 for n years. Therefore, we can substitute the expression for the P/A factor.

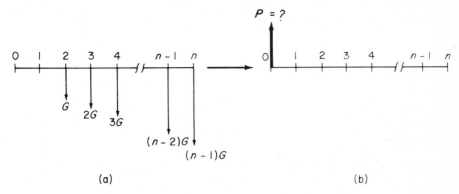

(a)   (b)

FIG. 6.4 Conversion diagram from a uniform gradient to a present worth.

$$P = \frac{G}{i}\left[\frac{(1 + i)^n - 1}{i(1 + i)^n}\right] - \frac{Gn}{i(1 + i)^n}$$

$$= \frac{G}{i}\left[\frac{(1 + i)^n - 1}{i(1 + i)^n} - \frac{n}{(1 + i)^n}\right] \qquad (6.3)$$

Equation (6.3) is the general equation which converts a uniform gradient of G for n years into a present worth at year zero; that is, the present-worth gradient factor converts Fig. 6.4(a) into Fig. 6.4(b). Keep in mind that the gradient starts in year 2. You will recognize the importance of this in the examples later in this chapter.

The equivalent uniform annual series of the gradient can be obtained simply by multiplying the present worth by the A/P factor. In this case, the gradient factor converts Fig. 6.5(a) into Fig. 6.5(b). You should realize that the annual series is

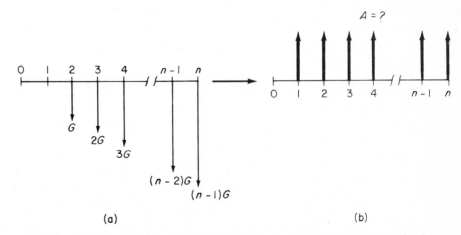

(a)   (b)

FIG. 6.5 Conversion diagram of a uniform gradient to an equivalent uniform annual series.

nothing but an A value equivalent to the gradient. Note from Fig. 6.5 that the gradient starts in year 2 and the A values occur from year 1 to year n inclusive.

Even though P in Fig. 6.4(b) is drawn as in income, it is actually the present worth of the disbursements shown in the gradient of Fig. 6.4(a). This convention is used to avoid drawing equivalent cash flows over actual cash flows in more complicated situations. In fact, this confusion would be present in Fig. 6.5 if the annual series were superimposed on the gradient.

The standard notation and equations used in working with uniform gradients, from Eq. (6.3), are:

$$P = G(P/G,i\%,n) \quad \text{and} \quad A = G(A/G,i\%,n)$$

where $(P/G, i\%, n) = \dfrac{1}{i}\left[\dfrac{(1 + i)^n - 1}{i(1 + i)^n} - \dfrac{n}{i(1 + i)^n}\right]$

The equivalent uniform capital recovery is

$$A = G(A/G, i\%, n) = G\left[\dfrac{1}{i} - \dfrac{n}{(1 + i)^n - 1}\right]$$

The $(A/G,i\%,n)$ factor is found by the relation

$$(A/G,i\%,n) = (P/G,i\%,n)(A/P,i\%,n)$$

Thus,

$$(P/G,i\%,n) = (A/G,i\%,n)(P/A,i\%,n)$$

Problem P6.3

6.3   Use of Gradient Tables

Appendix Tables A-25 and A-26 tabulate the present worth and annual series, respectively, of a uniform gradient of $1 for interest rates of 1–50% per year. Both are arranged and can be used in the same manner as the compound interest tables; that is, for a specified i and n, the correct gradient factor can be found at the intersection of the interest rate and n. Table 6.1 lists several examples of gradient factors taken from the tables.

Problems P6.4, P6.5

Table 6.1   EXAMPLES OF GRADIENT FACTORS

Value to be computed	Standard notation	i	Years, n	Table	Factor
P	$(P/G,5\%,10)$	5%	10	A-25	31.649
P	$(P/G,30\%,24)$	30	24	A-25	10.943
A	$(A/G,6\%,19)$	6	19	A-26	7.287
A	$(A/G,35\%,8)$	35	8	A-26	2.060

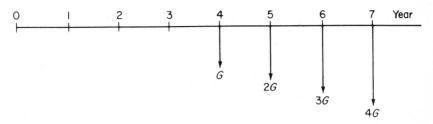

FIG. 6.6 Diagram of gradient, Example 6.2.

6.4 Location of Gradient Present Worth

In Sec. 6.2, Eq. (6.3), used for calculating the present worth of a uniform gradient, was derived. You will recall that the equation was derived for a present worth in year zero with the gradient starting in year 2 (see Fig. 6.2). Therefore, the present worth of a uniform gradient will always be located *two years before the gradient starts*. The examples that follow illustrate where the present worth of the gradient is located.

Example 6.2 For the cash-flow diagram shown in Fig. 6.6, locate the gradient present worth.

SOLUTION The present worth (P_G) of the gradient is shown in Fig. 6.7.

COMMENT In the derivation of the present worth of a gradient series, the present worth was located two years before the start of the gradient. Therefore, for the gradient in Fig. 6.6, the present worth would be located at the end of year 2. It is usually advantageous to renumber the cash-flow diagram so that the gradient year zero and the number of years (n) of the gradient can be determined. The best method for accomplishing this is to determine where the

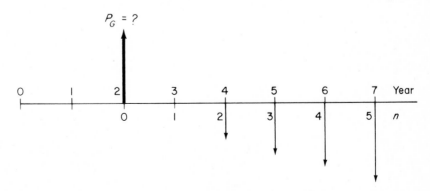

FIG. 6.7 Diagram locating the present worth of gradient in Fig. 6.6.

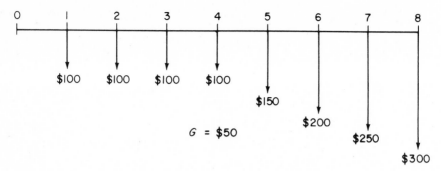

FIG. 6.8 Cash flow of gradient added to a base amount, Example 6.3.

gradient begins and label that time as year 2 and then work backward and forward. In this example, since the gradient started in year 4, gradient year 2 was placed below year 4 of the original diagram (Fig. 6.7). Year zero for the gradient was then located by moving back two years.   ////

Example 6.3 For the cash-flow diagram shown in Fig. 6.8, locate the present worth of the gradient.

SOLUTION The present worth is correctly located in Fig. 6.9.

COMMENT The gradient is $50 and begins between years 4 and 5 of the original cash-flow diagram. Therefore, year 5 of the original diagram represents year 2 of the gradient; the present worth of the gradient would then be located in year 3 (Fig. 6.9). If Fig. 6.8 is divided into two cash-flow diagrams, the location of the gradient becomes quite clear, as in Fig. 6.10.   ////

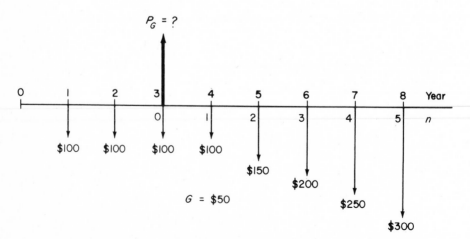

FIG. 6.9 Location of present worth of gradient in Fig. 6.8.

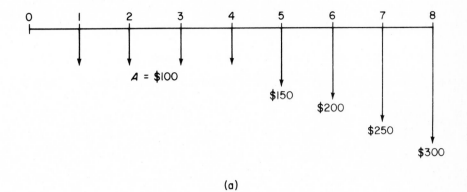

(a)

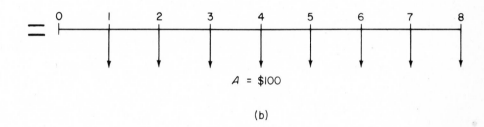

(b)

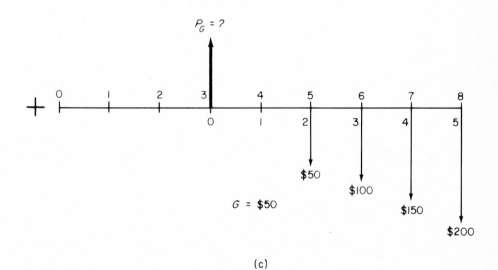

(c)

FIG. 6.10 Partitioned cash flow of Fig. 6.8, $(a) = (b) + (c)$.

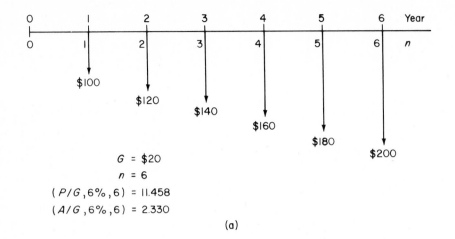

$$G = \$20$$
$$n = 6$$
$$(P/G, 6\%, 6) = 11.458$$
$$(A/G, 6\%, 6) = 2.330$$

(a)

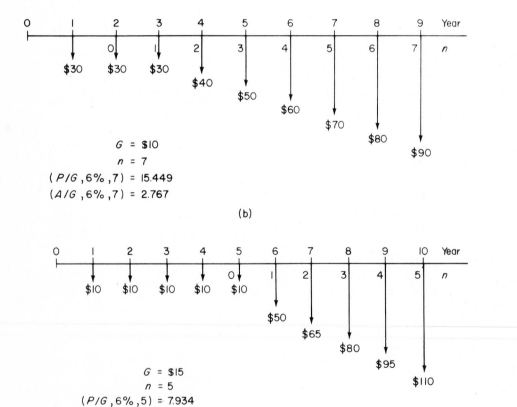

$$G = \$10$$
$$n = 7$$
$$(P/G, 6\%, 7) = 15.449$$
$$(A/G, 6\%, 7) = 2.767$$

(b)

$$G = \$15$$
$$n = 5$$
$$(P/G, 6\%, 5) = 7.934$$
$$(A/G, 6\%, 5) = 1.883$$

(c)

FIG. 6.11 Determination of G and n values used in the gradient factors.

To determine n when the gradient begins at a time after year 2, the same renumbering procedure used to determine where the present worth of the gradient is located is necessary. For the cash-flow diagrams shown in Fig. 6.11(a)–(c), the gradients (G), number of years (n), and gradient factors used to calculate the present worth and annual series of the gradients are shown on each cash-flow diagram, assuming an interest rate of 6% per year.

Example 6.9
Problems P6.6–P6.11

6.5 Present Worth and Equivalent Uniform Annual Series of Conventional Gradients

If the gradient begins in year 2, year zero of the gradient and year zero of the entire cash-flow diagram coincide, and the gradient is referred to as *conventional*. In this case the present worth or equivalent uniform annual series of the *gradient only* can be determined simply by multiplying the gradient by the appropriate gradient factor. The cash flow that forms the base amount of the gradient must be considered separately, as illustrated below.

Example 6.4 A couple plan to start saving money by depositing $500 into their savings account one year from now. They estimate that the deposits will increase by $100 per year for nine years thereafter. What would be the present worth of the investments if the interest rate is 5% per year?

SOLUTION The cash-flow diagram is shown in Fig. 6.12. Two computations must be made: the first to compute the present worth (P_A) of the base amount, and a second to compute the present worth (P_G) of the gradient. Then the total present worth, P_T equals P_A plus P_G since P_A and P_G occur in year zero. This

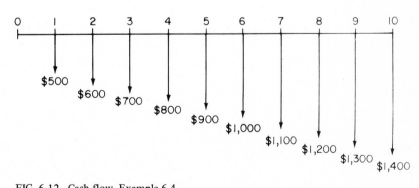

FIG. 6.12 Cash flow, Example 6.4.

is clearly illustrated by converting the cash-flow diagram as in Fig. 6.13. The present worth would be calculated as follows:

$$P_T = P_A + P_G$$
$$= 500(P/A,5\%,10) + 100(P/G,5\%,10)$$
$$= 500(7.7216) + 100(31.649)$$
$$= \$7,025.70$$

COMMENT   It is important to again emphasize that the gradient factor represents the present worth of the *gradient only*. Any other cash flow involved must be considered separately.   ////

Example 6.5   Work Example 6.4 solving for the equivalent uniform annual series.

SOLUTION   Here, too, it is necessary to consider the gradient and the other costs involved in the cash flow separately. From the diagrams shown in Fig. 6.13(b) and (c), the annual series would be

$$A = A_1 + A_G$$

where A_1 = equivalent annual series of the base amount $500
         A_G = equivalent annual series of the gradient = $100(A/G,5\%,10)$

Then,

$$A = 500 + 100(A/G,5\%,10) = 500 + 100(4.099)$$
$$= \$909.90 \text{ per year from years } 1\text{--}10$$

COMMENT   It is often helpful to remember that the present worth of the base amount and gradient can simply be multiplied by the appropriate A/P factor to get A. Here,

$$A = P_T(A/P,5\%,10) = 7,025.70(0.12951)$$
$$= \$909.90$$ ////

Problems P6.12–P6.14

6.6   Present Worth and Equivalent Annual Series of Shifted Gradients

When a gradient occurs in the middle of a series of payments it is referred to as a *shifted gradient*. The A/G factor *cannot* be used to find an equivalent A value for cash flows involving a shifted gradient. Consider the cash-flow diagram of Fig. 6.14. To find the equivalent annual disbursement, you must first find the present worth of the

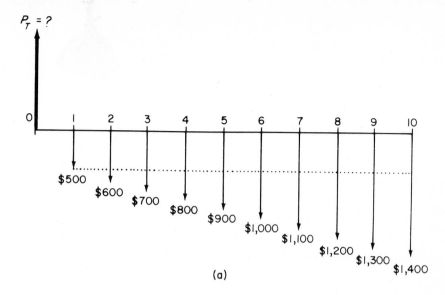

(a)

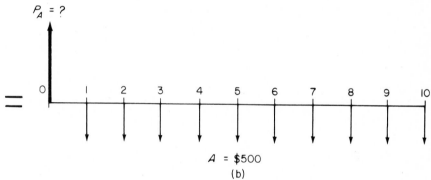

A = $500

(b)

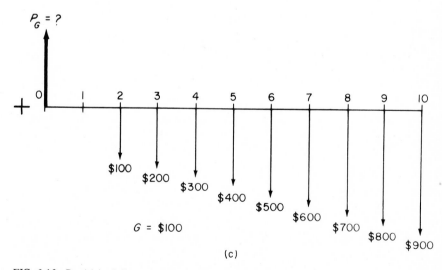

G = $100

(c)

FIG. 6.13   Partitioned diagram, Example 6.4, $(a) = (b) + (c)$.

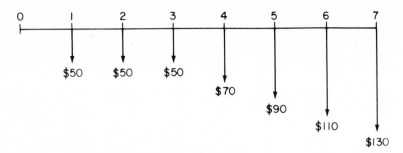

FIG. 6.14 Diagram illustrating a shifted gradient.

gradient, take this present worth back to year zero, then annualize the present worth from year zero with the A/P factor. If you used the annual-series gradient factor $(A/G,i\%,n)$, the gradient would be converted to an annual cost over years 3–7 only! For this reason, the first step in this type of problem is always to find the present worth of the gradient at actual year zero. The steps involved in handling problems of this type are illustrated in the example below.

Example 6.6 Compute the equivalent annual series for the payments of Fig. 6.14.

SOLUTION The solution steps are:

1 Consider the $50 base amount as an annual cost for all seven years (Fig. 6.15).

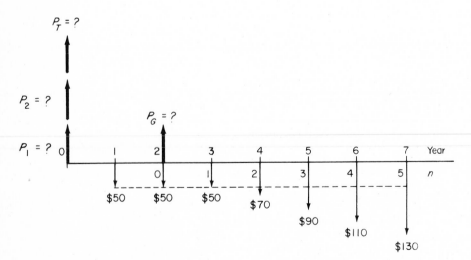

FIG. 6.15 Completed diagram, Example 6.6.

2 Find the present worth of the gradient (P_G) that occurs in year 2 as shown by the gradient-year time scale.

$$P_G = 20(P/G,i\%,5)$$

3 Bring the gradient present worth back to actual year zero.

$$P_1 = P_G(P/F,i\%,2)$$

4 Annualize the gradient present worth from year zero through year *n*.

$$A = P_1(A/P,i\%,7)$$

5 Finally, add the remaining annual costs to the gradient annual cost. Here the base amount is \$50 for all seven years and the annual series is

$$A = 20(P/G,i\%,5)(P/F,i\%,2)(A/P,i\%,7) + 50 \qquad ////$$

A total present worth (P_T) determination for the type of gradient above would involve the steps *1–3* in the example, plus the following:

4 Find the present worth (P_2) of the remaining costs.

$$P_2 = 50(P/A,i\%,7)$$

5 Add all present worths to obtain

$$P_T = P_1 + P_2 = 20(P/G,i\%,5)(P/F,i\%,2) + 50(P/A,i\%,7)$$

Example 6.10
Problems P6.15–P6.19

6.7 Decreasing Gradients

Example 6.1 involved the use of a decreasing uniform gradient, $G = \$7,500$ per year. The use of the gradient tables is the same for increasing and decreasing gradients, except that in the case of decreasing gradients the following are true:

1 The base amount is equal to the *largest* amount attained in the gradient series.
2 The gradient has a negative value; thus, the term $-G(A/G,i\%,n)$ or $-G(P/G,i\%,n)$ must be used in the computations.

The present worth of the gradient will still take place two years before the gradient starts and the *A* value will start at year 1 and continue through year *n*.

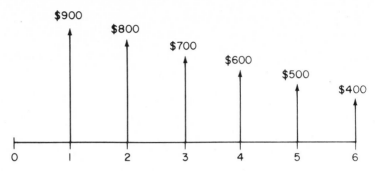

FIG. 6.16   Diagram including a decreasing gradient.

Example 6.7   Find the present worth of the receipts shown in Fig. 6.16 for $i =$ 7% per year.

SOLUTION   The cash flows of Fig. 6.16 may be separated as in Fig. 6.17. The dashed line in Fig. 6.17(a) indicates that the gradient is subtracted from an annual receipt of $900. The present worth (P_T) is computed as

$$P_T = P_A - P_G = 900(P/A,7\%,6) - 100(P/G,7\%,6)$$
$$= 900(4.7665) - 100(10.978)$$
$$= \$3,192.05 \qquad\qquad ////$$

Example 6.8   Compute an annual-series value for the cash flow of Example 6.7.

SOLUTION   The annual series is made up of two components: the base amount and the equivalent uniform-gradient amount. The annual receipt series $(A_1 = \$900)$ is the base amount, and the annual series (A_G), which is equivalent to the gradient, must be subtracted from A_1. Thus,

$$A = A_1 - A_G = 900 - 100(A/G,7\%,6)$$
$$= 900 - 100(2.303)$$
$$= \$669.70 \text{ per year for years 1-6} \qquad\qquad ////$$

Shifted decreasing gradients are handled in a fashion similar to shifted increasing gradients. For an example that combines conventional increasing and shifted decreasing gradients, see the Solved Examples.

Example 6.11
Problems P6.20–P6.25

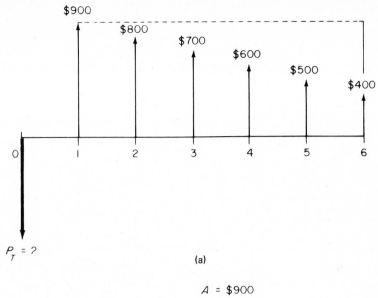

(a)

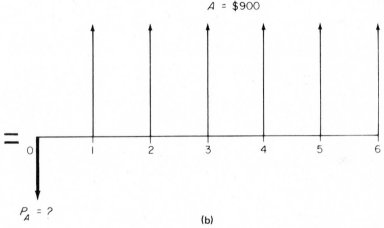

(b)

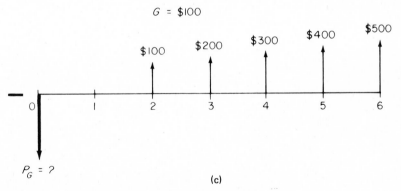

FIG. 6.17 Partitioned cash flow of Fig. 6.16, $(a) = (b) - (c)$.

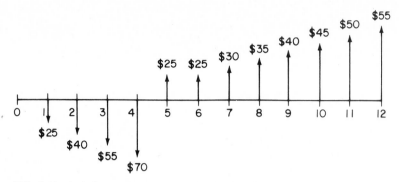

FIG. 6.18 Cash flow of two gradients, Example 6.9.

SOLVED EXAMPLES

Example 6.9 Determine the amount of the gradients, the location of the present worth of the gradients, and the n values of the cash flow of Fig. 6.18.

SOLUTION You should construct your own cash-flow diagram upon which you locate the gradient present worths and determine the n values. If we call the series from year 1 to year 4 G_1, the base amount is $25, G_1 is $15, n_1 equals 4 years, and P_{G1} occurs in year zero. For the second series, the base amount is also $25, but G_2 is $5, n_2 equals 7 years, and P_{G2} takes place in year 5.

COMMENT Even though Fig. 6.18 shows a series of disbursements and receipts, both gradients are increasing. Decreasing gradients are illustrated in Sec. 6.7.   ////

Sec. 6.4

Example 6.10 Using $i = 8\%$ for the cash flows of Fig. 6.19, compute (*a*) the equivalent annual series and (*b*) present worth.

SOLUTION

(*a*) The dashed lines of Fig. 6.19 should help you in the solution for present worth and equivalent annual series. For the annual series, if we use the steps outlined in Sec. 6.6:

  1 $A_1 = $60 for seven years
   A $= $40 base amount of gradient for four years
   $A_3 =$ equivalent series of base amount for seven years
     $= P_2(A/P,8\%,7)$

where $P_2 =$ present worth of $A = $40 series
     $= 40(P/A,8\%,4)(P/F,8\%,3)$
     $= $105.17

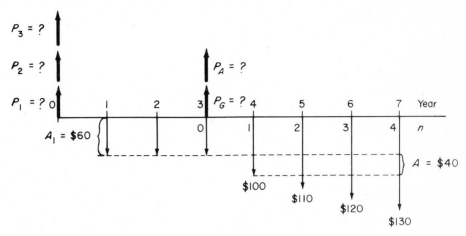

FIG. 6.19 Shifted gradient, Example 6.10.

Then, $A_3 = 105.17(A/P,8\%,7)$
$= \$20.20$

2 P_G = present worth of gradient in year 3
$= G(P/G,8\%,4) = 10(4.650)$
$= \$46.50$

3 P_1 = present worth of P_G in year zero
$= P_G(P/F,8\%,3) = 46.50(0.7938)$
$= \$36.91$

4 A_2 = equivalent seven-year A value of gradient
$= P_1(A/P,8\%,7) = 36.91(0.19207)$
$= \$7.09$

5 *Finally*, the equivalent annual series is

$$A = A_1 + A_2 + A_3 = 60.00 + 7.09 + 20.20$$
$$= \$87.29$$

(*b*) To find the present worth of the cash flows shown in Fig. 6.19, the present worth (P_1) of the gradient would be the same as calculated in step *3* above. The remaining steps yield:

4 The $40 series has a present worth (P_2) of

$$P_2 = 40(P/A,8\%,4)(P/F,8\%,3) = 40(3.3121)(0.7938)$$
$$= \$105.17$$

The $60 annual series has a present worth (P_3) of

$$P_3 = 60(P/A,8\%,7) = 60(5.2064)$$
$$= \$312.38$$

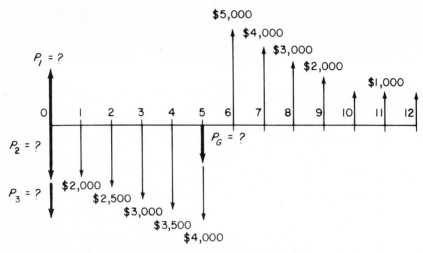

FIG. 6.20 An investment and withdrawal sequence, Example 6.11.

5 The total present worth (P_T) is

$$P_T = P_1 + P_2 + P_3 = 36.91 + 105.17 + 312.38$$
$$= \$454.46$$

which is equivalent to $87.29 per year, as in part (a). ////

Sec. 6.6

Example 6.11 Assume that you are planning to invest money at 7% per year shown by the increasing gradient of Fig. 6.20. In addition, you plan to withdraw according to the decreasing gradient shown. Find the present worth and equivalent annual series for the investment and withdrawal sequence.

SOLUTION For the investment sequence, G is $500, the base amount is $2,000, and n equals 5; for the withdrawal sequence, G is $-1,000, the base amount is $5,000, and n equals 5; there is also a two-year annual series with A equal to $1,000 in years 11 and 12. For the investment sequence,

$$P_I = \text{present worth of investments}$$
$$= 2,000(P/A,7\%,5) + 500(P/G,7\%,5)$$
$$= 2,000(4.1002) + 500(7.646)$$
$$= \$12,023.40$$

For the withdrawal sequence,

$$P_W = \text{present worth of withdrawal gradient}$$
$$+ \text{present worth of withdrawals in years 11 and 12}$$
$$= P_2 + P_3 = P_G(P/F,7\%,5) + P_3$$

$$P_W = [5,000(P/A,7\%,5) - 1,000(P/G,7\%,5)] (P/F,7\%,5)$$
$$+ 1,000(P/A,7\%,2)(P/F,7\%,10)$$
$$= [5,000(4.1002) - 1,000(7.646)] (0.7130)$$
$$+ 1,000(1.8080)(0.5084)$$
$$= \$10,084.80$$

Since P_I is actually a negative cash flow and P_W positive, the resultant present worth (P) is

$$P = P_W - P_I = 10,084.80 - 12,023.40$$
$$= \$-1,938.60$$

The A value may be computed using $P = \$-1,938.60$:

$$A = P(A/P,7\%,12)$$
$$= \$-244.07$$

COMMENT Thus, in present-worth equivalent, you will invest $1,938.60 more than you plan to withdraw. This is equivalent to an annual savings of $244.07 per year for the 12 years.   ////

Sec. 6.7

BIBLIOGRAPHY

Barish, pp. 66–70.
DeGarmo and Canada, pp. 128–132.
Emerson and Taylor, pp. 2-13 – 2-15.
Fabrycky and Thuesen, pp. 52–54.
Grant and Ireson, pp. 41–43.
Smith, pp. 47–50, 71–73.
Taylor, pp. 37–39.
Thuesen, Fabrycky, and Thuesen, pp. 60–63.

PROBLEMS

P6.1 The income from a newly purchased piece of construction equipment is expected to be $5,000 for the first year and to increase by $500 a year until the fifth year of ownership. However, starting in year 6 it is expected to bring in $800 less each year. The company plans to keep the equipment for a total of 12 years. Construct the cash-flow diagram of the revenues described.

P6.2   Construct the cash-flow diagram for the following deposits (year = k):

Year, k	Deposit
1	$ 60
2	60
3	60
4–7	$100 + 10(k-4)$

P6.3   Show that $A = G(A/G,i\%,n)$. Start with Eq. (6.3) of Sec. 6.2.

P6.4   Find the value of the factor to convert a 12-year gradient to an equivalent uniform annual series if interest is at 20% per year.

P6.5   Find the correct factor value for the following:

(a) $(P/G,10\%,8)$

(b) $(A/G,15\%,5)$

(c) $(A/G,17\%,13)$

(d) $(P/G,28\%,41)$

P6.6   A cash-flow sequence starts in year 1 at $200 and increases to $354 in year 8. Do the following: (a) construct the cash-flow diagram; (b) determine the amount of the gradient; (c) locate the gradient present worth on the diagram; and (d) determine the value of n for the gradient factor.

P6.7   For the data of problem P6.2, locate the gradient present worth and determine the gradient annual-series factor value for $i = 25\%$.

P6.8   Determine the n value for the cash flows of Fig. 6.8, using two different base amounts.

P6.9   A couple expect to borrow $500 each year for the next two years to cover Christmas expenses. Due to increasing costs, they expect to have to borrow $550 three years from now, $600 the next year, and $650 the following year. However, due to their children's ages, they hope to have to borrow only $300 per year after that. (a) Draw the cash-flow diagram of loans and (b) show where the gradient present-worth will occur; (c) also indicate the n value for the gradient series.

P6.10   The UR-OK Company is considering two types of machines, both of which will do the same job. The net cash flows for each are tabulated below. Locate the present worth of all gradients on a cash-flow diagram and determine the base amounts, gradient amounts, and n values for each machine.

	Cash flows			Cash flows	
Year	Machine A	Machine B	Year	Machine A	Machine B
0	$+2,000	$+2,000	7	$+3,000	$+2,500
1	2,000	2,000	8	3,500	2,500
2	2,000	2,500	9	4,000	2,500
3	2,500	3,000	10	4,500	3,000
4	2,500	3,500	11	3,000	3,500
5	2,500	4,000	12	3,000	4,000
6	2,500	2,500			

P6.11 For Examples 6.1 and 6.2 determine the values for G, n, $(P/G,4\%,n)$ and $(A/G,4\%,n)$.

P6.12 Consider the cash flow depicted in Fig. 6.11(*a*). For this situation, if $i = 10\%$ per year, compute the (*a*) present worth and (*b*) equivalent uniform annual series and construct its cash-flow diagram.

P6.13 An engineer has a second job of selling shoes at a neighborhood store. Her annual income has increased by $200 a year from a base of $500 four years ago. If she makes 5% on this extra income in a savings account, what is the equivalent future worth of her income after three more years of moonlighting, assuming her income continues to increase?

P6.14 Find the equivalent annual income for the situation presented in problem P6.13 from the time the engineer started the job to three years from now.

P6.15 Find the present worth in year zero of the *gradient only* depicted in Fig. 6.11(*b*) at $i = 6\%$.

P6.16 Find the equivalent annual series value for the cash flows depicted in Fig. 6.11(*c*) at $i = 6\%$.

P6.17 Compute the present worths for the two cash flows of problem P6.10 if $i = 5\%$ per year.

P6.18 Compute the present worth of the cash flows shown below at $i = 15\%$.

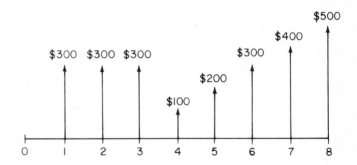

P6.19 A newspaper carrier's route has been increasing each year. The average annual net profits are given below.

Year	Net profit
1970	$400
1971	200
1972	300
1973	400
1974	500
1975	600
1976	700
1977	800

If money is worth 8%, compute the present worth of his income in 1969 two ways, using a different base amount of the gradient in each case. Compare the two answers to see how close they are. Hopefully they are equal. If not, try again!

P6.20 Compute the equivalent annual-cost series for $i = 6\%$ per year for a series of payments beginning at the end of the first year with $1,000 and decreasing by $100 per year to zero.

P6.21 Find the value of the factor to convert a 17-year decreasing gradient to a present worth if interest is at $9\frac{1}{2}\%$ per year.

P6.22 Compute the present worth and equivalent annual cost series at $i = 10\%$ per year for the cash flows below.

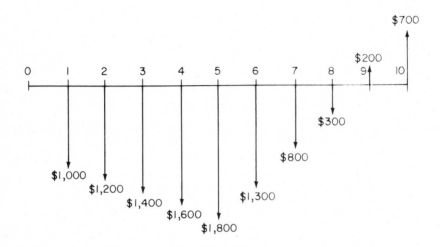

P6.23 Find the present worth of the cash flows shown below at 10% per year interest.

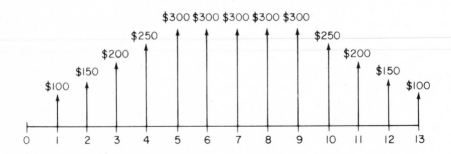

P6.24 The Chisel-Em Company plans to purchase a new piece of construction equipment now. Realizable income is $15,000 the first year, $12,000 in year 2, $9,000 in year 3, and so on. If the company plans to sell the equipment after seven years and interest is 15% per year, compute the present worth and equivalent annual series of the incomes.

P6.25 Compute the present worth and future worth of the cash flows shown below if $i = 6\%$ per year.

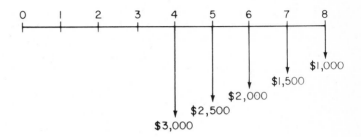

7

DEPRECIATION AND DEPLETION

The objective of this chapter is to acquaint you with the various methods of depreciating fixed assets. It will also teach you how to calculate yearly depreciation charges and book values using the three most common methods. In addition, you are introduced to the concept of depletion for reserves of natural resources.

CRITERIA

In order to complete this chapter, you must be able to do the following:

1 Define *depreciation*, *book value*, and *market value*.
2 State what is meant by additional first-year depreciation and investment tax credit.
3 State the procedure for calculating depreciation by the sinking-fund method.

   Given the initial cost, salvage value, and life of the asset(s) for criteria *4–8*, you must also be able to do the following:

4 Calculate the depreciation charge and book value for a specified year using the straight-line method.

5 Calculate the depreciation charge and book value for a specified year using the sum-of-year-digits method.
6 Calculate the depreciation charge and book value for a specified year using the declining-balance or the double declining-balance method.
7 Calculate the annual depreciation charge for a class of assets using the group method or composite method.
8 Determine which of two assets must pay extra income taxes, and the respective amount, using the straight-line method of depreciation.
9 Define *depletion* and compute the depletion charge using factor depletion, given the initial investment and resource capacity; or compute the depletion allowance given the depletion rate and annual gross incomes.

EXPLANATION OF MATERIAL

7.1 Depreciation Terminology

The dictionary definition of *depreciation* is "a decrease in value of property through wear, deterioration, or obsolescence." As stated in the definition, there are several reasons why an asset can decrease in value from its original worth. Thus, even though a machine might be in perfect mechanical condition, it may be worth considerably less than when it was new because of technological improvements in machine capability. Regardless of the reason for an asset's decrease in value, the depreciation must be taken into account in engineering-economy studies.

As you will see in Sec. 7.8, the major effect of depreciation in engineering-economy calculations is in income tax consideration. That is, income taxes are paid on net income *less* depreciation, thereby lowering the taxes paid. This allows the company to retain some of its income for replacement and additional investment.

Inasmuch as the word *value* was used in the definition of depreciation, perhaps it would be worthwhile to define what is meant by *value* in depreciation terminology. The *book value* of an asset refers to the difference between its original cost and the total amount of depreciation that has been charged to date. That is, book value represents the current worth of an asset as shown in the books of account. Since depreciation is charged once a year, the book value is computed at the end of the year and thus is in keeping with the end-of-year convention used previously. Book value is never taken into consideration in before-tax engineering-economy studies.

The *market value* of an asset refers to the amount of money that could be obtained for the asset if it were sold in the free market. In some cases, the market value bears very little relationship to the book value. For example, commercial buildings tend to increase in market value, while their book value decreases due to depreciation charges. It is the market value which must be taken into consideration in engineering-economy comparisons.

7.2 Additional First-Year Depreciation and Investment Tax Credit

In an effort to encourage more rapid economic growth, the federal government in recent years has permitted accelerated depreciation in order to provide more funds for industrial expansion. In addition to the accelerated depreciation methods that are detailed in the following sections, early write-off of depreciable assets can be increased even more through additional first-year depreciation and the investment tax credit.

Additional first-year depreciation allows the business firm to take an additional 20% depreciation deduction on assets costing up to $10,000 during the year an asset is purchased. If a joint return is filed (common in small businesses), the allowable asset first-cost increases to a maximum of $20,000. Thus, a small business may deduct up to $4,000 via the additional first-year allowance over and above the normal depreciation charges made.

A second method of accelerating depreciation is through the *investment tax credit*, which applies directly to taxes and can reduce the company's income tax by as much as 7% of the first cost of the asset in the year the asset is purchased. For example, if a company purchased a piece of equipment costing $8,000, it could deduct $8,000(0.07) = $560 from its taxes. This deduction can be made along with the additional first-year depreciation and the normal depreciation charges explained below. The investment tax credit is discussed in detail in Sec. 15.4.

Since there are certain conditions that preclude the use of either of these accelerated methods and other methods not mentioned here, none of the methods will be treated in detail in this chapter. Furthermore, in most cases, the use of these accelerated depreciation methods would not change the decision reached without their use.

Problems P7.1, P7.2

7.3 Sinking-Fund Depreciation

The sinking-fund method of depreciation, also known as annuity or compound-interest method, is one of the older methods of depreciation, almost never used today. In this method, a sinking fund is "established" which will amount to the first cost (less salvage value) of the asset at the time the asset is retired. In reality, an actual fund is never established, but the depreciation that is charged for any given year is equal to the amount the sinking fund would have increased in that year. The amount the fund increases in any given year is equal to the *uniform annual* "payment" plus the interest on all of the preceding payments and interest. Thus, in establishing a fund that would have an annual "payment" of $1,000 at an interest rate of 8%, the depreciation charge in the first year would be $1,000. In the second year, the charge would be $1,000 + $1,000(0.08) = $1,080. In the third year, the depreciation charge would be $1,000 + $2,080(0.08) = $1,166.40, and so on. It should be obvious that the depreciation charges increase rapidly in the later years of the asset's life. Since most business concerns prefer rapid write-off so that they will have the benefit of larger retained earnings in the early years and, therefore, for a longer period of time, the sinking-fund method is now primarily of historical significance.

The annual "deposit" that must be made into the sinking fund can be calculated by multiplying the $(A/F,i\%,n)$ factor by the first cost minus the salvage value of the asset. The depreciation charge for any given year can then be calculated by multiplying the $(F/P,i\%,n'-1)$ factor by the annual "deposit," where n' is equal to the number of years that have elapsed from the time the asset was purchased. The book value for any given year would be equal to the difference between the amount accumulated in the sinking fund for n years and the first cost of the asset. That is, if the asset described above had a first cost of $10,000, the book value after five years would be $10,000 − 1,000$(F/A,8\%,5)$ = $4,133.

Problem P7.3

7.4 Straight-Line Depreciation

The straight-line (SL) method of depreciation is one of the most common methods in use today. It derives its name from the fact that the book value of the asset decreases linearly with time, because the same depreciation charge is made each year. The yearly depreciation is determined by dividing the first cost of the asset minus its salvage value by the life of the asset. In equation form,

$$D = \frac{P - SV}{n} \qquad (7.1)$$

where D = annual depreciation
P = first cost of the asset
SV = salvage value of the asset
n = expected depreciable life of the asset

The first cost (P) includes purchase price, delivery cost, installation cost, and other equipment-related costs. The salvage value (SV) is a net realizable value after any dismantling or removal costs have been subtracted from actual monetary value.

Since the asset is depreciated by the same amount each year, the book value after m years of service (BV_m) will be equal to the first cost of the asset minus the annual depreciation times m. Thus,

$$BV_m = P - mD \qquad (7.2)$$

These calculations are illustrated in Example 7.1.

Example 7.1 If an asset has a first cost of $50,000 with a $10,000 salvage value after five years, (*a*) calculate the annual depreciation and (*b*) compute and plot the book value of the asset after each year using SL depreciation.

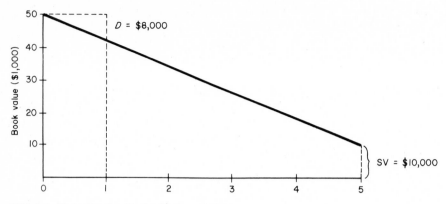

FIG. 7.1   Plot of book value for SL depreciation.

SOLUTION

(a) The depreciation per year can be found by using Eq. (7.1).

$$D = \frac{P - SV}{n} = \frac{50,000 - 10,000}{5}$$

$$= \$8,000 \text{ per year}$$

(b) The book value of the asset after each year can be found by Eq. (7.2).

$$BV_m = P - mD \quad (m = 1, 2, 3, 4, 5)$$

Thus,

$$BV_1 = 50,000 - 1(8,000) = \$42,000$$
$$BV_2 = 50,000 - 2(8,000) = \$34,000$$

and so on until

$$BV_5 = 50,000 - 5(8,000) = \$10,000 = SV$$

A plot of BV versus m is given in Fig. 7.1.

COMMENT   The simplicity of the SL depreciation method is evident in this example and is undoubtedly one of the major reasons for its wide acceptance. One further fact about SL depreciation you will be able to use later is the value $d = 1/n$, referred to as the *rate of depreciation*; that is, $1/n$ represents the fraction of the depreciable amount (i.e., $P - SV$) that is removed from the accounting records each year.   ////

Problems P7.4, P7.5

7.5 Sum-of-Year-Digits Depreciation

In Sec. 7.2, two methods that permitted rapid write-off of assets by allowing extra depreciation charges in the first year only were discussed. The sum-of-year-digits

(SYD) method of depreciation is a rapid write-off technique by which most of the value of the asset is written off in the first one-third of its life. That is, the depreciation charges are very high in the first few years, but decrease rapidly in later years of the asset's life. This procedure, therefore, is just the opposite of the sinking-fund method.

The mechanics of the method involve initially finding the sum of the year digits from 1 through n, the life of the asset. The number obtained in this manner represents the sum-of-year-digits. The depreciation charge for any given year is then obtained by multiplying the first cost of the asset less its salvage value $(P - SV)$ by the ratio of the number of years remaining in the life of the asset to the sum-of-year-digits. In equation form,

$$D_m = \frac{\text{depreciable years remaining}}{\text{sum-of-year-digits}} \text{ (first cost} - \text{salvage value)}$$

$$= \frac{n - m + 1}{SYD} (P - SV) \tag{7.3}$$

where D_m = depreciation charge for any given year, m
$\quad\quad SYD$ = sum-of-year-digits 1 to n

$$= \sum_{j=1}^{n} j = \frac{n(n + 1)}{2} \tag{7.4}$$

Note that the depreciable years remaining must include the year for which the depreciation charge is desired. That is why the 1 has been included in the numerator of Eq. (7.3). For example, if it is desired to determine the depreciation for the fourth year of an asset which has an eight-year life, the numerator of Eq. (7.3) would be $8 - 4 + 1 = 5$. If an asset has a ten-year life, using Eq. (7.4)

$$SYD = \sum_{j=1}^{10} j = 1 + 2 + \cdots + 10 = \frac{10(11)}{2}$$

$$= 55$$

The book value for any given year can be calculated without making the year-by-year depreciation determinations through the use of the following equation:

$$BV_m = P - \left[\frac{m(n - m/2 + 0.5)}{SYD} \right] (P - SV) \tag{7.5}$$

Example 7.2 illustrates the calculations for the SYD depreciation method.

Example 7.2 Calculate the depreciation charges for the first three years and the book value for year 3 for an asset which had a first cost of $25,000, a $4,000 salvage value, and a life of eight years.

SOLUTION   The sum-of-year-digits must be calculated first, using Eq. (7.4):

$$\text{SYD} = \frac{n(n+1)}{2} = \frac{8(9)}{2} = 36$$

The depreciation charges for each of the three years can now be calculated, using Eq. (7.3):

$$D_1 = \frac{(8-1+1)}{36}(25,000 - 4,000) = \$4,667$$

$$D_2 = \frac{7}{36}(21,000) = \$4,083$$

$$D_3 = \frac{6}{36}(21,000) = \$3,500$$

Note that $D_1 > D_2 > D_3$ to indicate that depreciation takes place at a decreasing rate. The book value for year 3, using Eq. (7.5), is

$$BV_3 = 25,000 - \left[\frac{3(8 - 3/2 + 0.5)}{36}\right](25,000 - 4,000)$$

$$= 25,000 - \left[\frac{3(7)}{36}\right](21,000)$$

$$= \$12,750 \qquad\qquad\qquad ////$$

Problems P7.6, P7.7

7.6   Declining-Balance Depreciation

The declining-balance method of depreciation, also known as the uniform- or fixed-percentage method, is another of the rapid write-off techniques. Very simply, the depreciation charge for any year is determined by multiplying a uniform percentage by the book value for that year. For example, if the uniform-percentage depreciation rate was 10%, then the depreciation write-off for any given year would be 10% of the book value for that year. Obviously, the depreciation charge is largest in the first year and decreases each succeeding year.

When the depreciation laws were liberalized in 1954 by the Internal Revenue Service, the maximum percentage depreciation that was permitted was *double the straight-line rate*, that is, $2/n$. When this rate is used, the method is known as the *double declining-balance* (DDB) method. Thus, if an asset has a useful life of 10 years, the straight-line rate would be $1/n = 1/10$ or 10% per year. The uniform rate of 20%, therefore, could be used in the DDB depreciation method. The general formula for calculating the maximum uniform depreciation rate per year is

$$\text{Maximum rate} = 2\left(\frac{1}{n}\right)100\% = \frac{200\%}{n}$$

Since the depreciation is determined by taking a fixed percentage of a decreasing number (i.e., the book value), the book value of the asset would never reach zero. Therefore, the tax laws permit reverting back to the straight-line method at any time during the life of the asset so that the firm can take advantage of the higher rate. Since the percentage that is used in the DDB method is twice the straight-line rate, it would be advantageous to change to the straight-line method when the asset's life is half over, if there is no salvage value. When salvage values are involved (which is usually the case, since almost everything has at least some scrap value), the most opportune time to switch would have to be determined by trial-and-error (see Solved Examples).

When DDB depreciation is used, the salvage value should *not* be subtracted from the first cost when calculating the depreciation charge. It is important that you remember this, since this procedure further increases the rate of write-off in the early years. However, even though salvage values are not considered in the depreciation calculations, an asset may not be depreciated below the amount that would be considered a reasonable salvage value. Generally, this is important only for short-lived assets ($n < 5$) or assets having large salvage values ($SV > 0.2P$).

The depreciation (D_m) for any given year (m) can be calculated for any value of depreciation rate (d) without making the intermediate calculations by using the formula

$$D_m = dP(1 - d)^{m-1} \qquad (7.6)$$

Similarly, whenever d is known,

$$D_m = BV_{m-1}(d) \qquad (7.7)$$

The book value for any year (BV_m) can be computed as

$$BV_m = P(1 - d)^m \qquad (7.8)$$

When DDB is used, $d = 2/n$ is substituted into the three equations above. Finally, since the SV is not used directly in the declining-balance method, a relation which will compute SV after n years is the same as BV_n, that is,

$$SV = BV_n = P(1 - d)^n \qquad (7.9)$$

With this information it is possible to compute d using Eq. (7.9) as

$$d = 1 - \left(\frac{SV}{P}\right)^{1/n} \qquad SV > 0 \qquad (7.10)$$

Naturally, in the case of DDB, $d = 2/n$ is known and Eq. (7.10) is not needed. Example 7.3 illustrates the calculations for the DDB method (see Solved Examples for an illustration where $d < 2/n$).

Example 7.3 Assume an asset has a first cost of $25,000 and a $4,000 salvage value after 12 years. Calculate its depreciation and book value for (a) year 1 and (b) year 4, and (c) the SV after 12 years using the DDB method.

SOLUTION   The first step is to calculate the DDB depreciation rate.

$$d = \frac{200\%}{12} = 16.67\% \text{ per year}$$

(a) For the first year, the depreciation and book value can be calculated using Eqs. (7.7) and (7.8) where $BV_0 = P$:

$$D_1 = 25,000(0.1667) = \$4,167.50$$

and

$$BV_1 = 25,000(1 - 0.1667)^1 = \$20,832.50$$

(b) From Eqs. (7.6) and (7.8) and $d = 0.1667$,

$$D_4 = \frac{2(25,000)}{12} (1 - 0.1667)^{4-1} = \$2,410.98$$

$$BV_4 = 25,000(1 - 0.1667)^4 = \$12,054.40$$

(c) Using Eq. (7.9), the salvage value at $n = 12$ is

$$SV = 25,000(1 - 0.1667)^{12} = \$2,802.57$$

Since the salvage value is anticipated to be \$4,000, the lower limit on BV is \$4,000.

COMMENT   The most important thing to remember about the DDB method is that the salvage value is not subtracted from the first cost when the yearly depreciation is calculated.   ////

Examples 7.9–7.11
Problems P7.8–P7.19

7.7   Group and Composite Methods of Depreciation

All previous methods of depreciation are called *unit depreciation* because they allow write-off on only one asset at a time. Due to the great number of assets in large companies, assets are usually grouped into classes and the depreciation charge is computed for the class, not each individual asset. An added advantage of classifying assets and depreciating all of them as a unit is that no capital gain or loss (Sec. 15.1) need be computed at the time of retirement of a single asset. The cost is simply removed from the asset class account and a gain or loss is determined at the time of retirement of the last asset in the class.

The *group depreciation* method requires computation of a single annual depreciation charge for a group of assets using their average life. Retirement after a number of years different than the average requires the prorating of depreciation charges to the remaining assets. Any method of unit depreciation can be used to compute the group depreciation charge.

Example 7.4 A purchase of 100 sewing machines has been made by the Zipper Underlings Company. Average useful life for the machines is five years and total purchase cost was $40,000. Compute the group depreciation charge for this collection of machines using straight-line depreciation.

SOLUTION If, by the absence of mention of a salvage value, we assume SV = 0, the annual straight-line depreciation charge by Eq. (7.1) is

$$D = \frac{P - SV}{n} = \frac{40,000}{5} = \$8,000$$

COMMENT Sale prior to or after five years would have to take account of the fact that annual unit depreciation is $80 per machine. See Solved Examples for further details. ////

Composite depreciation uses the same principle as group depreciation, but accounts for varying expected lives within the same asset class. A depreciation charge for assets with the same lives is calculated, and total depreciation is then found by adding the depreciation for all assets in the class. Finally, a composite life is computed as

$$\text{Composite life} = \frac{\text{total depreciable value}}{\text{total annual depreciation}}$$

The rate determined for the composite method is continued until the composite life is reached or until the lives in the class change significantly by the addition or retirement of assets. As in group depreciation, capital gains and losses on assets are not computed when an individual asset is retired.

Example 7.5 Four similar assets are purchased, the details of which are given in the first four columns of Table 7.1. Determine the annual depreciation charges and composite life for this class of asset using straight-line depreciation.

SOLUTION For each asset we compute the straight-line depreciation (Table 7.1). Total annual depreciation is $28,000 for the class (this is a depreciation rate of

Table 7.1 COMPOSITE METHOD OF DEPRECIATION

Asset	Life, n	Cost, P	Salvage, SV	Depreciable value	Annual depreciation
A	4	$ 50,000	$10,000	$ 40,000	$10,000
B	6	22,000	4,000	18,000	3,000
C	7	34,000	6,000	28,000	4,000
D	8	93,000	5,000	88,000	11,000
		$199,000		$174,000	$28,000

28,000/199,000 = 0.141 per year). The composite life is 174,000/28,000 = 6.21 years, which may be viewed as a weighted average life of assets in this class.   ////

Example 7.12
Problems P7.20, P7.21

7.8   Income Tax Calculations

As stated earlier in the chapter, depreciation is of concern in engineering economy primarily because of its effect on income taxes. In general, an income tax rate of 50% can be used in engineering-economy studies wherein income taxes are taken into consideration. This 50% rate is an approximation to the 48% corporate income tax on taxable income above $25,000, with some consideration for state or local income taxes. When the taxable income is not expected to greatly exceed $25,000 or when there is reason to believe that another income tax rate would be more applicable, then the appropriate income tax rate should be used. It would always be wise, of course, to check the validity of the 50% income tax rate in economy calculations, since income tax laws are varied and complicated. In this section, a 50% income tax rate and the straight-line method of depreciation will be assumed unless stated otherwise. For a complete treatment of tax considerations in engineering-economy studies, you should refer to Chaps. 15 and 16.

The general procedure for considering income taxes in engineering-economy calculations is to determine the difference in deductions between the two alternatives under consideration and add 50% of this difference to the alternative having the least deductions. This is done because the alternative having the least deductions will have to pay more income taxes; assuming the income would be unaffected by the alternative selected, the difference in income taxes would be 50% of the difference in deductions between the alternatives. When determining the deductions for each alternative, the first cost of each alternative must be amortized over the life of the asset using the SL method. All other annual costs are then added to this figure to obtain the total annual deductions. These calculations are illustrated in Example 7.6.

Example 7.6 Determine which of the following alternatives must pay extra income taxes and the amount of the tax, using SL depreciation and an income tax rate of 50%.

	Alternative A	Alternative B
First cost	$10,000	$15,000
Salvage value	1,000	2,000
Annual cost	1,500	600
Life, years	10	10

SOLUTION The annual SL depreciation for each alternative is

$$D_A = \frac{10,000 - 1,000}{10} = \$900 \text{ per year}$$

$$D_B = \frac{15,000 - 2,000}{10} = \$1,300 \text{ per year}$$

Adding the respective costs to the depreciation, the annual deduction for each alternative is

$$\text{Deductions}_A = 900 + 1,500 = \$2,400$$
$$\text{Deductions}_B = 1,300 + 600 = \$1,900$$

Since Alternative B has $2,400 - 1,900 = \$500$ less deductions than Alternative A, income taxes for Alternative B will be greater by $0.50(500) = \$250$ per year.

COMMENT Note that an interest rate was not necessary for this part of the problem. In order to compare the alternatives by one of the methods presented in the following chapters, however, the interest rate will have to be considered. ////

Problems P7.22–P7.26

7.9 Depletion Methods

Thus far we have computed depreciation for an asset which has a value that can be recovered by purchasing a replacement. *Depletion* is similar to depreciation; however, depletion is applicable to natural resources, which, when removed, cannot be "repurchased" as can a machine or building. You can see that a depletion method is applicable to natural deposits removed from mines, wells, quarries, forests, and the like.

There are two methods of depletion: *factor or cost depletion* and *depletion allowance*. Factor depletion is based on the level of activity or usage, not time, as in depreciation. The depletion factor (d_m) for year m is

$$d_m = \frac{\text{initial investment}}{\text{resource capacity}} \qquad (7.11)$$

and the annual depletion charge is d_m times the year's usage or activity volume. As for depreciation, accumulated depletion by the cost method cannot exceed total cost of the resource. Example 7.7 illustrates factor depletion.

Example 7.7 The Knotty Wood Company has purchased some forest acreage for \$350,000, from which an estimated 175 million board feet of lumber are recoverable. Determine the depletion charges if 15 million and 22 million board feet are removed in the first and second years, respectively.

SOLUTION Using Eq. (7.11) the depletion factor for each year $m = 1, 2, \ldots$ is

$$d_m = \frac{350,000}{175} = \$2,000 \text{ per million board feet}$$

Actual depletion charges are

$$\text{First year: } 2,000(15) = \$30,000 \qquad m = 1$$
$$\text{Second year: } 2,000(22) = \$44,000 \qquad m = 2$$

This will continue until $350,000 depletion is accumulated.

COMMENT Often the recoverable material estimate is altered once operation is begun, in which case d_m must be changed. The Solved Examples present an illustration of this case. ////

The second depletion method, that of *depletion allowance*, is a special consideration given when natural resources are exploited. A flat percentage of the resource's gross income may be depleted each year provided it does not exceed 50% of taxable income. Using depletion allowance, total depletion charges may exceed actual costs with no limitation. Since it is possible to use the depletion figure computed either by the factor or by the allowance method, the depletion-allowance method is usually chosen, due to the possibility of writing off more than the original cost of the venture. Below are listed some of the percentages for those activities that can use the depletion-allowance method, which is illustrated by Example 7.8.

Activity	Percentage of gross income
Oil and gas wells	22% (maximum)
Coal, sodium chloride	10
Gravel, sand, clay	5
Sulfur, cobalt, lead, nickel, zinc, etc.	22
Gold, silver, copper, iron ore	15

Example 7.8 A gold mine purchased for $750,000 has an anticipated gross income of $1.1 million per year for years 1–5, and $0.85 million per year after year 5. Compute annual depletion charges for the mine.

SOLUTION A 15% depletion allowance applies to the gold mine. Thus, assuming depletion charges do not exceed 50% of taxable income, depletion will be 0.15 (1.1 million) = $165,000 for years 1–5 and 0.15 (0.85 million) = $127,500 each year thereafter. At this rate, the cost of $750,000 will be recovered in approximately 4.5 years of operation. ////

Since there are several tax considerations to be made when depletion is used, the tax angles are detailed in Sec. 17.5. In 1975, the depletion allowance for oil and gas wells was altered due to the tax favoritism enjoyed by this industry.

A depreciation method similar to factor depletion, but applicable to depreciable assets, is the *unit-of-production* method. The asset can be depreciated using this method only if the rate of use or production is a measure of its rate of deterioration.

The depreciation factor is computed in a fashion similar to Eq. (7.11) with the salvage value removed from the initial investment. For example, oil-producing equipment at a lease site costing $300,000 with a salvage of $50,000 may be *depreciated* on a basis of the estimated 1.0 million barrels of oil to be extracted from the lease. Then, the depreciation factor is (300,000 − 50,000)/1,000,000 = $0.25 per barrel. This method is not commonly used due to the difficulty in finding a production unit which accurately measures the rate of asset deterioration.

Example 7.13
Problems P7.27–P7.29

SOLVED EXAMPLES

Example 7.9   The Dandy Company has just purchased an ore-crushing unit for $80,000. The unit has an anticipated life of ten years and a salvage of $10,000. Use the declining-balance method to compute (*a*) the depreciation and book value for each year and (*b*) the salvage value. Compare anticipated and computed salvage values.

SOLUTION

(*a*)   The depreciation rate from Eq. (7.10) is

$$d = 1 - \left(\frac{10,000}{80,000}\right)^{1/10} = 0.188$$

Note that $d = 0.188 < 2/n = 0.2$; thus, the write-off is not as accelerated as DDB. Table 7.2 presents D_m and BV_m values.

Table 7.2 D_m AND BV_m VALUES
USING DECLINING-
BALANCE DEPRECIATION

Year	D_m Eq. (7.7)	BV_m $=BV_{m-1} - D_m$
0	–	$80,000
1	$15,040	64,960
2	12,212	52,748
3	9,917	42,831
4	8,052	34,779
5	6,538	28,241
6	5,309	22,932
7	4,311	18,621
8	3,501	15,120
9	2,843	12,277
10	2,308	9,969

(b) Since $BV_{10} = \$9,969$, this is the model-computed salvage value, which compares with the anticipated SV of $10,000. Thus a loss of $2,308 - 2,277 = \$31$ in depreciation for year 10 will be experienced if the declining-balance method is used, since the BV cannot be less than reasonable SV = $10,000.

////

Sec. 7.6

Example 7.10 Graphically compare the rate of write-off for an asset with $P = \$80,000$, SV = $10,000, and $n = 10$ years using the following methods of depreciation: SL, SYD, DDB, and declining balance.

SOLUTION Since P, SV, and n are the same as those in Example 7.9, declining-balance results are presented in Table 7.2. Table 7.3 gives the results of the other three methods, where the general expression $BV_m = BV_{m-1} - D_m$ is used for book value. A plot of BV_m versus m (Fig. 7.2) indicates that the DDB method attempts to reduce book value below the SV value in year 10; thus, only $738 depreciation is allowed for this year. The declining balance plot between DDB and SYD is omitted to avoid confusion. Note that at year 5 the book value for DDB ($26,214) is approximately only 58% of the SL method ($45,000), attesting to the very rapid write-off of DDB. Under DDB, approximately $\frac{1}{2}P$ is removed from the books after only three years.

COMMENT If any accelerated method reduces book value to the SV value before year n, no additional depreciation is allowed. You can see that the SYD method is a fair approximation to the DDB method, thus accounting for some of its popularity.

////

Secs. 7.4, 7.5, 7.6

Table 7.3 COMPARISON OF DEPRECIATION METHODS
($P = \$80,000$, SV = $10,000, $n = 10$ years)

Year, m	Straight-line D_m (Eq. 7.1)	BV_m	Sum-of-year-digits D_m (Eq. 7.3)	BV_m	Double declining-balance D_m (Eq. 7.8)	BV_m
0	–	$80,000	–	$80,000	–	$80,000
1	$7,000	73,000	$12,727	67,273	$16,000	64,000
2	7,000	66,000	11,455	55,818	12,800	51,200
3	7,000	59,000	10,182	45,636	10,240	40,960
4	7,000	52,000	8,909	36,727	8,192	32,768
5	7,000	45,000	7,636	29,091	6,553	26,214
6	7,000	38,000	6,364	22,727	5,243	20,971
7	7,000	31,000	5,091	17,636	4,194	16,777
8	7,000	24,000	3,818	13,818	3,355	13,422
9	7,000	17,000	2,545	11,273	2,684	10,738
10	7,000	10,000	1,273	10,000	738*	10,000

*Due to Internal Revenue tax laws, only $738 can be claimed for depreciation.

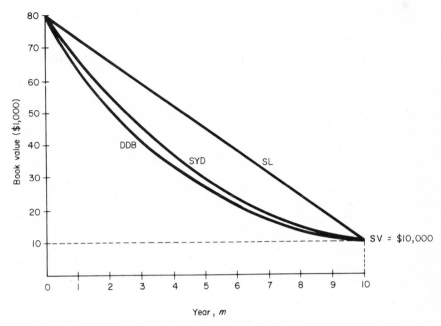

FIG. 7.2   Book-value comparison for several depreciation methods.

Example 7.11   Referring to the previous two solved examples and Table 7.2 and 7.3, when would you switch from one method of depreciation to the SL method?

SOLUTION   At any time that the annual depreciation for a method other than SL is *less* than the corresponding SL depreciation, a switch is advisable.

(*a*)   *Declining-balance* (Table 7.2): In year 5 the depreciation $6,538 < $7,000, the annual uniform SL depreciation. So, switch to SL in year 5.

(*b*)   *Sum-of-year-digits* (Table 7.3): In year 6, $6,364 < $7,000.

(*c*)   *Double declining-balance* (Table 7.3): In year 5, $6,553 < $7,000. As stated in Sec. 7.6, the switch is advisable at about half the life of the asset.   ////

Sec. 7.6

Example 7.12   Reconsider the 100 sewing machines that were group depreciated in Example 7.4. Assume 30 of these were disposed of after four years, 40 more were retired after five years, and the remaining 30 retired after six years. Compute the annual group depreciation charges for each year.

SOLUTION   From Example 7.4, annual write-off is $8,000 per year, or $80 per machine-year. Therefore, depreciation for the first four years is $8,000; for the fifth

Table 7.4 SUMMARY OF GROUP DEPRECIATION
METHOD

Year	In-service machines	Annual depreciation	Accumulated depreciation	Class book value
1	100	$8,000	$ 8,000	$32,000
2	100	8,000	16,000	24,000
3	100	8,000	24,000	16,000
4	100	8,000	32,000	8,000
5	70	5,600	37,600	2,400
6	30	2,400	40,000	0

year, with 70 machines in operation, depreciation is 70(80) = $5,600. Results are detailed in Table 7.4.

COMMENT Using *unit depreciation*, the sale after four years would require reporting of a 30(80) = $2,400 capital loss, since disposal at SV = 0 did not allow recovery of this amount of depreciation. This fact need not be recognized with group depreciation. Similarly, with unit depreciation there could be no write-off in year 6, as has been done here. ////

Sec. 7.7

Example 7.13 Consider again the forest acquisition discussed in Example 7.7. If after two years of operation total recoverable board feet is reestimated at 225 million, compute the new d_m ($m = 3, 4, \ldots$).

SOLUTION After two years, a total of $74,000 had been depleted; thus, the new d_m value must be based on the remaining $350,000 - 74,000 = $276,000 of undepleted investment. Additionally, with the new estimate of 225 million, a total of $225 - 15 - 22 = 188$ million board feet remain. Now, for $m = 3, 4, \ldots$,

$$d_m = \frac{\$276,000}{188 \text{ million}} = \$1,468 \text{ per million board feet}$$

////

Sec. 7.9

BIBLIOGRAPHY

Barish, pp. 78–86, 89–92.
DeGarmo and Canada, pp. 150–152, 159–163, 172–174.
Emerson and Taylor, pp. 10-4 – 10-13, 12-9 – 12-12, 14-7 – 14-9.
Fabrycky and Thuesen, pp. 157–165, 168, 178–180.
Grant and Ireson, pp. 152–169, 373.
Ostwald, pp. 98–102.
Park, pp. 102–109.

Reisman, pp. 246–258.
Riggs, pp. 190–204.
Smith, pp. 195–208, 211–214.
Taylor, pp. 78–80, 212–219, 329–336.
Thuesen, Fabrycky, and Thuesen, pp. 254–265, 266–267.

PROBLEMS

P7.1   A janitorial supply company has just purchased a new van for $6,500. If the sole owner of the company computes a $1,300 depreciation for the first year of ownership, what additional amount can be written off using the allowance for additional first-year depreciation?

P7.2   A cosmetic company has purchased a new piece of mixing equipment for $5,800. What amount of additional depreciation and tax credit is allowed in the first year of ownership?

P7.3   An asset was purchased in 1973 for $20,000. If expected life were ten years and salvage $2,500, use the sinking-fund depreciation method at $i = 12\%$ to compute the (a) depreciation for year 2 and (b) book value after nine years.

P7.4   Smoothline Construction, Inc., has just purchased a Glop-a-de-glop machine for $275,000. A $75,000 installation charge is required to use the machine. The expected life is 30 years with a salvage value of 10% of the purchase price. Use the straight-line (SL) method of depreciation to determine (a) first cost, (b) salvage value, (c) annual depreciation, and (d) book value after 20 years.

P7.5   A machine costing $12,000 has a life of eight years with a $2,000 salvage value. Calculate the (a) depreciation charge and (b) book value of the machine for each year using the straight-line method.

P7.6   Work problem P7.5 using the sum-of-year-digits (SYD) method.

P7.7   Earth-moving equipment having a first cost of $82,000 is expected to have a life of 18 years. The salvage value at that time is expected to be $15,000. Calculate the depreciation charge and book value for years 2, 7, 12, and 18 using the sum-of-year-digits method.

P7.8   What is the basic difference between the declining-balance and the double declining-balance methods of depreciation?

P7.9   Work problem P7.5 using the double declining-balance (DDB) method and plot the book value for the SL, SYD, and DDB depreciation methods.

P7.10 A building costing $320,000 is expected to have a 35-year life with a 25% salvage value. Calculate the depreciation charge for years 4, 9, 18, and 26 using the (a) straight-line method, (b) sum-of-year-digits method, and (c) double declining-balance method.

P7.11 Calculate the book value of the building in problem P7.10 for year 13 using the (a) straight-line method, (b) sum-of-year-digits method, and (c) double declining-balance method.

P7.12 An asset has a first cost of $45,000, a life of 15 years, and a $5,000 salvage value. What is the first year that the depreciation charge is larger by the sum-of-year-digits method than it is by the double declining-balance method?

P7.13 Using the data from problem P7.12, determine the first year that the book value will be lower by the sum-of-year-digits method than it is by the double declining-balance method.

P7.14 An asset having a first cost of $23,000 is expected to have a life of 12 years with a $2,000 salvage value. In what year does the depreciation charge by the straight-line method first exceed the depreciation charge allowed by the (*a*) sum-of-year-digits method? (*b*) double declining-balance method?

P7.15 Work problem P7.7 using (*a*) the double declining-balance method and (*b*) the declining-balance method with a depreciation rate from Eq. (7.10). Compare the book values.

P7.16 The Zap Electric Company owns a building with a first cost of $155,000 and a $65,000 salvage value after 25 years. (*a*) Determine the number of years the building should be depreciated by the sum-of-year-digits method until it would be more advantageous to use the straight-line method. (*b*) What is the book value of the building at the time the switch should be made?

P7.17 Work problem P7.16 using the double declining-balance method instead of the sum-of-year-digits method.

P7.18 A building having a first cost of $450,000 is expected to have a life of 30 years. At what salvage value would the depreciation calculated by the double declining-balance method just exceed that calculated by the sum-of-year-digits method (*a*) in year 3 and (*b*) in year 15?

P7.19 Rework problem P7.18, except use the double declining-balance method and the straight-line method.

P7.20 Western Tool and Die owns a number of three different models of lathes. The values of P, SV, and n for each lathe are given below.

Model number	Number owned	First cost	Expected life	Salvage value
A	30	$37,000	8	$5,000
B	10	4,800	12	0
C	6	20,000	10	2,500

One-half of the lathes of each model type will be sold after eight years and the remainder will be replaced after 12 years. Compute the total annual straight-line depreciation charges using (*a*) the group depreciation method and (*b*) the composite depreciation method. (*c*) Compare the total annual depreciation charges for each method, and compute the composite life based on the depreciation when all assets are owned by the company.

P7.21 Rework problem P7.20 using the same data. However, calculate the *actually realized* annual depreciation by assuming that the company uses a life of eight years for all models and the salvage values are one-half of those listed in problem P7.20.

P7.22 What is the present worth of the savings in extra income taxes realized by making the switch to the straight-line method in problem P7.16 assuming an income tax rate of 50% and an interest rate of 10%?

P7.23 An asset having a first cost of $138,000 was purchased by the Not-Too-Quik Company (NTQ Co.) five years ago. In order to reduce bookkeeping to a minimum, the company decided to use the straight-line method of depreciation instead of the double declining-balance method. If the asset has a life of 11 years and a $28,000 salvage value, (a) how much extra money did the company actually pay in income taxes, assuming an income tax rate of 50%? (b) What is the present-worth difference at the time of purchase in income taxes for the two depreciation methods if money is worth 10%?

P7.24 Work problem P7.23, except compare the sum-of-year-digits method with the straight-line method.

P7.25 A company is considering the purchase of one of the following machines:

	Machine A	Machine B
First cost, P	$10,000	$15,000
Salvage value, SV	1,000	2,000
Annual operating cost, AOC	2,000	1,000
Life, n	10	10

Assuming an income tax rate of 50%, determine which asset would involve extra income taxes and the amount of this extra tax if straight-line depreciation is used.

P7.26 Determine which of the following assets would require the payment of extra income taxes and the amount of the tax if sum-of-year-digits depreciation is used.

	Machine C	Machine D
First cost, P	$28,000	$45,000
Salvage value, SV	2,000	3,000
Annual operating cost, AOC	2,500	1,000
Overhaul every 3 years	4,000	–
Life, n	9	9

P7.27 A coal mining company has owned a mine for the past five years. During this time the following tonnage of ore has been removed each year: 40,000; 52,000; 58,000; 60,000; and 56,000 tons. The mine is estimated to contain a total of 2.0 million tons of coal and the mine had an initial cost of $3.5 million. If the company had a gross income for this coal of $15 per ton for the first two years and $18 per ton for the last three years, (*a*) compute the depletion charges each year using the larger of the values for the two accepted depletion methods and (*b*) compute the percent of the initial cost that has been written off in these five years.

P7.28 If the mine operation explained in problem P7.27 is reevaluated after the first three years of operation and estimated to contain a remaining 1.5 million tons, answer the two questions posed in P7.27.

P7.29 Assume that the earth-moving equipment described in problem P7.7 is to be depreciated by the unit-of-production method. Total tons moved in a lifetime is based on an annual average of 150,000 tons per year. If the tonnage moved in the first three years is 200,000, 250,000, and 175,000 tons, respectively, compute the depreciation charge and book value for each year.

LEVEL II

At this point you have all the computational skills necessary to perform an engineering-economy analysis. The next three chapters give you a chance to learn three of the most basic methods of alternative evaluation. All methods must give identical decisions as to which alternative is best when they are used to compare the same alternatives.

Chapter	Subject
8	Present worth, capitalized cost
9	Annual cost
10	Rate of return

8

PRESENT-WORTH AND
CAPITALIZED-COST EVALUATION

The objective of this chapter is to teach you how to compare alternatives on the basis of present-worth or capitalized-cost determinations. As you will see in later chapters, the present-worth method is probably the most versatile of all alternative evaluation procedures.

CRITERIA

In order to complete this chapter, you must be able to do the following:

1 Select the better of two alternatives using present-worth calculations for alternatives that have equal lives, given the costs and their respective dates, the lives and salvage values of the alternatives, and the interest rate.

2 Same as *1* above, except for alternatives that have different lives.

3 Define *capitalized cost* and calculate the capitalized cost of a series of disbursements, given the disbursements, their respective dates, and the interest rate.

4 Select the better of two alternatives on the basis of capitalized cost, given the disbursements, dates, lives, and salvage values for each alternative, and the interest rate.

EXPLANATION OF MATERIAL

8.1 Present-Worth Comparison of Equal-Lived Alternatives

The present-worth (PW) method of alternative evaluation is very popular because future expenditures or receipts are transformed into *equivalent dollars now*. In this form, it is very easy, even for a person unfamiliar with economic analysis, to see the economic advantage of one alternative over one or more other alternatives.

Since the present worth is always less than the future worth (in terms of dollars, not equivalently) of a disbursement or receipt when the interest rate is greater than zero, the present-worth amount is known as the *discounted cash flow*. Similarly, the interest rate used in present-worth calculations is sometimes referred to as the *discount rate*, particularly in financial institutions. These terms will not be used in this text except when they more appropriately reflect everyday usage.

The comparison of alternatives having equal lives by the present-worth method is straightforward. If both alternatives would be used in identical capacities, they are termed *equal-service* alternatives and the annual incomes will have the same numerical value. Therefore, the cash flow will involve disbursements only, in which case it is generally convenient to omit the minus sign from the disbursements. Then the alternative with the *lowest* present-worth value should be selected. On the other hand, when disbursements *and* incomes must be considered, if the above sign convention is used, the alternative selected would be the one with the *highest* present-worth value *provided incomes exceed disbursements*, and vice versa. While it does not matter which sign is used for disbursements, it is important to be consistent in assigning the proper sign to each cash-flow element. For single-project evaluation, $PW < 0$ indicates a net loss at a certain rate of return and $PW > 0$ implies a net gain greater than the stated rate of return.

It should be pointed out that a present-worth analysis can be conducted when multiple alternatives are under consideration, using the same procedures presented here for two alternatives. This is one of the advantages of this method over the rate-of-return method, as you will see in Chaps. 10 and 17.

Example 8.1 Make a present-worth comparison of the equal-service machines for which the costs are shown below, if $i = 10\%$.

	Type A	Type B
First cost, P	$2,500	$3,500
Annual operating cost, AOC	900	700
Salvage value, SV	200	350
Life, years	5	5

SOLUTION The cash-flow diagram is left to the reader. The present worth of each machine is calculated as follows:

$$P_A = 2,500 + 900(P/A,10\%,5) - 200(P/F,10\%,5) = \$5,788$$
$$P_B = 3,500 + 700(P/A,10\%,5) - 350(P/F,10\%,5) = \$5,936$$

Type A should be selected, since $P_A < P_B$.

COMMENT Note the minus sign on the salvage value, since it is a negative cost. Also, when alternatives are evaluated by the present-worth method it is common to use PW rather than P. In this case, then, $PW_A = \$5,788$ and $PW_B = \$5,936$.   ////

Example 8.5
Problems P8.1–P8.8

8.2 Present-Worth Comparison of Different-Lived Alternatives

When using the present-worth method for comparing alternatives that have different lives, the procedure of the previous section is used with the following exception: *The alternatives must be compared over the same number of years.* That is, the cash flow for one "cycle" of an alternative must be duplicated for the least common multiple of years, so that service is compared over the same total life for each alternative. For example, if it is desired to compare alternatives which have lives of three years and two years, respectively, the alternatives must be compared over a period of six years, with reinvestment assumed at the end of each life cycle. It is important to remember that when an alternative has a terminal salvage value, this must also be included and shown as an income on the cash-flow diagram at the time reinvestment is made. Example 8.2 is a present-worth comparison of alternatives having different lives.

Example 8.2 A plant superintendent is trying to decide between the machines detailed below.

	Machine A	Machine B
First cost	$11,000	$18,000
Annual operating cost	3,500	3,100
Salvage value	1,000	2,000
Life, years	6	9

Determine which one should be selected on the basis of a present-worth comparison using an interest rate of 15%.

SOLUTION Since the machines have different lives, they must be compared over their least common multiple of years, which is 18 years in this case. The

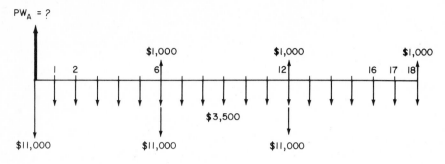

Machine A

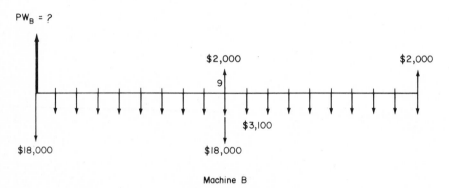

Machine B

FIG. 8.1 Cash-flow diagram for unequal life assets, Example 8.2.

cash-flow diagram is shown in Fig. 8.1. Thus,

$$\begin{aligned}
PW_A &= 11{,}000 + 11{,}000\,(P/F,\,15\%,6) - 1{,}000\,(P/F,\,15\%,6) \\
&\quad + 11{,}000\,(P/F,\,15\%,12) - 1{,}000\,(P/F,\,15\%,12) \\
&\quad - 1{,}000\,(P/F,\,15\%,18) + 3{,}500\,(P/A,15\%,18) \\
&= \$38{,}559 \\
PW_B &= 18{,}000 + 18{,}000\,(P/F,\,15\%,9) - 2{,}000\,(P/F,\,15\%,9) \\
&\quad - 2{,}000\,(P/F,\,15\%,18) + 3{,}100\,(P/A,\,15\%,18) \\
&= \$41{,}384
\end{aligned}$$

Machine A should be selected since $PW_A < PW_B$.

COMMENT Note that the salvage value of each machine must be recovered *after each life cycle* of the asset. In this example, the salvage value of Machine A was recovered in years 6, 12, and 18 while that for Machine B was recovered in years 9 and 18.   ////

Examples 8.6, 8.7
Problems P8.9–P8.16

8.3 Capitalized-Cost Calculations

Capitalized cost refers to the present-worth value of a project that is assumed to last forever. Certain public works projects such as dams and irrigation systems fall into this category. In addition, permanent university or charitable organization endowments must be handled by capitalized-cost methods.

In general, the procedure that should be followed in calculating the capitalized cost or initial cost of a permanent endowment is as follows:

1 Draw a cash-flow diagram showing all nonrecurring (one-time) expenditures or receipts and at least two cycles of all recurring (periodic) expenditures or receipts.

2 Find the present worth of all nonrecurring expenditures (receipts).

3 Find the equivalent uniform annual cost (EUAC) through one cycle of all recurring expenditures and uniform annual cost series.

4 Divide the EUAC obtained in step *3* by the interest rate to get the capitalized cost of the EUAC.

5 Add the value obtained in step *2* to the value obtained in step *4*.

The purpose for beginning the solution by drawing a cash-flow diagram should be evident from previous chapters. However, the cash-flow diagram is probably more important in this calculation than anywhere else, because it facilitates the differentiation between nonrecurring and periodic expenditures. In step *2*, the present worth of all nonrecurring expenditures (receipts) should be determined. Since the capitalized cost is the *present worth* of a perpetual project, the reason for this step should be obvious. In step *3* the EUAC (which has been called A thus far) of all recurring and uniform annual expenditures should be calculated. This is done to compute the present worth of a perpetual annual cost (capitalized cost) using

$$\text{Capitalized cost} = \frac{\text{EUAC}}{i} \qquad (8.1)$$

This can be illustrated by considering the time value of money. If \$100 is deposited into a savings account at 6% interest compounded annually, the maximum amount of money that could be withdrawn at the end of every year for *eternity* is \$6, or the amount equal to the interest that accumulated in that year. This would leave the original \$100 deposit to earn interest so that another \$6 would be accumulated in the next year. Mathematically, the amount of money that could be accumulated and withdrawn each year is

$$A = Pi \qquad (8.2)$$

Thus, for the example,

$$A = 100(0.06) = \$6 \text{ per year}$$

The capitalized-cost calculation proposed in step *4* is the reverse of the one just made; that is, Eq. (8.2) is solved for P:

$$P = \frac{A}{i} \qquad (8.3)$$

For the example just cited, if it were desired to withdraw $6 every year for eternity at an interest rate of 6%, from Eq. (8.3),

$$P = \frac{6}{0.06} = \$100$$

After the present worths of all cash flows have been obtained, the total capitalized cost is simply the sum of these present worths. Capitalized-cost calculations are illustrated in Example 8.3.

Example 8.3 Calculate the capitalized cost of a project that has an initial cost of $150,000 and an additional investment cost of $50,000 after ten years. The annual operating cost will be $5,000 for the first four years and $8,000 thereafter. In addition, there is expected to be a recurring major rework cost of $15,000 every 13 years. Assume $i = 5\%$.

SOLUTION The format outlined above will be used:

1 Draw cash flows for two cycles (Fig. 8.2).
2 Find the present worth (P_1) of the nonrecurring costs of $150,000 now and $50,000 in year 10:

$$P_1 = 150,000 + 50,000(P/F,5\%,10) = \$180,695$$

3 Convert the recurring cost of $15,000 every 13 years to an EUAC (A_1) for the first 13 years:

$$A_1 = 15,000(A/F,5\%,13) = \$847$$

4 The capitalized cost for the annual-cost series can be computed in two ways. These are (*a*) consider a series of $5,000 from now to infinity and find the present worth of $8,000 − $5,000 = $3,000 from year 5 on, or (*b*) find the present worth of $5,000 for four years and the present worth of $8,000 from year 5 to infinity. Using the first method, the annual cost (A_2) is $5,000 and the

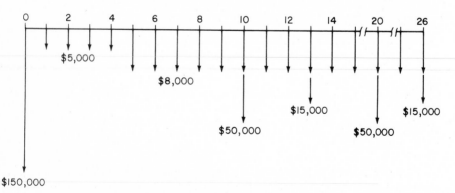

FIG. 8.2 Diagram used to compute capitalized cost, Example 8.3.

present worth (P_2) of \$3,000 from year 5 to infinity, using Eq. (8.3) and the P/F factor, is

$$P_2 = \frac{3,000}{0.05} \ (P/F, 5\%,4) = \$49,362$$

The two annual costs are converted to a capitalized cost (P_3):

$$P_3 = \frac{A_1 + A_2}{i} = \frac{847 + 5,000}{0.05} = \$116,940$$

5 The total capitalized cost (P_T) can now be obtained by addition:

$$P_T = P_1 + P_2 + P_3 = \$346,997$$

COMMENT In calculating P_2, $n = 4$ was used in the P/F factor because the present worth of the annual \$3,000 cost is computed in year 4, since P is always one year ahead of the first A. You should rework the problem using the second method suggested for calculating P_2. ////

Problems P8.17–P8.22

8.4 Capitalized-Cost Comparison of Two Alternatives

When two or more alternatives are compared on the basis of their capitalized cost, the procedure of Example 8.3 is followed. Since the capitalized cost represents the present total cost of financing and maintaining a given alternative forever, the alternatives will automatically be compared for the same number of years (i.e., infinity). The alternative with the smaller capitalized cost will represent the most economical one. As in present-worth and all other alternative evaluation methods, it is only the differences in cash flow between the alternatives which must be considered. Therefore, whenever possible, the calculations should be simplified by eliminating the elements of cash flow which are common to both alternatives. Example 8.4 shows the procedure for comparing two alternatives on the basis of their capitalized cost.

Example 8.4 Two sites are currently under consideration for a bridge to cross the Ohio River. The north site would connect a major state highway with an interstate loop around the city and would alleviate much of the local through traffic. The disadvantages of this site are that the bridge would do little to ease local traffic congestion during rush hours, and the bridge would have to stretch from one hill to another to span the widest part of the river, railroad tracks, and local highways below. This bridge would therefore be a suspension bridge. The south site would require a much shorter span allowing for construction of a truss bridge, but would require new road construction.

The suspension bridge would have a first cost of \$30 million with annual inspection and maintenance costs of \$15,000. In addition, the concrete deck will have to be resurfaced every ten years at a cost of \$50,000. The truss bridge and approach roads are expected to cost $12 million and will have annual maintenance costs of $8,000. The bridge will have to be painted every three years at a cost of $10,000. In addition, the bridge will have to be sandblasted and painted every ten years at a cost of $45,000. The cost of purchasing right-of-way is expected to be $800,000 for the suspension bridge and $10.3 million for the truss bridge. Compare the alternatives on the basis of their capitalized cost if the interest rate is 6%.

SOLUTION Construct the cash-flow diagrams before you attempt to solve the problem. You should do this *now*.

Capitalized cost of suspension bridge

$$P_1 = \text{present worth of initial cost} = 30.0 + 0.8 = \$30.8 \text{ million}$$

The recurring operating cost is $A_1 = \$15,000$, while the annual equivalent of the resurface cost is

$$A_2 = 50,000(A/F,6\%,10) = \$3,794$$

$$P_2 = \text{capitalized cost of recurring costs} = \frac{A_1 + A_2}{i}$$

$$= \frac{15,000 + 3,794}{0.06}$$

$$= \$313,233$$

Finally, the total capitalized cost (P_S) is

$$P_S = P_1 + P_2 = \$31,113,233 \quad (\$31.1 \text{ million})$$

Capitalized cost of truss bridge

$$P_1 = 12.0 + 10.3 = \$22.3 \text{ million}$$
$$A_1 = \$8,000$$
$$A_2 = \text{annual cost of painting} = 10,000(A/F,6\%,3)$$
$$\quad = \$3,141$$
$$A_3 = \text{annual cost of sandblasting} = 45,000(A/F,6\%,10)$$
$$\quad = \$3,414$$

$$P_2 = \frac{A_1 + A_2 + A_3}{i} = \$242,583$$

The total capitalized cost (P_T) is

$$P_T = P_1 + P_2 = \$22,542,583 \quad (\$22.5 \text{ million})$$

Since $P_T < P_S$, the truss bridge should be constructed. ////

Example 8.8
Problems P8.23, P8.24

SOLVED EXAMPLES

Example 8.5 A traveling saleswoman expects to purchase a new car this year. She has collected or estimated the following data: first cost is $4,800; trade-in value will be $500 after four years; annual maintenance and insurance costs are $350; and annual income due to ability to travel is $1,500. Will the woman be able to make a rate of return of 20% on her investment?

SOLUTION Compute the PW value of the investment at $i = 20\%$. (A cash-flow diagram will aid you.)

$$PW = -4,800 - 350(P/A,20\%,4) + 1,500(P/A,20\%,4)$$
$$+ 500(P/F,20\%,4)$$
$$= \$-1,582$$

Indeed, she would not make 20%, since the PW is much less than zero.

COMMENT If the PW value had been greater than zero, an excess of 20% would be returned. In Chap. 10, calculations similar to those above will be made to determine the rate of return on project investments.   ////

Sec. 8.1

Example 8.6 A cement plant plans to open a new rock pit. Two plans have been devised for movement of raw material from the quarry to the plant. Plan A requires the purchase of two earth movers and construction of an unloading pad at the plant. Plan B calls for construction of a conveyor system from the quarry to the plant. The costs for each plan are itemized in Table 8.1. Which plan should be selected if money is presently worth 15%?

SOLUTION Evaluation will take place over 24 years, since we plan to use present-worth (PW) analysis. Reinvestment in the two movers will occur in years 8 and 16 while the unloading pad must be repurchased in year 12. No reinvestment is necessary for Plan B. You are advised to construct your own cash-flow diagram for each plan to better follow the PW analysis.

Table 8.1 DETAILS OF PLANS TO MOVE ROCK FROM QUARRY TO CEMENT PLANT

	Plan A		Plan B
	Mover	Pad	Conveyor
P	$45,000	$28,000	$175,000
AOC	6,000	300	2,500
SV	5,000	2,000	10,000
n, years	8	12	24

To simplify computations, we can use the fact that Plan A will have an extra AOC in the amount of $6,300 − $2,500 = $3,800 per year.

PW of Plan A

$$PW_A = PW_{movers} + PW_{pad} + PW_{AOC}$$
$$PW_{movers} = 2(45,000)[1 + (P/F,15\%,8) + (P/F,15\%,16)]$$
$$- 2(5,000)[(P/F,15\%,8) + (P/F,15\%,16) + (P/F,15\%,24)]$$
$$= \$124,355$$
$$PW_{pad} = 28,000[1 + (P/F,15\%,12)] - 2,000[(P/F,15\%,12)$$
$$+ (P/F,15\%,24)]$$
$$= \$32,790$$
$$PW_{AOC} = 3,800(P/A,15\%,24) = \$24,448$$
$$PW_A = \$181,593$$

PW of Plan B

$$PW_B = PW_{conveyor} = 175,000 - 10,000(P/F,15\%,24)$$
$$= \$174,651$$

Since $PW_B < PW_A$, the conveyor should be constructed. ////

Sec. 8.2

Example 8.7 A restaurant owner is trying to decide between two different garbage disposals. A regular steel (RS) disposal has an initial cost of $65 and a life of four years. The alternative is a corrosion-resistant disposal constructed primarily of stainless steel (SS). The initial cost of the SS disposal is $110, but it is expected to last ten years. Because the SS disposal has a slightly larger motor, it is expected to cost about $5 per year more to operate than the RS disposal. If the interest rate is 6%, which disposal should be selected, assuming both have a negligible salvage value?

SOLUTION The cash-flow diagram (Fig. 8.3) uses a comparison period of 20 years with reinvestment in year 10 for the SS disposal and in years, 4, 8, 12, and 16 for the RS disposal. Inflation is not considered in the reinvestments. The present-worth calculations are as follows:

$$PW_{RS} = 65 + 65(P/F,6\%,4) + 65(P/F,6\%,8) + 65(P/F,6\%,12)$$
$$+ 65(P/F,6\%,16) = \$215$$
$$PW_{SS} = 110 + 110(P/F,6\%,10) + 5(P/A,6\%,20) = \$229$$

The regular steel disposal should be purchased since $PW_{RS} < PW_{SS}$.

COMMENT In the solution presented, the extra operating cost of $5 per year was regarded as an expense for the SS disposal. However, the same decision would have been reached if the $5 per year had been shown as an income for the RS disposal, but the present worths of both would have been lower by $5(P/A,6\%,20). This

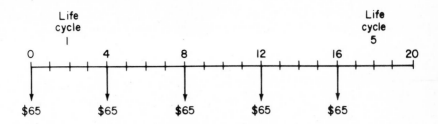

Regular steel disposal

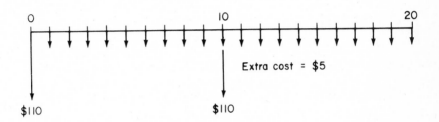

Stainless steel disposal

FIG. 8.3 Present-worth comparison of two unequal life assets, Example 8.7.

illustrates that unless the absolute money values are sought, it is only important to consider *differences* in cash flow for alternative evaluation.　　　　////

Sec. 8.2

Example 8.8 A city engineer is considering two alternatives for the local water supply. The first alternative would involve construction of an earthen dam on a nearby river, which has a highly variable flow. The dam would serve as a reservoir so that the city would have a dependable source of water indefinitely. The initial cost of the dam is expected to be $8 million and will require annual upkeep costs of $25,000. The dam is expected to last indefinitely.

Alternatively, the city can drill wells as needed and construct pipelines for transporting the water to the city. The engineer estimates that an average of ten wells will be required initially at a cost of $45,000 per well, including the pipeline. The average life of a well is expected to be five years with an annual operating cost of $5,000 per well. If the city uses an interest rate of 5%, determine which alternative should be selected on the basis of their capitalized costs.

SOLUTION The cash-flow diagram is left to the reader. The capitalized cost of the dam is calculated as follows:

$$PW_{dam} = 8,000,000 + \frac{25,000}{0.05} = \$8,500,000$$

The capitalized cost of the wells can be calculated by first converting the recurring costs and annual operating costs to an EUAC and then dividing by the interest rate. Thus,

$$EUAC_{wells} = EUAC \text{ of investment} + \text{annual operating costs}$$
$$= 45,000(10)(A/P,5\%,5) + 5,000(10)$$
$$= \$153,941$$

The capitalized cost of the wells, using Eq. (8.3), is

$$PW_{wells} = \frac{153,941}{0.05} = \$3,078,820$$

The wells should be constructed instead of the dam.

COMMENT The capitalized cost of the wells could also have been obtained by using the A/F factor for calculating the EUAC. The value obtained should then be divided by i and added to the initial investment cost. Thus,

$$EUAC_{wells} = \frac{450,000(A/F,5\%,5) + 50,000}{0.05}$$
$$= \$2,628,820$$
$$PW_{wells} = 2,628,820 + 450,000 = \$3,078,820 \qquad ////$$

Sec. 8.4

BIBLIOGRAPHY

Barish, pp. 139–146.
DeGarmo and Canada, pp. 223–225, 137–138.
Emerson and Taylor, pp. 7-4 – 7-11, 9-7 – 9-8.
Fabrycky and Thuesen, pp. 91–98, 111–115.
Grant and Ireson, pp. 88–108.
Riggs, pp. 180–181, 239–240.
Smith, pp. 100–103, 106–108, 111–116.
Taylor, pp. 100–107.
Thuesen, Fabrycky, and Thuesen, pp. 103–108.

PROBLEMS

P8.1 Two machines are under consideration by a metal fabricating company. Machine A will have a first cost of $15,000, an annual maintenance and operation cost of $3,000, and a $3,000 salvage value. Machine B will have a first cost of $22,000, an annual cost of $1,500, and a $5,000 salvage value. If both machines are expected to last for ten years, determine which machine should be selected on the basis of their present-worth values using an interest rate of 12%.

P8.2 A public utility is trying to decide between two different sizes of pipe for a new water main. A 10-inch line will have an initial cost of $35,000 while a 12-inch line will cost $55,000. Since there is less head loss through the 12-inch pipe, the pumping cost for the larger line is expected to be $3,000 per year less than for the 10-inch line. If the pipes are expected to last for 20 years, which size should be selected if the interest rate is 6%? Use a present-worth analysis.

P8.3 A couple are trying to decide between purchasing a house and renting one. They can purchase a new house with a down payment of $5,000 and a monthly payment of $150. Taxes and insurance are expected to amount to $60 per month. In addition, they expect to paint the house every four years at a cost of $300. Alternatively, they can rent a house for $125 per month and a $300 deposit, which will be returned when they vacate the house. The utilities are expected to average $35 per month whether they purchase or rent. If they expect to be able to sell the house for $3,000 more than they paid for it in six years, should they buy a house or rent one, if the interest rate is a nominal 12% per year? Use a present-worth analysis.

P8.4 A consulting engineer is trying to determine which of two methods should be specified for screening sewage. A manually cleaned bar screen will have an initial installed cost of $400. The labor cost for cleaning is expected to be $800 the first year, $850 the second year, $900 the third year, and increase by $50 each year. An automatically cleaned bar screen will have an initial cost of $2,500 with an annual power cost of $150. In addition, the motor will have to be replaced every two years at a cost of $40 per motor. General maintenance is expected to cost $100 the first year and increase by $10 per year. If the screens are expected to last for ten years, which method should be selected if the interest rate is 6%? Use the present-worth method.

P8.5 A consulting engineering firm is trying to decide between purchasing cars and leasing. It estimates that medium-size cars will cost $5,300 and will have a probable trade-in value in four years of $1,100. The annual cost of such items as fuel and repairs is expected to be $750 the first year and will increase by $50 per year. Alternatively, the company can lease the same cars for $1,500 per year. Since some maintenance is included in the rental price, the annual maintenance and operation expenses are expected to be $100 lower if the cars are leased. If the company's minimum rate of return is 10%, which alternative should be selected?

P8.6 A building contractor is trying to determine if it would be economically feasible to install rainwater drains in a large shopping center currently under construction. Since the project is being built in the arid Southwest, the total annual amount of rainfall is slight, but the rain that does occur is in the form of brief but heavy thundershowers. The thundershowers tend to cause erosion of soil in the project site, which was formed by filling in a large arroyo. In the three years required for construction, 12 heavy thundershowers are expected. If no drains are installed, the cost of refilling the washed-out area is expected to be $800 per thunderstorm. Alternatively, a corrugated steel drainpipe could be installed which will prevent the soil erosion. The installation cost of the pipe would be $1.50 per foot, with a total length of 5,000 feet required. After the three-year construction period, some of the pipe could be recovered with an estimated value of $3,000. Assuming the thunderstorms occur at three-month intervals, which alternative should be selected, if the interest rate is a nominal 20% per year compounded quarterly?

P8.7 A southwestern university is considering installing electric valves with automatic timers on some of their sprinkler systems. They estimate that they would need 45 valves and timers at a cost of $65 per set. The initial installation cost is expected to be $2,000. At the present time, there are four employees who are in charge of maintaining these lawns. These employees spend 25% of their time in watering and they earn $8,000 per year each. The present cost of watering these lawns is $1,200 per year. If the automatic system is installed, the manpower cost for watering could be reduced by 80% and the water bill would be reduced by 35%. However, extra maintenance on the automatic system is expected to cost $250 per year. If the timers and valves are expected to last for eight years, which system should be used if the interest rate is 6%? Use a present-worth analysis.

P8.8 A manufacturing company is in need of 10,000 square feet of storage space for three years. The company is considering the purchase of an acre of land for $8,000 and erecting a temporary, metal structure at a cost of $7 per square foot. At the end of the three-year use period, the company expects to be able to sell the land for $9,000 and the building for $12,000. Alternatively, the company can lease storage space for $0.15 per square foot per month payable at the beginning of each year. If the company's minimum attractive rate of return is 10%, which type of storage space should be used? Use the present-worth method of analysis.

P8.9 Machines that have the following costs are under consideration for a continuous production process.

	Machine G	Machine H
First cost	$62,000	$77,000
Annual operating cost	15,000	21,000
Salvage value	8,000	10,000
Life, years	4	6

Using an interest rate of 15%, determine which alternative should be selected on the basis of a present-worth analysis.

P8.10 Rework problem P8.9 assuming Machine G requires an extensive overhaul at the end of two years costing $10,000.

P8.11 Which screen should be selected in problem P8.4 if the manually cleaned screen will last 20 years and the automatically cleaned screen will last only 10 years?

P8.12   A production plant manager has been presented with two proposals for automating an assembly process. Proposal A involves an initial cost of $15,000 and an annual operating cost of $2,000 per year for the next four years. Thereafter, the operating cost is expected to increase by $100 per year. This equipment is expected to have a ten-year life with no salvage value. Proposal B requires an initial investment of $28,000 and an annual operating cost of $1,200 per year for the first three years. Thereafter, the operating cost is expected to increase by $120 per year. This equipment is expected to last for 20 years and will have a $2,000 salvage value. If the company's minimum attractive rate of return is 10%, which proposal should be accepted on the basis of a present-worth analysis?

P8.13   Which alternative should be selected in problem P8.12 if the company's minimum attractive rate of return is a nominal 12% compounded semi-annually?

P8.14   An environmental engineer is trying to decide between two operating pressures for a wastewater irrigation system. If a high pressure system is used, fewer sprinklers and less pipe will be required, but the pumping cost will be higher. The alternative is to use lower pressure with more sprinklers. The pumping cost is estimated to be $0.10 per psi of pressure per million gallons of wastewater. If a pressure of 80 psi is used, 25 sprinklers will be required at a cost of $22 per unit. In addition, 4,000 feet of aluminum pipe will be required at a cost of $2.80 per foot. If a lower pressure of 50 psi is used, 85 sprinklers and 13,000 feet of pipe will be required. The aluminum pipe is expected to last ten years and the sprinklers four years. If the volume of wastewater is expected to be 120 million gallons per year, which pressure should be selected if the company's minimum attractive rate of return is 20%?

P8.15   The owner of the Good Flick Drive-In Theatre is considering two proposals for upgrading the parking ramps. The first proposal involves asphalt paving of the entire parking area. The initial cost of this proposal would be $15,000, and it would require annual maintenance of $150 beginning three years after installation. The owner expects to have to resurface the theatre in 15 years. Resurfacing will cost only $8,000 since grading and surface preparation are not necessary. Alternatively, gravel can be purchased and spread in the drive areas and grass planted in the parking areas. The owner estimates that 40 tons of gravel will be needed per year at a cost of $15 per ton. In addition, a riding lawn mower, which will cost $500 and have a life of ten years, will be needed. The cost of labor for spreading gravel, cutting grass, etc., is expected to be $700 the first year and $750 the second, and will increase by $50 per year thereafter. The owner figures that a gravel surface would not be used for more than 30 years. If the interest rate is 7%, which alternative should be selected? Use a present-worth analysis.

P8.16   An automobile owner is trying to decide between purchasing four new radial tires or having the worn-out tires recapped. Radial tires for the car will cost $55 each and will last 42,000 miles. The old tires can be recapped for $15 each, but they will last for only 12,000 miles. Since this is a "second" car, it probably will register only 6,000 miles per year. If the radial tires are purchased, the gasoline mileage will increase by 10%. If the cost of gasoline is assumed to be $0.60 per gallon and the car gets 20 miles per gallon, what type of tires should be purchased if the interest rate is 6%? Use the present-worth method and assume the salvage value of the tires is zero.

P8.17 A local planning commission has estimated the first cost of a new city-owned amusement park to be $35,000. They expect to improve the park by adding new rides every year for the next five years at a cost of $6,000 per year. Annual operating costs are expected to be $12,000 the first year, and these will increase by $2,000 per year until year 5. After that time, the operating expenses will remain at $20,000 per year. The city expects to receive $11,000 in profits the first year, $14,000 the second, and amounts increasing by $3,000 per year until year 8, after which the net profit will remain the same. Calculate the capitalized cost of the park if the interest rate is 6%.

P8.18 How much additional uniform annual cost can the city incur for the amusement park in problem P8.17 to break even?

P8.19 What is the capitalized cost of $75,000 now, $60,000 five years from now, and a uniform annual amount of $700 per year for year 10 and every year thereafter, if the interest rate is 8%?

P8.20 What is the capitalized cost of $200,000 now, $300,000 four years from now, $50,000 every five years, and a uniform annual amount of $8,000 beginning 15 years from now, if the interest rate is 6%?

P8.21 A wealthy alumnus of a small university wants to establish a permanent fund for tuition scholarships. He wants to support three students for the first five years after the fund is established and five students thereafter. If tuition is expected to cost $400 per year, how much money must the alumnus donate now if the university can earn 6% on the fund?

P8.22 If the tuition in problem P8.21 increases by $20 per year for the first 20 years, how much money must the alumnus donate?

P8.23 A city planning commission is considering two proposals for a new civic center. Proposal F would require an initial investment of $10 million now and an expansion cost of $4 million ten years from now. The annual operating cost is expected to be $250,000 per year. Income from conventions, shows, etc., is expected to be $190,000 the first year and to increase by $20,000 per year for four more years and then remain constant until year 10. In year 11 and thereafter income is expected to be $350,000 per year. Proposal G would require an initial investment of $18 million now and an annual operating cost of $300,000 per year. However, income is expected to be $260,000 the first year and increase by $30,000 per year to year 7. Thereafter income will remain at $440,000 per year. Determine which proposal should be selected on the basis of capitalized cost if the interest rate is 6% per year.

P8.24 Rework problem P8.12, comparing the alternatives on the basis of their capitalized cost.

EQUIVALENT UNIFORM ANNUAL COST EVALUATION

The objective of this chapter is to teach you the primary methods of calculating the equivalent uniform annual cost (EUAC) of an asset and how to select the better of two alternatives on the basis of an annual-cost comparison. The alternative selected using EUAC must be the same as that chosen using present-worth or any other evaluation method; that is, all methods should result in identical decisions, just in a different manner.

CRITERIA

In order to complete this chapter you must be able to do the following:

1 State why the EUAC must be calculated for only one cycle of each alternative, when the alternatives have different lives.
2 Calculate the EUAC of an asset having a salvage value, using the *salvage sinking-fund* method, given the asset initial cost, salvage value, life, and the interest rate.
3 Same as 2 above, except using the *salvage present-worth* method.

4 Same as 2 above, except using the *capital-recovery-plus-interest* method.
5 Select the better of two alternatives on the basis of their EUAC, given their initial costs, salvage values, lives, amount and time of the operating costs, and the interest rate.
6 Calculate the EUAC of a perpetual investment, given the initial cost of the asset, amount and timing of disbursements, and the interest rate.

EXPLANATION OF MATERIAL

9.1 Study Period for Alternatives Having Different Lives

The EUAC (equivalent uniform annual cost) is another method that is commonly used for compared alternatives. As illustrated in Chap. 5, the EUAC means that all disbursements (irregular and uniform) must be converted to an equivalent uniform annual cost, that is, a year-end amount which is the *same each year*. The major advantage of this method over all of the other methods is that it is not necessary to make the comparison over the same number of years when the alternatives have different lives. When the EUAC method is used, the equivalent uniform annual cost of the alternative must be calculated for *one life cycle only*. Why? Because, as its name implies, the EUAC is an equivalent annual cost over the life of the project. If the project is continued for more than one cycle, the equivalent annual cost for the next cycle and all succeeding cycles would be exactly the same as for the first, assuming all cash flows were the same for each cycle. The EUAC for one cycle of an alternative, therefore, represents the equivalent uniform annual cost of that alternative *forever*.

Problem P9.1

9.2 Salvage Sinking-Fund Method

When an asset of a given alternative has a terminal salvage value (SV), there are several ways by which the EUAC can be calculated. This section presents the salvage sinking-fund method, probably the simplest of the three discussed in this chapter. This is the method that is used hereafter in this book. In the salvage sinking-fund method, the initial cost (P) is first converted to an equivalent uniform annual cost using the A/P (capital-recovery) factor. The salvage value, after conversion to an equivalent uniform cost via the A/F (sinking-fund) factor, is *subtracted* from the annual-cost equivalent of the first cost. The calculations can be represented by a general equation:

$$\text{EUAC} = P(A/P,i\%,n) - \text{SV}(A/F,i\%,n) \qquad (9.1)$$

Naturally, EUAC is nothing more than an A value, but it is referred to as EUAC here. The calculations are illustrated in Example 9.1.

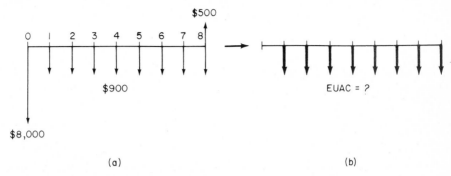

FIG. 9.1 (*a*) Diagram for machine costs and (*b*) conversion to an EUAC.

Example 9.1 Calculate the EUAC of a machine that has an initial cost of $8,000 and a salvage value of $500 after eight years. Annual operating costs (AOC) for the machine are estimated to be $900, and the interest rate of 6% is applicable.

SOLUTION The cash-flow diagram (Fig. 9.1) requires us to compute

$$EUAC = A_1 + A_2$$

where A_1 = annual cost of initial investment less salvage value, Eq. (9.1)
          A_2 = annual maintenance cost = $900

$$A_1 = 8,000(A/P,6\%,8) - 500(A/F,6\%,8) = \$1,238$$

$$EUAC = 1,238 + 900 = \$2,138$$

COMMENT Since the maintenance cost was already expressed as an annual cost over the life of the asset, no conversions were necessary. The simplicity of the salvage sinking-fund method should be obvious from the straightforward calculations shown in this example. The steps in this method are the following:

1 Annualize the initial investment cost over the life of the asset using the A/P factor.
2 Annualize the salvage value using the A/F factor.
3 Subtract the annualized salvage value from the annualized investment cost.
4 Add the uniform annual costs to the value from step *3*.   ////

Example 9.6
Problems P9.2, P9.3

9.3 Salvage Present-Worth Method

The salvage present-worth method is the second method by which investment costs having salvage values can be converted into an EUAC. The present worth of the salvage value is subtracted from the initial investment cost, and the resulting difference is annualized for the life of the asset. The general equation is

$$\text{EUAC} = [P - \text{SV}(P/F,i\%,n)]\,(A/P,i\%,n) \qquad (9.2)$$

The steps that must be followed in this method are the following:

1   Calculate the present worth of the salvage value via the P/F factor.
2   Subtract the value obtained in step *1* from the initial cost, P.
3   Annualize the resulting difference over the life of the asset using the A/P factor.
4   Add the uniform annual costs to the result of step *3*.

Example 9.2   Compute the EUAC for the machine detailed in Example 9.1 using the salvage present-worth method.

SOLUTION   Using the steps outlined above and Eq. (9.2),

$$\text{EUAC} = [8{,}000 - 500(P/F,6\%,8)]\,(A/P,6\%,8) + 900 = \$2{,}138 \qquad ////$$

Problem P9.4

9.4   Capital-Recovery-Plus-Interest Method

The final procedure that will be presented here for calculating the EUAC of an asset having a salvage value is the capital-recovery-plus-interest method. The general equation for this method is

$$\text{EUAC} = (P - \text{SV})(A/P,i\%,n) + \text{SV}(i) \qquad (9.3)$$

In subtracting the salvage value from the investment cost *before* multiplying by the A/P factor, it is recognized that the salvage value will be recovered. However, the fact that the salvage value will not be recovered for n years must be taken into account by adding the interest (SVi) lost during the asset's life. Failure to include this term would assume that the salvage value was obtained in year zero instead of year n. The steps to be followed for this method are as follows:

1   Subtract the salvage value from the initial cost.
2   Annualize the resulting difference with the A/P factor.
3   Multiply the salvage value by the interest rate.
4   Add the values obtained in steps *2* and *3*.
5   Add the uniform annual costs to the result of step *4*.

Example 9.3   Use the values of Example 9.1 to compute the EUAC using the capital-recovery-plus-interest method.

SOLUTION From Eq. (9.3) and the steps above,

$$EUAC = (8,000 - 500)(A/P,6\%,8) + 500(0.06) + 900 = \$2,138 \qquad ////$$

While it makes no difference which method is used to compute the EUAC, it would be good procedure hereafter to use only the one method you prefer in order to avoid unnecessary errors caused by mixing the methods. We shall use the salvage sinking-fund method (Sec. 9.2).

Problem P9.5

9.5 Comparing Alternatives by EUAC

The equivalent-uniform-annual-cost method of comparing alternatives is probably the simplest of the alternative evaluation techniques presented in this book. Selection is made on the basis of EUAC with the alternative having the lowest cost being the most favorable. Obviously, nonquantifiable data must also be considered in arriving at the final decision, but in general, the alternative having the lowest EUAC would be selected.

Perhaps the most important rule to remember when making EUAC comparisons is that *only one cycle* of the alternative must be considered. This assumes, of course, that the costs in all succeeding periods will be the same. While it is true that the cost of an asset today will probably be much lower than the cost of the same asset ten years from today, because of inflation, it must be remembered that, in general, the costs of the other alternatives would increase as well. Inasmuch as the analytical methods presented here are mainly for the purpose of making comparisons and not for determining actual costs, the same conclusions would be reached at any future date as long as all costs increased proportionately. Obviously, when information is available which would indicate that the costs of certain assets will increase or decrease considerably, because of technical improvements or increased competition, these factors must be taken into consideration in arriving at a final decision.

Example 9.4 The following costs are proposed for two equal-service tomato-peeling machines in a food canning plant:

	Machine A	Machine B
First cost	$26,000	$36,000
Annual maintenance cost	800	300
Annual labor cost	11,000	7,000
Extra income taxes	–	2,600
Salvage value	2,000	3,000
Life, years	6	10

If the minimum required rate of return is 15%, which machine should be selected?

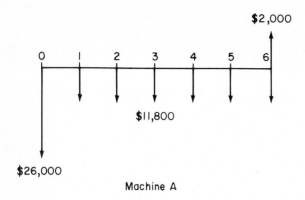

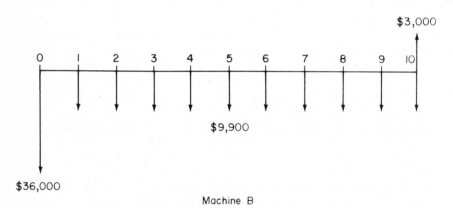

FIG. 9.2   Cash flows for two alternative tomato-peeling machines, Example 9.4.

SOLUTION   The cash-flow diagram for each alternative is shown in Fig. 9.2. The EUAC of each machine using the salvage sinking-fund method, Eq. (9.1), is calculated as follows:

EUAC$_A$ = 26,000(A/P,15%,6) − 2,000(A/F,15%,6) + 11,800 = $18,442
EUAC$_B$ = 36,000(A/P,15%,10) − 3,000(A/F,15%,10) + 9,900 = $16,925

Select Machine B, since EUAC$_B$ < EUAC$_A$.   ////

Example 9.7
Problems P9.6–P9.19

9.6   EUAC of a Perpetual Investment

It is sometimes necessary to compare alternatives that can be expected to have a perpetual life, such as flood-control dams or irrigation projects. For this type of analysis, it is important to recognize that the annual cost of a perpetual *initial* investment is simply the annual interest on the initial lump sum. That is, if the federal government were to invest $10,000 in a certain public works project, the EUAC of the investment would be $10,000(0.04) = $400, if the interest rate were 4%. This calculation is easy to understand when it is realized that the government could either spend $10,000 now or $400 per year forever. By spending $10,000 now, the government is losing the $400 per year interest it would make on the $10,000 if it were kept in a bank at 4%. On the other hand, the federal government could let an investor pay the $10,000 for the public works project and then pay the investor the $400 per year lost interest. From another point of view, a person receiving $10,000 now would have an amount of money equivalent to what another person who was to receive $400 per year forever would if the interest rate were 4%, since both persons would receive only $400 per year perpetually.

Costs recurring at regular or irregular intervals are handled exactly as in conventional EUAC problems. That is, all other costs must be converted into equivalent uniform annual costs for *one cycle*. They are thus automatically annual costs forever, as discussed in Sec. 9.1. Example 9.5 illustrates EUAC calculations for a perpetual project.

Example 9.5   The U.S. Bureau of Reclamation is considering two proposals for increasing the capacity of the main canal in their Lower Valley irrigation system. Proposal A would involve dredging the canal in order to remove sediment and weeds which have accumulated during previous years' operation. Since the capacity of the canal will have to be maintained near its design peak flow because of increased water demand, the Bureau is planning to purchase the dredging equipment and accessories for $65,000. The equipment is expected to have a ten-year life with a $7,000 salvage value. The annual labor and operating costs for the dredging operation is estimated to be $22,000. In order to control weeds in the canal itself and along the banks, herbicides will be sprayed during the irrigation season. The yearly cost of the weed-control program, including labor, is expected to be $12,000.

Proposal B would involve lining the canal with concrete at an initial cost of $650,000. The lining is assumed to be permanent, but minor maintenance will be required every year at a cost of $1,000. In addition, lining repairs will have to be made every five years at a cost of $10,000. Compare the two alternatives on the basis of equivalent uniform annual cost using an interest rate of 5%.

SOLUTION   The cash-flow diagram is left to the reader. The EUAC of each proposal is determined as follows:

Proposal A

EUAC of dredging equipment:	
$65,000(A/P,5\%,10) - 7,000(A/F,5\%,10)$	$ 7,861
Annual cost of dredging	22,000
Annual cost of weed control	12,000
	$41,861

Proposal B

EUAC of initial investment: $650,000(0.05)	$32,500
Annual maintenance cost	1,000
Lining repair cost: $10,000(A/F,5%,5)	1,810
	$35,310

Proposal B should be selected.

COMMENT For Proposal A, it was necessary to consider only one cycle. No calculations were necessary for the dredging and weed-control costs since they were already expressed as annual costs. For Proposal B, the EUAC of the initial investment was obtained by multiplying by the interest rate, which is nothing more than Eq. (8.3), that is, $P = A/i$, solved for A and renamed EUAC.

If nonrecurring single or series costs were involved, these should be converted to a present worth and then multiplied by the interest rate. Note the use of the A/F (sinking-fund) factor for the lining repair cost. The A/F factor is used instead of the capital-recovery (A/P) factor because the lining repair cost began in year 5 instead of year zero and continued indefinitely at five-year intervals.   ////

Examples 9.8, 9.9
Problems P9.20–P9.26

SOLVED EXAMPLES

Example 9.6 A drugstore chain has just purchased a fleet of five pickup trucks to be used for delivery in a particular city. Initial cost was $4,600 per truck and the expected life and salvage value is five years and $300, respectively. The combined insurance, maintenance, gas, and lubrication costs are expected to be $650 the first year and increase by $50 per year thereafter, while delivery service will bring an extra $1,200 per year for the company. If a return of 10% is required, use the EUAC method to determine if the purchase should have been made.

SOLUTION The cash-flow diagram is shown in Fig. 9.3. If we compute the EUAC by the salvage sinking-fund method, we can first compute

$$A_1 = \text{annual cost of fleet purchase, Eq. (9.1)}$$
$$= -5(4,600)(A/P,10\%,5) + 5(300)(A/F,10\%,5)$$
$$= \$-5,822$$

The minus signs are used for costs since incomes and disbursements are involved. The annual disbursement and income can be combined into an annual-income equivalent (A_2) so that the net income conveniently follows a decreasing gradient.

$$A_2 = 550 - 50(A/G,10\%,5) = \$460$$

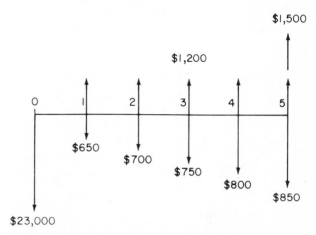

FIG. 9.3   Diagram used to compute EUAC, Example 9.6.

Now the EUAC is equal to the *algebraic* sum of the annual disbursement and annual income.

$$\text{EUAC} = -5{,}822 + 460 = \$-5{,}362$$

Since EUAC < 0, a return less than 10% will be made, and therefore, the purchase is not justified.

COMMENT   Try one of the EUAC methods of the Secs. 9.3 and 9.4 to solve the problem. They all work!   ////

Secs. 9.2, 9.3, 9.4

Example 9.7   Compare the two plans proposed in Example 8.6 using the EUAC method.

SOLUTION   Even though the two component parts of Plan A, movers and pad, have different lives, the EUAC analysis must be conducted for only one life cycle. For the salvage sinking-fund method, Eq. (9.1):

$$\text{EUAC}_A = \text{EUAC}_{\text{movers}} + \text{EUAC}_{\text{pad}} + \text{EUAC}_{\text{AOC}}$$

where $\text{EUAC}_{\text{movers}} = 90{,}000(A/P,15\%,8) - 10{,}000(A/F,15\%,8) = \$19{,}328$
$\qquad \text{EUAC}_{\text{pad}} = 28{,}000(A/P,15\%,12) - 2{,}000(A/F,15\%,12) = \$5{,}096$
$\qquad \text{EUAC}_{\text{AOC}} = \$3{,}800$
Then, $\text{EUAC}_A = 19{,}328 + 5{,}096 + 3{,}800$
$\qquad\qquad = \$28{,}224$

$\qquad \text{EUAC}_B = \text{EUAC}_{\text{conveyor}}$
$\qquad\qquad = 175{,}000(A/P,15\%,24) - 10{,}000(A/F,15\%,24) = \$27{,}146$

As was also shown in the present-worth analysis of Example 8.6, select Plan B.

COMMENT You should recognize a fundamental relation between the PW and EUAC values for the two examples discussed here. If you have the PW of a given plan, you can get EUAC by EUAC $= \text{PW}(A/P,i\%,n)$ or with an EUAC, PW $=$ EUAC$(P/A,i\%,n)$. The question is: What value does n assume? What would you use? We vote for the least-common-multiple value used in the present-worth method, since this method of evaluation must take place over an equal time period for each alternative. Therefore, the present-worth values are

$$\text{PW}_A = \text{EUAC}_A\,(P/A,15\%,24) = \$181,588$$
$$\text{PW}_B = \text{EUAC}_B\,(P/A,15\%,24) = \$174,652$$

as found in Example 8.6.   ////

Sec. 9.5

Example 9.8 If an investor deposits $1,000 now, $3,000 three years from now, and $600 per year for five years starting four years from now, how much money can be withdrawn every year forever beginning 12 years from now, if the rate of return on the investment is 8%?

SOLUTION The cash-flow diagram is shown in Fig. 9.4. The uniform amount of money that can be withdrawn every year forever is equal to the amount of interest that accumulates each year on the principal amount. To solve the problem, therefore, it is necessary to determine the total amount that would be accumulated in year 11 (*not* year 12) and then multiply by the interest rate (i) to obtain A. The future amount in year 11 would be

$$F_{11} = 1,000(F/P,8\%,11) + 3,000(F/P,8\%,8)$$
$$+ 600(F/A,8\%,5)(F/P,8\%,3)$$
$$= \$12,319$$

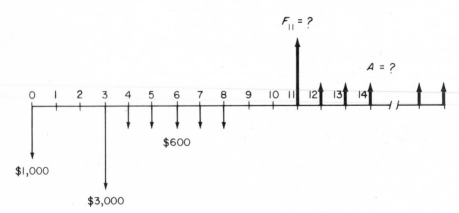

FIG. 9.4   Diagram to determine perpetual annual withdrawal, Example 9.8.

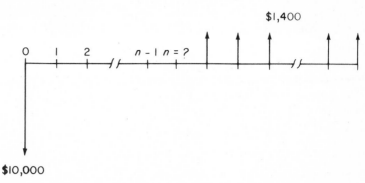

FIG. 9.5   Diagram to determine n for a perpetual withdrawal, Example 9.9.

The perpetual withdrawal can now be found by multiplying F_{11} (which is now a P value with respect to the perpetual withdrawal) by the interest rate.

$$A = Pi = 12,319(0.08) = \$986 \qquad ////$$

Sec. 9.6

Example 9.9   If an investor deposits $10,000 now at an interest rate of 7% per year, how many years must the money accumulate before the investor can withdraw $1,400 per year forever?

SOLUTION   The cash-flow diagram is shown in Fig. 9.5. The first step is to find the total amount of money that must be accumulated in year n (P_n), which is one year prior to the first withdrawal, to permit the perpetual $1,400 per year withdrawal using

$$P_n = \frac{A}{i} = \frac{1,400}{0.07} = \$20,000$$

When $20,000 is accumulated, the investor can withdraw $1,400 per year forever. The next step is to determine when the initial $10,000 deposit will accumulate to $20,000. This can be done with the F/P factor:

$$20,000 = 10,000(F/P,7\%,n)$$

By interpolation, $(F/P,7\%,n) = 2.0000$ when $n = 10.24$ years. $////$

Sec. 9.6

BIBLIOGRAPHY

Barish, pp. 129–133.
DeGarmo and Canada, pp. 221–223.
Emerson and Taylor, pp. 7-4 – 7-11.
Grant and Ireson, pp. 66–84.

Park, pp. 132–150.
Riggs, pp. 177–180.
Smith, pp. 94–95, 100–107, 112–114.
Taylor, pp. 76–77, 85–86.
Thuesen, Fabrycky, and Thuesen, pp. 108–112.

PROBLEMS

P9.1 Why must the least common multiple of years be used for assets that have different lives when making a present-worth comparison while an EUAC analysis requires annualization of costs over only one life cycle of each asset?

P9.2 If a woman purchased a new car for $6,000 and sold it three years later for $2,000, what was the equivalent uniform annual cost if she spent $750 per year for upkeep and operation? Use an interest rate of 5% and the salvage sinking-fund method.

P9.3 An entrepreneur purchased a dump truck for the purpose of offering a short-haul earth-moving service. He paid $14,000 for the truck and sold it five years later for $3,000. His operation and maintenance expense while he owned the truck was $3,500 per year. In addition, he had the truck engine overhauled for $220 at the end of the third year. Calculate his equivalent uniform annual cost using the salvage sinking-fund method, if the interest rate was 7%.

P9.4 Work the following problems by the salvage present-worth method: (*a*) P9.2 and (*b*) P9.3.

P9.5 Work the following problems by the capital-recovery-plus-interest method: (*a*) P9.2 and (*b*) P9.3.

P9.6 The manager in a canned-food processing plant is trying to decide between two labeling machines. Their respective costs are as follows:

	Machine A	Machine B
First cost	$15,000	$25,000
Annual operating cost	1,600	400
Salvage value	3,000	6,000
Life, years	7	10

(*a*) Determine which machine should be selected using a minimum attractive rate of return of 12% and an EUAC analysis. (*b*) If a present-worth analysis were used, over how many years would you make the comparison?

P9.7 Compare the following machines on the basis of their equivalent uniform annual cost. Use $i = 8\%$.

	New machine	Used machine
First cost	$44,000	$23,000
Annual operating cost	7,000	9,000
Annual repair cost	210	350
Overhaul every 2 years	–	1,900
Overhaul every 5 years	2,500	–
Salvage value	4,000	3,000
Life, years	15	8

P9.8 A moving and storage company is considering two possibilities for warehouse operations. Proposal 1 would require the purchase of a fork lift for $5,000 and 500 pallets that cost $5 each. The average life of a pallet is assumed to be two years. If the fork lift is purchased, the company must hire an operator for $9,000 annually and spend $600 per year in maintenance and operation. The life of the fork lift is expected to be 12 years with a $700 salvage value.

Alternatively, Proposal 2 requires that the company hire two men to operate power-driven hand trucks at a cost of $7,500 per man. One hand truck will be required at a cost of $900. The hand truck will have a life of six years with no salvage value. If the company's minimum attractive rate of return is 12%, which alternative should be selected?

P9.9 The supervisor of a country club swimming pool is trying to decide between two methods of adding chlorine. If gaseous chlorine is added, a chlorinator, which has an initial cost of $800 and a useful life of five years, will be required. The chlorine will cost $200 per year and the labor cost will be $400 per year.

Alternatively, dry chlorine can be added manually at a cost of $500 per year for chlorine and $800 per year for labor. If the interest rate is 6% per year, which method should be used?

P9.10 A carpenter is trying to determine how much insulation should be put into a ceiling. The higher the R rating of the insulation, the better the insulation. The choices are limited to either R-11 or R-19 insulation. The R-11 insulation costs $0.10 per square foot while the R-19 costs $0.15 per square foot. The annual saving in heating and cooling costs is estimated to be $25 per year greater with R-19 than with R-11. If the house has 2,500 square feet and the owner expects to keep the house for 25 years, which insulation should be installed at an interest rate of 10%?

P9.11 In problem P9.10, how much would the saving have to be per year in order for the R-19 insulation to be just as economical as the R-11?

P9.12 A meat-packing plant manager is trying to decide between two different methods for cooling cooked hams. The spray method involves spraying water over the hams until the ham temperature is reduced to 95°F. With this method, approximately 20 gallons of water are required for each ham. Alternatively, an immersion method can be used in which only four gallons of water are required per ham. However, this method will require an initial extra investment of $2,000 and extra overhaul expenses of $100 per year with the equipment expected to last ten years. The company cooks 10 million hams per year and pays $0.25 per 1,000 gallons for water. The company must also pay $0.09 per 1,000 gallons for wastewater discharged. If the company's minimum attractive rate of return is 15%, which method of cooling should be used?

P9.13 Work problem P8.2 using the EUAC method of evaluation.

P9.14 Work problem P8.3 using the EUAC method.

P9.15 Work problem P8.4 using the EUAC method.

P9.16 Rework the following problems using EUAC analysis: (*a*) P8.9 and (*b*) P8.10.

P9.17 Two environmental chambers (A and B) are being considered for a government project, which is to last for 6 years. Pertinent data are listed below.

	Chamber A	Chamber B
First cost	$4,000	$2,500
Annual operating cost	400	300
Salvage value	1,000	−100
Estimated life, years	3	2

(*a*) What chamber should be selected if money is worth 5%? (*b*) What must the difference in annual operating cost be to make the equivalent annual cost of both chambers equal?

P9.18 Compare the two plans below at $i = 15\%$.

		Plan B	
	Plan A	Machine 1	Machine 2
First cost	$10,000	$30,000	$5,000
Annual operating cost	500	100	200
Salvage value	1,000	5,000	−200
Life, years	40	40	20

P9.19 The Mighty Mouse Company is considering the purchase of a trap system to rid the plant of stray cats. Compare the two systems below at 10% interest.

	Scram-um	Catch-um
First cost	$25,000	$50,000
Annual operating cost	500	200
Salvage value	1,000	500
Life, years	20	40

P9.20 Calculate the perpetual equivalent uniform annual cost of $14,000 now, $55,000 six years from now, and $5,000 per year thereafter if the interest rate is 8% per year.

P9.21 Rework problem P9.20 if the interest rate is a nominal 8% per year compounded semiannually.

P9.22 An alumnus of Watsa Matta U. desires to establish a permanent university scholarship in his name. He plans to donate $20,000 per year for ten years starting one year from now and $100,000 when he dies. If the alumnus dies 15 years from now, how much money can be given to each of five students beginning one year from now and continuing forever if the interest rate is 8% per year?

P9.23 The first cost of a small dam is expected to be $3 million. The annual maintenance cost is expected to be $10,000 per year and a $35,000 outlay will be required every five years. If the dam is expected to last forever, what will be its equivalent uniform annual cost at an interest rate of 5% per year?

P9.24 A city that is attempting to attract a professional football team is planning to build a new football stadium costing $12 million. Annual upkeep is expected to amount to $25,000 per year. In addition, the artificial turf will have to be replaced every ten years at a cost of $150,000. Painting every five years will cost $65,000. If the city expects to maintain the facility indefinitely, what will be its equivalent uniform annual cost? Assume $i = 6\%$.

P9.25 Calculate the equivalent uniform annual cost of the park described in problem P8.17.

P9.26 Rework problem P8.23, comparing the alternatives on the basis of their equivalent uniform annual costs.

10

RATE–OF–RETURN EVALUATION

This chapter presents the method used to determine the rate of return of a project and the method by which two alternatives can be compared by the rate-of-return method. As with the present-worth and EUAC analyses of the last two chapters, this type of economic evaluation also results in selection of the same alternative.

CRITERIA

To complete this chapter you must be able to do the following:

1 Prepare a tabulation of cash flow for two alternatives having the same or different lives, given the details of each alternative.
2 Calculate the rate of return for a single project using the present-worth method, given the amounts and dates of project incomes and disbursements.
3 Same as 2, except using the EUAC method.
4 State how you would determine which alternative should be selected on the basis of the calculated incremental rate of return on the extra investment using the present-worth method.

5 Select the better of two alternatives on the basis of present-worth incremental-investment analysis by computing the breakeven rate of return, given the initial cost, life, and salvage value of each alternative, dates and amounts of disbursements, and the minimum acceptable rate of return.

6 Same as 5, except using EUAC analysis.

EXPLANATION OF MATERIAL

10.1 Tabulation of Cash Flow

The preparation of a cash-flow tabulation was discussed briefly in Chap. 2 with respect to net cash flow for a single alternative. In this chapter, it will be necessary to prepare a cash-flow tabulation for each of the two alternatives as well as the net cash flow that results when the two alternatives are compared. The column headings for a cash-flow tabulation involving two alternatives are shown in Table 10.1. If the alternatives have equal lives, the years column will go from zero to n, the life of the alternatives. If the alternatives have unequal lives, the years column will go from zero to the least common multiple of the two lives. The use of the least-common-multiple rule is necessary because rate-of-return analysis on the net cash-flow values must always be done over the same number of years for each alternative (as is the case with present-worth comparisons). If the least common multiple of lives is tabulated, reinvestment in each alternative is shown at appropriate times (as was done in Chap. 8 for cash flow in present-worth analysis).

As is shown in Sec. 10.5, a cash-flow tabulation is part of the procedure for selecting one of two alternatives on the basis of incremental rate of return. Since the cash-flow tabulation is an important part of the alternative evaluation process, a standardized format for the tabulation will simplify interpretation of the final results. Therefore, in this chapter, the alternative with the *higher initial investment* will always be regarded as *alternative B*. That is,

$$\text{Net cash flow} = \text{cash flow}_B - \text{cash flow}_A$$

Table 10.1 FORMAT FOR CASH–FLOW TABULATION

Year	Cash flow Alternative A (1)	Alternative B (2)	Net cash flow (3) = (2) − (1)
0			
1			
2			
.			
.			
.			

Table 10.2 CASH-FLOW TABULATION FOR EXAMPLE 10.1

Year	Cash flow		Net cash flow (new − old)
	Old mill	New mill	
0	$ −15,000	$ −21,000	$ −6,000
1-25	−8,200	−7,000	+1,200
25	+750	+1,050	+300
	$−219,250	$−194,950	$+24,300

The following examples demonstrate cash-flow tabulation for equal-life and unequal-life projects.

Example 10.1 A tool and die company is considering the purchase of an additional milling machine. The company has the opportunity to buy a slightly used machine for $15,000 or a new one for $21,000. Because the new machine is a more sophisticated model with some automatic features, its operating cost is expected to be $7,000 per year, while the old machine is expected to cost $8,200 per year. The machines are expected to have a 25-year life with 5% salvage values. Tabulate the net cash flow of the two alternatives.

SOLUTION Net cash flow is tabulated in Table 10.2. The salvage values in year 25 are separated from ordinary cash flow for clarity. Note that a sign must be included to indicate a disbursement (minus) or an income (plus).

COMMENT Note that when the cash-flow columns are subtracted, the difference between the totals of the two alternatives should equal the total of the net cash-flow column. This will provide a check of your addition and subtraction in preparing the tabulation.

When disbursements are the same for a number of consecutive years, it saves time to make a single cash-flow listing, as is done for years 1-25 of the example. However, be careful to remember that several years were combined when adding to get the column totals. ////

Example 10.2 The Fresh-Pak Tomato Cannery has under consideration two different types of conveyors. Type A has an initial cost of $7,000 and a life of eight years. The initial cost of Type B is $9,500 and it is expected to last 12 years. The operating cost for Type A is expected to be $900 while the cost for Type B is expected to be $700. If the salvage values are $500 and $1,000 for Type A and Type B conveyors, respectively, (a) tabulate the net cash flows using the least common multiple of lives for present-worth analysis and (b) tabulate the cash flows of each asset for an EUAC analysis.

Table 10.3 CASH FLOW FOR 24 YEARS
FOR EXAMPLE 10.2(a)

Year	Cash flow Type A	Cash flow Type B	Net cash flow
0	$ −7,000	$ −9,500	$−2,500
1-7	−900	−700	+200
8	⎧ −7,000 ⎨ −900 ⎩ +500	−700	+6,700
9-11	−900	−700	+200
12	⎧ ⎨ −900 ⎩	−9,500 −700 +1,000	−8,300
13-15	−900	−700	+200
16	⎧ −7,000 ⎨ −900 ⎩ +500	−700	+6,700
17-24	−900	−700	+200
24	+500	+1,000	+500
	$−41,100	$−33,800	$+7,300

SOLUTION

(a) The least common multiple of years between 8 and 12 is 24 years. The cash-flow tabulation must, therefore, be prepared over 24 years (Table 10.3).

(b) Cash-flow tabulation for respective asset life (Table 10.4) is similar to that of part (a), except that tabulation stops at n for each alternative, since only one cycle of each alternative is required for the EUAC method of analysis. ////

Problems P10.1–P10.4

Table 10.4 CASH FLOW FOR RESPECTIVE
ASSET LIFE FOR EXAMPLE 10.2(b)

Year	Cash flow Type A	Cash flow Type B
0	$−7,000	$−9,500
1-7	−900	−700
8	−900 + 500	−700
9-11	–	−700
12	–	−700 + 1,000

10.2 Rate-of-Return Calculations by the Present-Worth Method

In Sec. 3.7 the method for calculating the rate of return on an investment was illustrated when only one factor was involved. In this section a method for calculating the rate of return on an investment when several factors are involved is demonstrated. To help you understand rate-of-return calculations more clearly, it would be helpful for you to remember that the basis for engineering-economy calculations is *equivalence*, or time value of money. In previous chapters, we have shown that a present sum of money is equivalent to a larger sum of money at some future date when the interest rate is greater than zero. In rate-of-return calculations, the objective is to find the interest rate at which the present sum and future sum are equivalent; in other words, the calculations that will be made here are simply the reverse of calculations made in previous chapters, where the interest rate was known.

The backbone of the rate-of-return method is a rate-of-return equation, which is simply an expression equating a present sum of money to the present worth of future sums. For example, if you invest $1,000 now and are promised $500 three years from now and $1,500 five years from now, the rate-of-return equation would be

$$1,000 = 500(P/F,i\%,3) + 1,500(P/F,i\%,5) \qquad (10.1)$$

where the value of i to make the equality correct is to be computed (see Fig. 10.1). If the $1,000 is moved to the right side of Eq. (10.1), we have

$$0 = -1,000 + 500(P/F,i\%,3) + 1,500(P/F,i\%,5) \qquad (10.2)$$

Equation (10.2) is the form that will be followed in setting up all rate-of-return calculations. The equation must then be solved for i by trial and error to obtain $i = 16.95\%$. Since there is always some income and some disbursements involved in any project, some value of i can be found; however, only if the total amount of incomes is greater than the total amount of disbursements will the rate of return be greater than zero.

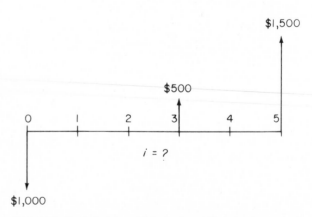

FIG. 10.1   Cash flow for which a value of i is to be determined.

It should be evident that rate-of-return calculations are merely the reverse of present-worth calculations. That is, if the above interest rate were given (16.95%) and it were desired to find the present worth of $500 three years from now and $1,500 five years from now, the equation would be

$$P = 500(P/F,16.95\%,3) + 1,500(P/F,16.95\%,5) = \$1,000$$

which can be arranged as follows:

$$0 = -1,000 + 500(P/F,16.95\%,3) + 1,500(P/F,16.95\%,5) \qquad (10.3)$$

Equation (10.3) is in the same form as Eq. (10.2), illustrating that rate-of-return and present-worth equations are set up in exactly the same fashion. The only difference lies in what is given and what is sought. The general procedure used to make a rate-of-return calculation is the following:

1 Draw a cash-flow diagram.
2 Set up the rate-of-return equation in the form

$$0 = \pm P \pm \sum_{j=1}^{n} F(P/F,i\%,j) \pm A(P/A,i\%,n) \qquad (10.4)$$

3 Select values of i by trial and error until the equation is balanced.

In a trial-and-error solution, it is advantageous to get fairly close to the correct answer on the first trial. If the cash flows could be combined in such a manner that the income and disbursements could be used to compute a *single factor* such as P/F, P/A, etc., as in Chap. 3, it would be possible to look up the interest rate corresponding to the value of that factor for n years. The problem, then, is to combine the cash flows into the format of only one of the standard factors. This can be done through the following procedure:

1 Convert all *disbursements* to either single amounts (P or F) or uniform amounts (A) by neglecting the time value of money. For example, if it were desired to convert an A into an F value, simply multiply the A by the number of years (n). The movement of cash flows should minimize the error caused by neglecting the time value of money.
2 Convert all *incomes* to either single or uniform values as in step *1*.
3 Having reduced all disbursements and incomes to either a P/F, P/A, or A/F format, use the interest tables to find the approximate interest rate at which the P/F, P/A, or A/F value, respectively, is satisfied for the proper n value. The rate obtained is a good "ball park" figure to use in the first trial.

It is important to recognize that the rate of return obtained in this manner is only an *estimate* of the actual rate of return because the time value of money is neglected. This procedure is illustrated in Example 10.3.

Example 10.3 If $5,000 is invested now in common stock that is expected to yield $100 per year for ten years and $7,000 at the end of ten years, what is the rate of return?

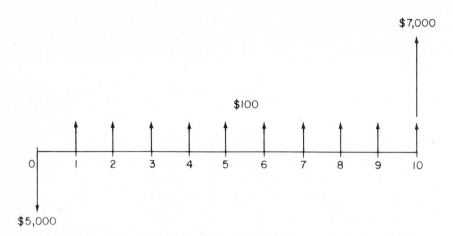

FIG. 10.2 Cash flow for a stock investment, Example 10.3.

SOLUTION The cash-flow diagram (Fig. 10.2) indicates that the rate of return can be calculated using Eq. (10.4) in the form

$$0 = -5{,}000 + 100(P/A, i\%, 10) + 7{,}000(P/F, i\%, 10)$$

In order to estimate the interest rate for the first trial, all income will be regarded as a single F in year 10 so that the P/F factor can be used. The P/F factor is selected since most of the cash flow already fits this factor and errors created by neglecting the time value of money will be minimized. Thus,

$$P = \$5{,}000$$
$$F = 10(100) + 7{,}000 = 8{,}000$$
$$n = 10$$

Now we can state that

$$5{,}000 = 8{,}000(P/F, i\%, 10)$$
$$(P/F, i\%, 10) = 0.625$$

The interest rate is between 4% and 5%. Therefore, try $i = 5\%$.

$$0 = -5{,}000 + 100(P/A, 5\%, 10) + 7{,}000(P/F, 5\%, 10)$$
$$0 \neq \$69.46$$

We are too large on the positive side. Therefore, try $i = 6\%$.

$$0 = -5{,}000 + 100(P/A, 6\%, 10) + 7{,}000(P/F, 6\%, 10)$$
$$0 \neq -\$355.19$$

Since the interest rate of 6% is too high, we now interpolate (Sec. 3.5) using Eq. (3.10):

$$c = \frac{a}{b}(d) = \frac{(69.46 - 0)}{69.46 - (-355.19)}(1.0) = 0.16$$

$$i = 5.00 + 0.16 = 5.16\%$$

COMMENT Note that 5% was used for the first trial rather than 4%. The higher value was used because, by assuming that the ten $100 amounts were equivalent to a single $1,000 in year 10, the approximate rate from the P/F factor was *lower* than the true value. This is due to the neglect of the time value of money. Therefore, the first-trial i value used was above that indicated by the P/F factor in order to improve the "first shot" accuracy. ////

Examples 10.7, 10.8
Problems P10.5–P10.15

10.3 Rate-of-Return Calculations by the EUAC Method

Just as i can be found using the present worth of a project, it can likewise be found by an EUAC formulation of the general form

$$0 = \pm P(A/P,i\%,n) \pm A \qquad (10.5)$$

where P is the present worth of all nonuniform amounts and A is an equivalent uniform annual series for n years. As before, a trial-and-error solution for i is necessary.

Example 10.4 Use the EUAC method to find the rate of return for the data in Example 10.3.

SOLUTION To use Eq. (10.5), we must first compute an equivalent A for the receipts (Fig. 10.2). This is computed as

$$A = 100 + 7,000(A/F,i\%,10)$$

The EUAC formulation for the rate of return is

$$0 = -5,000(A/P,i\%,10) + 100 + 7,000(A/F,i\%,10)$$

Results are as follows: $i = 5\%$, $0 \neq \$+9.02$ and $i = 6\%$, $0 \neq \$-48.26$. Interpolation gives equation satisfaction at $i = 5.16\%$, as before. ////

Thus, for rate-of-return calculations, you can choose either method, present-worth or EUAC. It is generally better to get accustomed to using only one of the two methods in order to avoid unnecessary errors.

Problems P10.5–P10.15

10.4 Interpretation of Rate of Return on Extra Investment

The first step in calculating the rate of return on the extra investment between two alternatives is the preparation of a cash-flow tabulation similar to that of Tables 10.2 or 10.3. When the evaluation is conducted by the present-worth method, the least common multiple of lives must be used for the study period. The net cash-flow column then reflects the *extra investment* that would be required if the alternative with the larger first cost is selected. Thus, in Example 10.1 (Sec. 10.1), the new milling machine would require an extra investment of $6,000, as shown in the last column of Table 10.2. On the other hand, if the new machine is purchased instead of the old one, there would be a "savings" of $1,200 per year for 25 years, plus $300 in year 25 as a result of the difference in salvage values. Therefore, spending the extra $6,000 for the new machine now will result in a "savings" of $1,200 per year for 25 years, plus an additional $300 in year 25. The decision of whether to buy the old or the new milling machine can be made on the basis of the profitability of investing the extra $6,000 in the new machine. If the present worth of the savings is greater than the present worth of the extra investment using the company's minimum acceptable rate of return (MARR), then the extra investment should be made (i.e., the higher first-cost proposal should be accepted). On the other hand, if the present worth of the savings is less than the present worth of the extra investment, then the lower first-cost proposal should be accepted.

You should note (Table 10.2) that if the new machine is selected, there will be a net savings of $24,300. Keep in mind that this figure does not take into account the time value of money, since this total was obtained by adding the values for the various years without using the time factors and, therefore, cannot be used as a basis for the decision. The totals at the bottom of the table serve only as a check against the additions and subtractions for the individual years. In fact, the $24,300 is the present worth of net cash flow at $i = 0\%$. Right? Sure, think about it!

The rationale for making the decision is the same as if only *one alternative* were under consideration, that alternative being the one represented by the difference (net cash-flow) column in the cash-flow tabulation. When viewed in this manner, it is obvious that, unless this investment would yield a rate of return greater than the MARR, the investment should not be made (meaning that the lower-priced alternative should be selected *to avoid this extra investment*). On the other hand, if the rate of return on the difference investment is greater than MARR, the investment should be made (meaning that the higher-priced alternative should be selected).

Problem P10.16

10.5 Alternative Evaluation by Incremental-Investment Analysis

The information presented in the previous sections of this chapter serves as a basis for alternative evaluation by the incremental-investment analysis using present-worth computations. The basic procedure for the incremental-investment analysis method is the following:

1 Prepare a tabulation of cash flow and net cash flow, keeping in mind that the least common multiple of years must be used when the alternatives have different lives (present-worth method).
2 Draw a cash-flow diagram of net cash flows.
3 Set up the rate-of-return relation, Eq. (10.4), from the difference (net cash-flow) column of the cash-flow tabulation.
4 Calculate the rate of return on the extra investment by trial and error.
5 Select the better alternative in accordance with the rationale presented in Sec. 10.4.

The procedure for conducting an incremental-investment rate-of-return analysis is illustrated in Example 10.5.

Example 10.5 A manufacturer of boys' pants is considering purchasing a new sewing machine, which can be either semiautomatic or fully automatic; the estimates for each are as follows:

	Semiautomatic	Fully automatic
First cost	$8,000	$13,000
Annual disbursements	3,500	1,600
Salvage value	0	2,000
Life, years	10	5

Determine which machine should be selected if the MARR is 15%.

SOLUTION The cash-flow diagram for each machine is left to you, but the tabulation of net cash flow is shown in Table 10.5. The rate-of-return equation is

Table 10.5 CASH–FLOW TABULATION FOR EXAMPLE 10.5

	Cash flow		
Year	Semiautomatic (1)	Fully automatic (2)	Difference (3) = (2) − (1)
0	$ −8,000	$−13,000	$ −5,000
1–5	−3,500	−1,600	+1,900
5	—	$\left\{ \begin{array}{r} +2,000 \\ -13,000 \end{array} \right\}$	−11,000
6–10	−3,500	−1,600	+1,900
10	—	+2,000	+2,000
	$−43,000	$−38,000	$ +5,000

formulated from the difference column of Table 10.5. This can be done more easily after drawing a cash-flow diagram for the difference column (Fig. 10.3). The rate of return can be interpolated from the equation

$$0 = -5,000 + 1,900(P/A, i\%, 10) - 11,000(P/F, i\%, 5)$$
$$+ 2,000(P/F, i\%, 10)$$

Solving and interpolating between 12 and 15% yields

$$i = 12.7\%$$

Since the rate of return on the extra investment is less than the 15% minimum attractive rate, the lower-cost, or semiautomatic, machine should be purchased.

COMMENT If the rate of return on the incremental investment had been equal to or greater than 15%, the fully automatic machine should be purchased. To estimate the interest rate for the first trial, the $-11,000 (Fig. 10.3) could be assumed to occur at year zero so that $P = \$-16,000$. The $2,000 future amount could be converted to a uniform annual amount ($2,000/10 = \$200$) and added to the $1,900 to obtain a total A of $2,100. The resulting P/A factor for $n = 10$ years reveals that i is between 5% and 6%. This estimate is not close to the actual i because of the large error introduced by the movement of the $11,000 cash flow to year zero.   ////

Actually the rate of return obtained above can easily be interpreted as the *breakeven* value of i, that is, the rate of return at which either alternative might be selected. If the i found by the rate-of-return equation, Eq. (10.4), is greater than the MARR, the other alternative is selected. As an illustration, the breakeven rate in Example 10.5 is $i = 12.7\%$. Figure 10.4 is a general plot of present-worth values for different rates of return with costs considered as positive and incomes as negative costs. At values of $i < 12.7\%$, the present worth for the fully automatic machine is less than that of the semiautomatic, that is, $PW_{fully} < PW_{semi}$. For values of $i > 12.7\%$,

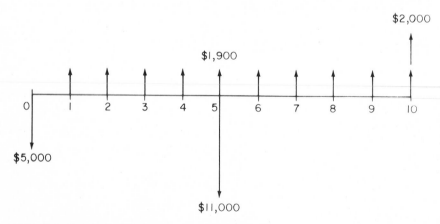

FIG. 10.3 Cash flow of difference column, Table 10.5.

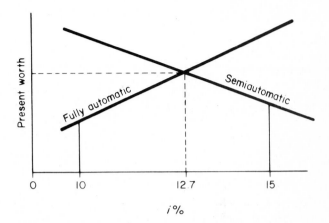

FIG. 10.4 Breakeven chart of present worth versus rate of return, Example 10.5.

$PW_{fully} > PW_{semi}$. Thus if MARR = 10%, select the fully automatic machine, whereas if MARR = 15%, as in Example 10.5, select the semiautomatic, because in each case the lower present-worth value is desired.

Example 10.9
Problems P10.17–P10.21

10.6 Alternative Evaluation by the EUAC Method

Even though the use of present-worth computations to obtain i for alternative evaluation is recommended, whether you use the present-worth or EUAC method, the conclusions must be identical. However, on some problems you might find EUAC computationally simpler. You should remember that the incremental rate of return must always be computed for the least common multiple of the lives if present worth is used. But for EUAC this is not necessary, so that it is often simpler to use the general form

$$0 = EUAC_B - EUAC_A \qquad (10.6)$$

to find the values of i for unequal-life alternatives. The EUAC computations in Eq. (10.6) utilize the cash flow for each alternative. The net cash flow can be used in EUAC comparisons *only* when the alternatives have *equal* lives. The technique for using EUAC to compare assets having unequal lives is demonstrated in the Solved Examples below.

The form of the rate-of-return equation for EUAC and *equal* alternative lives using net cash flow (difference) values and the salvage sinking-fund method (Sec. 9.2) is

$$0 = \pm P(A/P,i\%,n) - SV(A/F,i\%,n) \pm A \qquad (10.7)$$

Example 10.6 Compare the equal-life milling machines of Example 10.1 using the EUAC to compute the rate of return. Assume MARR = 15%.

SOLUTION Since the lives are equal, Eq. (10.7) can be applied to the net cash-flow column of Table 10.2 to get

$$0 = -6,000(A/P,i\%,25) + 300(A/F,i\%,25) + 1,200$$

Thus,

$$i = 19.79\%$$

The extra investment is justified for the new milling machine (as it would be with the present-worth method).

COMMENT The value $i = 19.79\%$ is the breakeven rate of return. A graph similar to Fig. 10.4 may be constructed in which EUAC would replace present worth. For values of MARR > 19.79%, the old milling machines should be purchased; for values of MARR < 19.79%, as is the case here, select the new milling machine.   ////

Example 10.10
Problems P10.21–P10.26

SOLVED EXAMPLES

Example 10.7 Assume a couple invest $10,000 now and $3,000 three years from now, and will receive $500 one year from now, $600 two years from now, and amounts increasing by $100 per year for a total of ten years. They will also receive lump-sum payments of $5,000 in five years and $2,000 in ten years. Calculate the rate of return on their investment.

SOLUTION The cash-flow diagram shown in Fig. 10.5 allows us to form the relation

$$0 = -10,000 - 3,000(P/F, i\%,3) + 500(P/A, i\%,10) + 100(P/G, i\%,10)$$
$$+ 5,000(P/F, i\%,5) + 2,000(P/F, i\%,10)$$

Solving and interpolating between $i = 4\%$ and $i = 5\%$, we find $i = 4.36\%$.

COMMENT Note that the single values at years 3, 5, and 10 were handled separately so that the P/A and P/G factors could be used. Use the procedure of Sec. 10.2 to estimate the interest rate by the P/A factor for $n = 10$ years. Assume the gradient term is an A with an average value of $500. Your estimate should show that i is in the neighborhood of 5% to 6%.   ////

Sec. 10.2

Example 10.8 Assume a series of cash flows as shown below for an ongoing project. (Data are adapted from an article by McLean[1] and some results of Barrish.[2])

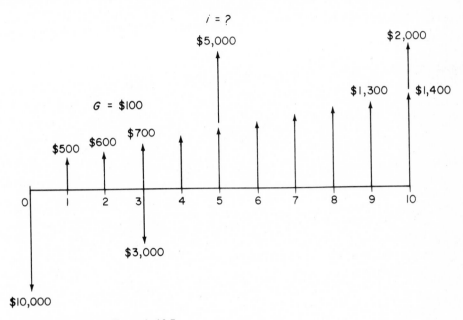

FIG. 10.5 Cash flow, Example 10.7.

Year	Net cash flow	Year	Net cash flow
1	$ 200	6	$500
2	100	7	400
3	50	8	300
4	−1,800	9	200
5	600	10	100

The negative net cash flow in year 4 is the result of a major alteration to the project. Compute the rate of return for the project.

SOLUTION The rate of return can be determined using the present-worth equation

$$0 = 200(P/F,i\%,1) + 100(P/F,i\%,2) + \ldots + 100(P/F,i\%,10) \qquad (10.8)$$

Tabulation of the results of Eq. (10.8) at several values of i yield the following curious values, which are plotted in Fig. 10.6.

i	Results of Eq. (10.8)
10%	$+198
20	+42
30	−2
40	−8
50	+1

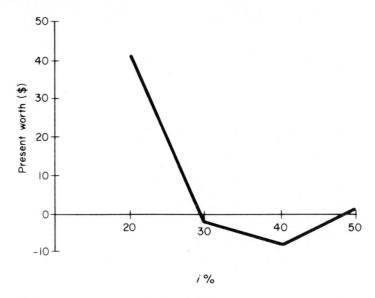

FIG. 10.6 Plot of present worth versus i, Example 10.8.

As you can see, there are two values of i which satisfy Eq. (10.8), at approximately $i = 28\%$ and $i = 49\%$. This phenomenon is referred to as the multiple rate-of-return problem. What is the true rate of return? There are several ways to deal with the dilemma. Grant and Ireson[3] present a method using an auxiliary interest rate. Usually, the two return values are far enough separated that the larger can be discarded as extraneous; that is, the use of reason assists in a solution to the dilemma.

COMMENT Two comments are in order. First, note that from Fig. 10.6, if the required MARR is such that $28\% \leqslant \text{MARR} \leqslant 49\%$, the proposal is not advantageous, but for $28\% > \text{MARR} > 49\%$, it is worthwhile, since the present worth is greater than zero. Second, the number of i values that can be found for a given series of net cash flows is equal to the number of times that the present worth of the net cash-flow values, Eq. (10.8), *changes sign*. Thus, if two switches in the sign of the present worth are observed (Fig. 10.6), there are two i values which can be found.   ////

Sec. **10.2**

Example 10.9 Determine which milling machine should be purchased if the explanation of Example 10.1 is followed. Assume a MARR of 15%.

SOLUTION Since the alternatives are *equal lived*, the net cash-flow values of Table 10.2 can be used directly to obtain

$$0 = -6{,}000 + 1{,}200(P/A,i\%,25) + 300(P/F,i\%,25)$$

An interpolated value of $i = 19.79\%$ can be computed. Try it yourself *now*, using the procedure of Sec. 10.2 to estimate the interest rate for the first trial. Since $i = 19.79\% > \text{MARR} = 15\%$, the purchase of the new milling machine is justified.

COMMENT You should realize that you *can* compare two alternatives, A and B, using the actual cash flows with the help of the general relation

$$0 = \text{PW}_\text{B} - \text{PW}_\text{A}$$

where alternative B has the larger first cost. In the milling-machine trade-off analysis, this approach results in the following:

$$\text{PW}_\text{new} = -21,000 - 7,000(P/A, i\%, 25) + 1,050(P/F, i\%, 25)$$
$$\text{PW}_\text{old} = -15,000 - 8,200(P/A, i\%, 25) + 750(P/F, i\%, 25)$$

Then

$$\begin{aligned} 0 &= \text{PW}_\text{new} - \text{PW}_\text{old} \\ &= (-21,000 + 15,000) + (-7,000 + 8,200)(P/A, i\%, 25) \\ &\quad + (1,050 - 750)(P/F, i\%, 25) \\ &= -6,000 + 1,200(P/A, i\%, 25) + 300(P/F, i\%, 25) \end{aligned}$$

Note that the reduced form is identical to that used in the solution to this example (above). This method is not advised for present-worth analysis since assets with different lives require reinvestment for comparison purposes. Try, for example, the data of Example 10.2. You will find it gets messy, thus resulting in possible mistakes when the rate of return is computed. ////

Sec. 10.5

Example 10.10 Compare the sewing machines of Example 10.5 using the EUAC method with a MARR of 15%.

SOLUTION Present-worth analysis in Example 10.5 showed that the semi-automatic machine should be purchased. For the EUAC we can use the salvage sinking-fund method, Eq. (9.1), and the general EUAC form, Eq. (10.6), since lives are unequal. Then we have

$$0 = \text{EUAC}_\text{fully} - \text{EUAC}_\text{semi}$$
$$\text{EUAC}_\text{fully} = P(A/P, i\%, n) - \text{SV}(A/F, i\%, n) + A_1$$

where A_1 = equivalent annual disbursements

From Table 10.5,

$$\text{EUAC}_\text{fully} = -13,000(A/P, i\%, 5) + 2,000(A/F, i\%, 5) - 1,600$$
$$\text{EUAC}_\text{semi} = -8,000(A/P, i\%, 10) - 3,500$$

Then

$$0 = -13,000(A/P, i\%, 5) + 2,000(A/F, i\%, 5) + 8,000(A/P, i\%, 10) + 1,900$$

At $i = 12\%$, $0 \neq \$+24.33$; for $i = 15\%$, $0 \neq \$-87.52$. Interpolation yields $i = 12.7\%$ (as for the present-worth method) and the semiautomatic should be purchased.

COMMENT It is important to remember that if an EUAC analysis is to be made on the *net cash flow*, the cash-flow tabulation must be extended for the least common multiple of lives (ten years in this example), as in the present-worth method.   ////

Sec. 10.6

REFERENCES

1. **J. G. McLean**, How to Evaluate New Capital Investments, *Harvard Business Review,* November–December, 1958, pp. 59–69.
2. **N. N. Barish**, "Economic Analysis for Engineering and Managerial Decision-making," pp. 169–171, McGraw-Hill Book Company, New York, 1962.
3. **E. L. Grant** and **W. G. Ireson**, "Principles of Engineering Economy," 5th ed., pp. 152–160, The Ronald Press Company, New York, 1970.

BIBLIOGRAPHY

Barish, pp. 147–175.
DeGarmo and Canada, pp. 220–221, 234–240.
Emerson and Taylor, pp. 7-4 - 7-11.
Fabrycky and Thuesen, pp. 102–110.
Grant and Ireson, pp. 109–116, 117–129.
Park, pp. 37–48.
Riggs, pp. 181–184.
Smith, pp. 127–134, 140–143.
Taylor, pp. 125–158.
Thuesen, Fabrycky, and Thuesen, pp. 113–121.

PROBLEMS

P10.1 Prepare a tabulation of cash flow for the following alternatives:

	Alternative R	Alternative S
First cost	$8,000	$12,000
Annual operating cost	900	1,400
Salvage value	1,000	2,000
Life, years	5	10

P10.2 Prepare a tabulation of cash flow for the following alternatives:

	Alternative A	Alternative B
First cost	$15,000	$11,000
Annual operating cost	2,300	2,600
Annual income	4,000	3,100
Salvage value	2,000	1,500
Life, years	3	2

P10.3   Two different machines are being considered for a certain process. Machine X has a first cost of $12,000 and annual operating expenses of $3,000 per year. It is expected to have a life of 12 years with no salvage value. Machine Y can be purchased for $21,000, and it will have an annual operating cost of $1,200 per year. However, a factory overhaul will be required every four years at a cost of $2,500. It will have a useful life of 12 years with a $1,500 salvage value. Prepare a tabulation of net cash flow for the two alternatives.

P10.4   The owners of a new home are trying to decide between two types of landscape. They can purchase grass seed for $2.50 per pound and plant it themselves at no cost. They estimate that they will need 1 pound for 250 square feet. The front yard is $100' \times 30'$ and the back yard $100' \times 20'$. They will also have to purchase three tons of compost when the lawn is seeded and one ton every three years thereafter at a cost of $20 per ton. If they elect to plant grass, they will also install a sprinkler system with a $350 installation cost, a $15 per year maintenance cost, and a life of 18 years. The cost of such items as water, fertilizer, and chemicals is expected to be $120 per year.

Alternatively, they can have desert landscaping installed in the front and back yards at a cost of $2,800. If they elect this alternative, the annual cost of water, etc., is expected to be only $25 per year. However, they will have to replace the grass barrier under the gravel every six years at a cost of $350. Prepare a tabulation of cash flows and net cash flow for the two alternatives if they are to be compared for an 18-year period.

P10.5   A real estate investor purchased a piece of property for $6,000 and sold it 17 years later for $21,000. The property taxes were $80 the first year and $90 the second year, and increased by $10 per year until the property was sold. What was the rate of return on the investment?

P10.6   If the property taxes in problem P10.5 increased by $10 per year for the first six years and then increased by $20 per year thereafter, what was the rate of return on the investment?

P10.7   A family purchased a rundown house for $25,000 with the idea of making major improvements and then selling for a profit. In the first year that they owned the house, they spent $5,000 on improvements. They spent $1,000 the second year and $800 the third year. In addition, they paid property taxes of $500 per year for three years, and then sold the house for $35,000. What rate of return did they make on their investment?

P10.8   The Karry-Mor Trucking Company purchased a new dump truck for $54,000. The total operating expenses were $36,000 the first year, $39,000 the second, and amounts increasing by $3,000 per year thereafter. The income for the first year was $66,000 and decreased by $500 per year thereafter. If the company kept the truck for ten years and then sold it for $15,000, what was the rate of return on the investment?

P10.9 If a company spends $6,000 now and $900 per year for 17 years with the first disbursement five years from now, what rate of return would the company receive if the income during the 21 years was $5,000 at the end of year 3 and $1,200 per year thereafter?

P10.10 A real estate investor purchased a building for $130,000 and received $2,000 per month in rent. The taxes were $2,500 per year and maintenance costs $700 every three years. If the building was sold for $180,000 12 years after it was bought, what nominal annual and effective monthly rate of return was made on the investment?

P10.11 Rework problem 10.5 assuming the property taxes increased by $50 per year.

P10.12 An investor purchased three types of stocks (identified here as A, B, and C). The investor purchased 200 shares of A at $13 per share, 400 shares of B at $4 a share, and 100 shares of C at $18 per share. The dividends were $0.50 a share from Stock A for three years and then the stock was sold for $15 a share. There were no dividends from Stock B but the stock was sold for $5.50 a share two years after it was purchased. Stock C resulted in dividends of $2.10 a share for 10 years, but due to a depressed market at the time it was sold, the stock sold for only $12 a share. Calculate the rate of return on each stock as well as the overall rate of return on the stock investments.

P10.13 Firewood can be purchased during July for $55 a cord. If the purchaser waits until November, the cost of the same kind of wood is $70 a cord. What rate of return would the purchaser receive if the firewood was purchased in July instead of November?

P10.14 A careful shopper is trying to decide between buying an artificial Christmas tree and continuing to buy cut trees. The artificial tree costs $34 and can be used for eight years, at which time it will be thrown away for no salvage value.

Alternatively, the shopper can continue to buy cut trees at a cost of $8 now, $9 next year, $10 two years from now and so on for eight years. If the artificial tree is purchased, what rate of return is made on the investment?

P10.15 A homeowner who plans to build a large brick patio is expecting the price of bricks to increase considerably in the next two years. The cost is expected to increase from the present price of $60 per 1,000 bricks to $75 per 1,000 next year and $95 the following year. If 3,000 bricks are purchased now instead of 1,000 now and 1,000 each year for the following two years, what will be the rate of return on the investment?

P10.16 After preparing a tabulation of cash flow, what would you know immediately if the total of the difference (net cash-flow) column was a negative amount?

P10.17 Determine which alternative should be selected in problem P10.1 if the company's MARR is 15%. Use the incremental-investment method.

P10.18 Determine which alternative should be selected in problem P10.2 on the basis of incremental rate of return if the company's MARR is 20%.

P10.19 Determine which alternative should be selected in problem P10.3 if the company's MARR is 12%. Use the incremental-investment method.

P10.20 Determine which alternative should be selected in problem P10.4 if the purchaser's MARR is 6%. Use the incremental-investment analysis.

P10.21 A food-processing company is considering plant expansion. Under the current set-up, the company can increase profits $25,000 per year by extending the workday two hours through overtime. No investment will be required for this alternative. On the other hand, if additional cookers and freezers are added at a cost of $175,000, the company's profits will increase by $50,000 per year. If the company expects to use the current process for ten years, would the plant expansion be justified if the company's MARR is 25%? Solve this problem two ways.

P10.22 Which alternative should be selected in problem P10.3 if EUAC rate-of-return analysis and a MARR of 12% is used? Compare your answer and values with those of problem P10.19.

P10.23 Select the more economic alternative in problem P10.4 by EUAC rate-of-return analysis if the homeowner's MARR is 6% and the initial cost of desert landscaping will be $1,700 rather than $2,800.

P10.24 The engineer at the Smoke Ring Cigar Company wants to do a rate-of-return analysis using annual costs of two wrapping machines. The details below are available; however, the engineer does not know what value to use for a MARR figure since some projects are evaluated at 8% and some at 10%. Determine whether this difference in MARR would change the decision of which machine to buy. Use rate-of-return on the incremental-investment method.

	Machine A	Machine M
First cost	$10,000	$9,000
Annual labor cost	5,000	5,000
Annual maintenance cost	500	300
Salvage value	1,000	1,000
Life, years	6	4

P10.25 Would the answer to problem P10.24 change if the lives of both machines were six years?

P10.26 A family needs a new roof on their home. They have estimates from Roofer A and Roofer B. Roofer A wants $1,400 for shingle roofing, material, and labor. If the tin on the eves and valleys and old boards are replaced, an extra $100 cost is incurred. The roof itself will reasonably last 15 years if $20 per year for the first five years and $5 per year more each year after year 5 is spent on preventive maintenance. The replacement of the tin and board will increase the life to 18 years. Roofer B will put a gravel roof on the house for $1,200. It is estimated that the same annual maintenance expenditure as with shingles will give the roof an expected life of 12 years. Use rate-of-return analysis to compare (*a*) shingles without tin and board replacement with gravel and (*b*) shingles with the replacements and gravel. Assume a MARR of 6%.

LEVEL III

Knowledge of the material in the next three chapters gives you a thorough understanding of engineering-economy analysis. Besides learning two more alternative evaluation methods, you learn to make an economic comparison of an asset that is currently owned with one being considered as its replacement. The procedure to determine minimum cost life is studied from a replacement and retirement point of view. Additionally, a general discussion of breakeven analysis is included. Finally, there is a treatment of bonds—their types and economic consideration.

Chapter	Subject
11	Benefit/cost ratio and service-life evaluation
12	Replacement, retirement, and breakeven analysis
13	Bonds

11
BENEFIT/COST RATIO AND SERVICE–LIFE EVALUATION

The objectives of this chapter are to teach you how to compare two alternatives on the basis of a benefit/cost ratio or service-life analysis. These methods are sometimes regarded as *supplementary* tools, since they are usually used in conjunction with present-worth, annual-cost, or rate-of-return analysis.

CRITERIA

To complete this chapter, you must be able to do the following:

1 State the definition used to classify specified expenditures or savings as benefits, costs, or disbenefits.
2 Determine whether a project should be undertaken by comparing its benefits and costs, given values for the benefits, disbenefits, and costs, and the interest rate.
3 Select the better of two alternatives on the basis of a benefit/cost analysis, given the initial cost, life, salvage value, and disbursements for each alternative and the required rate of return.
4 State the procedure for selecting the best alternative from three or more projects using a benefit/cost ratio analysis.

5 State the purpose of and the mathematical model used in service-life computation.
6 Determine whether a single asset should be purchased using service-life computation, given the asset's first cost, salvage value, cash flows, an estimated usable life
value, and a required rate of return.
7 Determine which of two alternatives should be selected by computing a breakeven
service life, given alternative first costs, annual cash flows, salvage values, a required
rate of return, and an estimated usable project life.

EXPLANATION OF MATERIAL

11.1 Classification of Benefits, Costs, and Disbenefits

The method for selecting alternatives that is most commonly used by federal agencies
for analyzing the desirability of public works projects is the benefit/cost ratio (B/C
ratio). As its name suggests, the B/C method of analysis is based on the ratio of the
benefits to costs associated with a particular project. Therefore, the first step in a B/C
analysis is to determine which of the elements are benefits and which are costs. In
general, *benefits* are advantages, expressed in terms of dollars, which happen to the
owner. On the other hand, when the project under consideration involves disadvantages to the owner, these are known as *disbenefits*. Finally, the *costs* are the
anticipated expenditures for construction, operation, maintenance, etc. Since B/C
analysis is common to economy studies by federal agencies, it is helpful to think of the
owner as the *public* and the one who incurs the costs as the *federal government*. The
determination of whether an item is to be considered as a benefit, disbenefit, or cost,
therefore, depends upon *who is affected* by the consequences. Some examples of each
are illustrated in Table 11.1.

While the examples presented in this chapter are straightforward with regard
to identification of benefits, disbenefits, or costs, it should be pointed out that in
actual situations, judgments must sometimes be made which are subject to

Table 11.1 EXAMPLES OF BENEFITS, DISBENEFITS,
AND COSTS

Item	Classification
Expenditure of $11,000 for new interstate highway	Cost
$50,000 annual income to local residents from tourists because of new reservoir and recreation area	Benefit
$150,000 per year upkeep cost for irrigation canals	Cost
$25,000 per year loss by farmers because of highway right-of-way	Disbenefit

interpretation, particularly when it is necessary to determine whether an element of cash flow is a disbenefit or a cost. In other instances, it is not possible to simply place a dollar value on all benefits, disbenefits, or costs that are involved. These nonquantifiable considerations must be included in the final decision, as they are in other methods of analysis. In general, however, dollar values are available, or obtainable, and the results of a proper B/C analysis would agree with the methods studied in preceding chapters (such as present worth, equivalent uniform annual cost, or rate of return on incremental investment). James and Lee have presented many of the categories that must be considered in determining benefits, costs, and disbenefits for water resources projects.[1]

Problems P11.1, P11.2

11.2 Benefits, Disbenefits, and Cost Calculations

Before a B/C ratio can be computed, all of the benefits, disbenefits, and costs that are to be used in the calculation must be converted to common dollar units, as in present-worth calculations, or dollars per year, as in annual-cost comparisons. It is irrelevant whether the present-worth or annual-cost method is used so long as the procedures learned in Chaps. 8 and 9 are followed. With the use of either the present-worth or EUAC values, the B/C ratio can be calculated as

$$B/C = \frac{\text{benefits} - \text{disbenefits}}{\text{costs}} \qquad (11.1)$$

Note that the *disbenefits are subtracted from the benefits*, not added to the costs. It is important to recognize that the B/C ratio could change considerably if disbenefits are regarded as costs. For example, if the numbers 10, 8, and 8 are used to represent benefits, disbenefits, and costs, respectively, the correct procedure would result in a B/C ratio of $(10 - 8)/8 = 0.25$, while the incorrect procedure would yield a B/C ratio of $10/(8 + 8) = 0.625$, which is over twice as large. Clearly then, the method by which disbenefits are handled is very important. When the proper procedure is followed, a B/C ratio greater than or equal to 1.0 indicates that the project under consideration is economically advantageous.

An alternative method that can be used to evaluate the feasibility of federal projects is to subtract the costs from the benefits, that is, B − C. In this case, if B − C is greater than or equal to zero, the project is acceptable. This method has the obvious advantage of eliminating the discrepancies noted above when disbenefits are regarded as costs, since B represents the *net benefits*. Thus, for the numbers 10, 8, and 8 the same result is obtained regardless of how disbenefits are treated.

Subtracting disbenefits: B − C = [(10 − 8) − 8] = −6
Adding disbenefits to costs: B − C = [10 − (8 + 8)] = −6

Before calculating the B/C ratio, check to be sure that the proposal with the higher EUAC is the one that yields the higher benefits *after the benefits and costs have been expressed in common units.* Thus, a proposal having a higher initial cost may actually have a lower EUAC or present worth when all other costs are considered. Example 11.1 illustrates this point.

Example 11.1 Alternative routes are being considered by the State Highway Department for location of a new highway. Route A, costing $4,000,000 to build, will provide annual benefits of $125,000 to local businesses. Route B would cost $6,000,000 but will provide $100,000 in benefits. The annual cost of maintenance is $200,000 for A and $120,000 for B, respectively. If the life of each road is 20 years and an interest rate of 8% is used, which alternative should be selected on the basis of a benefit/cost analysis?

SOLUTION The benefits in this example are $125,000 for Route A and $100,000 for Route B. The EUAC of each alternative is as follows:

$$EUAC_A = 4,000,000(A/P, 8\%, 20) + 200,000 = \$607,400$$
$$EUAC_B = 6,000,000(A/P, 8\%, 20) + 120,000 = \$731,100$$

Route B has a *higher* EUAC than Route A by $123,700 per year, and *less benefits* than A. Therefore, there would be no need to calculate the benefit/cost ratio for Route B, since this alternative is obviously inferior to Route A. Furthermore, if the decision had been made that either Route A or B *must* be accepted (which would be the case if there were no other alternatives), then no other calculations would be necessary and Route A would be accepted.   ////

Example 11.5
Problems P11.3–P11.7

11.3 Alternative Comparison by Benefit/Cost Analysis

In computing the benefit/cost ratio by Eq. (11.1) for a given alternative, it is important to recognize that the benefits and costs used in the calculation represent the *differences* between two alternatives. This will always be the case, since sometimes doing nothing is an acceptable alternative. Thus, when it seems as though only one proposal is involved in the calculation, such as whether or not a flood-control dam should be built to reduce flood damage, it should be recognized that the construction proposal is being compared against another alternative–the "do nothing" alternative. Although this is also true for the other alternative evaluation techniques previously presented, it is emphasized here because of the difficulty often present in determining the benefits and costs between two alternatives when only costs are involved. See Example 11.2 for an illustration.

Once the B/C ratio on differences is computed, a B/C $\geqslant$ 1.0 means that the extra benefits of the higher-cost alternative justify this higher cost. If B/C $<$ 1.0, the extra cost is not justified and the lower-cost alternative is selected. Note that this lower-cost project may be the "do nothing" alternative, if the B/C analysis is for only one project.

Example 11.2 Two routes are under consideration for a new interstate highway. The northerly route (N) would be located about five miles from the central business district and would require longer travel distances by local commuter traffic. The southerly route (S) would pass directly through the downtown area and, although its construction cost would be higher, it would reduce the travel time and distance for local commuters. Assume the costs for the two routes are as follows:

	Route N	Route S
Initial cost	$10,000,000	$15,000,000
Maintenance cost per year	35,000	55,000
Road-user cost per year	450,000	200,000

If the roads are assumed to last 30 years with no salvage value, which route should be accepted on the basis of a benefit/cost analysis using an interest rate of 5%?

SOLUTION Since most of the costs are already annualized, the EUAC method will be used to obtain the equivalent annual cost. The *costs* to be used in the B/C ratio are the initial cost and maintenance cost:

$$\text{EUAC}_N = 10,000,000(A/P, 5\%, 30) + 35,000 = \$685,500$$
$$\text{EUAC}_S = 15,000,000(A/P, 5\%, 30) + 55,000 = \$1,030,750$$

The *benefits* in this example are represented by the road-user costs, since these are costs "to the public." The benefits, however, are not the road-user costs themselves but the *difference* in road-user costs if one alternative is selected over the other. In this example, there is a $450,000 - $200,000 = $250,000 per year benefit if Route S is chosen instead of Route N. Therefore, the benefit (B) of Route S over Route N is $250,000 per year. On the other hand, the costs (C) associated with these benefits are represented by the difference between the annual costs of Routes N and S. Thus,

$$C = \text{EUAC}_S - \text{EUAC}_N = \$345,250 \text{ per year}$$

Note that the route that costs more (Route S) is the one that provides the benefits. Hence, the B/C ratio can now be computed by Eq. (11.1).

$$\text{B/C} = \frac{250,000}{345,250} = 0.724$$

The B/C ratio of less than 1.0 indicates that the extra benefits associated with Route S are less than the extra costs associated with this route. Therefore, Route N would be selected for construction. Note that there is no "do nothing" alternative in this case, since one of the roads *must* be constructed.

COMMENT If there had been disbenefits associated with each route, the difference between the disbenefits would have to be added or subtracted from the net benefits ($250,000) for Route S, depending on whether the disbenefits for Route S were less than or greater than the disbenefits for Route N. That is, if the disbenefits for Route S were less than those for Route N, the difference between the two would have to be added to the $250,000 benefit for Route S, since the disbenefits involved would also favor Route S. However, if the disbenefits for Route S were greater than those for Route N, their difference should be subtracted from the benefits associated with Route S, since the disbenefits involved would favor Route N instead of Route S. Example 11.6 illustrates the calculations when disbenefits must be considered.   ////

Example 11.6
Problems P11.8–P11.12

11.4   Benefit/Cost Analysis for Multiple Alternatives

When only one alternative must be selected from three or more mutually exclusive (stand-alone) alternatives, a multiple alternative evaluation is required. In this case, it is necessary to conduct an analysis on the *incremental* benefits and costs similar to the method used in Chap. 10 for incremental rates of return. The "do nothing" alternative may be one of the considerations.

There are two situations which must be considered with regard to multiple alternative analysis by the benefit/cost method. In the first case, if funds are available so that *more than one* alternative can be chosen from among several, it is necessary only to compare the alternatives against the "do nothing" alternative. The alternatives are referred to as *independent* in this situation. For example, if several flood-control dams could be constructed on a particular river and adequate funding is available for all dams, the B/C ratios should be those associated with a particular dam versus no dam. That is, the result of the calculations could show that three dams along the river would be economically justifiable on the basis of reduced flood damage, recreation, etc., and, therefore, should be constructed.

On the other hand, when only *one* alternative can be selected from among several, it is necessary to compare the alternatives against each other rather than against the "do nothing" alternative. The exact procedure for doing this is discussed in Chap. 17. However, it is important for you to understand at this time the difference between the procedure to be followed when multiple projects are mutually exclusive and when they are not. In the case of mutually exclusive projects, it is necessary to compare them against each other, while in the case of projects that are not mutually exclusive (independent projects), it is necessary only to compare them against the "do nothing" alternative.

Problem P11.13

11.5   Purpose and Formulas of Service-Life Analysis

Basically, service-life analysis is used to determine the number of years an asset must be retained and used to recover its initial cost with a stated return, given the annual cash flow and salvage value. The analysis should be performed using after-tax cash-flow values (CF) (Chaps. 15 and 16), so that the results are more realistic. To find the economic service life of an asset, the following model is utilized:

$$0 = -P + \sum_{j=1}^{n'} (CF)_j (P/F, i\%, j) \qquad (11.2)$$

where $(CF)_j$ = net cash flow at the end of year j ($j = 1, 2, \ldots, n'$). For a given interest rate (i), the value of n' is sought. After n' years (not necessarily an integer), the cash flows will recover the first cost (P) and a return of $i\%$. A common, but incorrect, industrial practice is to determine n' at $i = 0\%$, that is, with no return accounted for; this is illustrated in the Solved Examples section. In this case Eq. (11.2) becomes

$$0 = -P + \sum_{j=1}^{n'} (CF)_j \qquad (11.3)$$

which is used to compute no-interest service life, more commonly called *payback* or *payout period*. If the cash flow (CF) is the same for each year, Eq. (11.3) is usually solved for n' directly.

$$n' = \frac{P}{CF} \qquad (11.4)$$

For a brief look at payback and some of its fallacies see Solved Examples, *after* you read the next section.

Problem P11.14

11.6   Use of Service Life to Determine Required Life

Equation (11.2) can be used to find the number of years necessary to recover the first cost at a stated rate of return. If the service life (n') is less than the time you would expect to be able to employ or retain the asset, it should be bought. If n' is greater than the expected usable life, the asset should not be bought, since there will not be enough time to recover the investment plus the stated return during the usable life.

Example 11.3   A semiautomatic assembly machine can be purchased for $18,000 with a salvage value of $3,000 and an annual cash flow of $3,000. If a return of 15% is required and the company would never expect such a machine to be used for more than ten years, should it be purchased?

SOLUTION Of course, there are several ways to answer this question—present-worth, EUAC, or rate-of-return analysis. But, let's use the service-life approach. Using Eq. (11.2), we have

$$0 = -18,000 + \sum_{j=1}^{n'} (CF)_j (P/F, 15\%, j)$$

We assume the salvage value of $3,000 is correct regardless of how long the asset is retained. We, therefore, can modify the above relation as follows:

$$0 = -18,000 + CF(P/A, 15\%, n) + SV(P/F, 15\%, n)$$

where $SV(P/F,15\%,n)$ is the present worth of the salvage after n years and the P/A factor has been used where possible. At $n = 15$ years we have

$$P = -18{,}000 + 3{,}000(P/A, 15\%,15) + 3{,}000(P/F, 15\%,15)$$
$$= \$-89.10$$

For $n = 16$, the result is $\$+183.30$. Interpolation indicates that in $n' = 15.3$ years the first cost plus 15% will be recovered. Since a fair estimate of usability is ten years, the machine should not be purchased.

COMMENT The salvage value and cash flows will be allowed to vary in the material of Chap. 12.   ////

Example 11.7
Problems P11.15–P11.20

11.7 Comparison of Two Alternatives Using Service-Life Computation

If capital is tight and the future uncertain (about available money *and* proposed investments), a breakeven (or equivalent-point) service life of two proposals may be computed for use in decision-making. Still, other evaluation methods, such as present-worth, should be pursued, because service-life analysis is considered only a supplementary tool. If a firm is short of capital and requires quick recovery of investment capital, the service-life computations can indicate the speed with which the project will "pay for itself." Therefore, capital recovery being important, service life at a stated rate of return is found by equating alternative present-worth or EUAC values and finding n' by trial and error. Depending on how many years the purchase will *reasonably* be used, the proposal with the smaller present-worth or EUAC value is selected. The method is the same as that used in rate-of-return breakeven analysis (Sec. 10.5), but with the value of n sought here.

Example 11.4 A dirt-moving company requires the service of dirt-moving equipment. The service may be acquired by purchasing a mover for $\$25{,}000$ having a negligible salvage value, $\$5{,}000$ annual operating cost, and a $\$12{,}000$ overhaul cost in year 10. Alternatively, the company may lease the mover at a total cost of $\$10{,}000$ per year. If all other costs are equal and service is needed for 12 years at a 12% rate of return, use service-life analysis to determine whether the mover should be purchased or leased.

SOLUTION We use the relation $\text{EUAC}_{\text{buy}} = \text{EUAC}_{\text{lease}}$ and find the breakeven n value (n').

$$\text{EUAC}_{\text{buy}} = 25{,}000(A/P, 12\%,n) + 5{,}000$$
$$+ 12{,}000(P/F, 12\%, 10)(A/P, 12\%,n)$$
$$\text{EUAC}_{\text{lease}} = \$10{,}000$$

The last term of EUAC_{buy} is used only when $n \geqslant 10$. Then when $n < 10$, equating EUAC relations gives

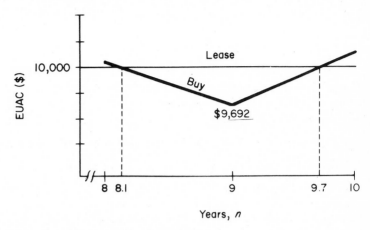

FIG. 11.1   Breakeven chart, Example 11.4.

$$25,000\,(A/P,\,12\%,n) + 5,000 = 10,000$$

$$(A/P,\,12\%,n) = \frac{5,000}{25,000} = 0.20000$$

At $n = 8$ years, $(A/P,12\%,8) = 0.20130$ and at $n = 9$ years, $(A/P,12\%,9) = 0.18768$. Interpolation gives $n' = 8.1$ years. However, if $n = 10$, $\text{EUAC}_{\text{buy}} = \$10,108$ due to the overhaul cost. This indicates that there is a second value of n' at which the buy and lease alternatives are equal. Computation of EUAC_{buy} for $n = 9$ yields a value of $\$9,692$. Interpolation indicates $n' = 9.7$ years. Figure 11.1 presents the entire picture of the service-life computations. Thus, for 12 years of service lease the mover, since here $\text{EUAC}_{\text{buy}} > \text{EUAC}_{\text{lease}}$.

COMMENT   If the service is needed for less than 8.1 or more than 9.7 years, the lease decision would be made, because $\text{EUAC}_{\text{lease}} < \text{EUAC}_{\text{buy}}$.

The possibility that lease costs could be reduced after 8.1 years or that mover costs would decrease, can be taken into account in this method if such decreases are predictable; but in the usual situation, they are not.   ////

It is important to realize that in the previous example there are *two* values of n' at which the alternatives are economically indifferent. If the *second* point had not been determined, obviously the wrong decision would be made for a 12-year expected life. This example illustrates the importance of determining *all* of the *equivalence points* in service-life analysis, if more than one is present. This complete analysis challenges the common objection that the service-life type of analysis neglects the economic life of the alternatives. Unless all equivalence points are computed, the cash flows occurring after the initial service-life value are overlooked, thereby introducing the possibility of selecting the wrong alternative. Therefore, when the service-life method is used correctly, it results in the same decision as other analysis methods.

Example 11.8
Problems P11.21–P11.24

SOLVED EXAMPLES

Example 11.5 The Wartol Foundation, a nonprofit educational research organization, is contemplating an investment of $1,500,000 in grants to develop new ways to teach people the rudiments of a profession. The grants would extend over a ten-year period and would create an estimated savings of $500,000 per year in professors' salaries, student tuition, and other expenses. The foundation uses a rate of return of 6% on all grant investments. In this case the program would be an addition to ongoing and planned activities, thus an estimated $200,000 a year would have to be released from other programs to support the educational research. Use (a) B/C ratio and (b) B − C analysis to determine the advisability of the program over a ten-year period.

SOLUTION The following definitions are in order:

Benefit: $500,000 per year
Cost: $1,500,000(A/P, 6%,10) = $203,805 per year
Disbenefit: $200,000 per year

(a) Using B/C ratio analysis, Eq. (11.1), yields

$$B/C = \frac{500,000 - 200,000}{203,805} = 1.47$$

The project is justified, since B/C > 1.0.
(b) If B is the net benefit, then

$$B - C = (500,000 - 200,000) - 203,805 = \$96,195$$

Again, since (B − C) > 0, investment is wise.

COMMENT In (a), if you happened to incorrectly add the disbenefits to costs you would have

$$B/C = \frac{500,000}{203,805 + 200,000} = 1.238$$

which still justifies the investment. But, the $200,000 is not a direct cost to this program and should be subtracted from B, not added to C. ////

Sec. 11.2

Example 11.6 Assume the same situation as in Example 11.2 for the routing of a new interstate highway. The B/C analysis showed that the northerly route, N, was to be constructed. However, this route will go through an agricultural region and the local farmers have complained about the great loss in revenue they and the economy will suffer. Likewise, the downtown merchants have complained about Route S because of the loss in revenue due to reduced merchandising ability, parking problems, etc. To consider these eventualities, the state highway department has undertaken a study and predicted that the loss to state agriculture for Route N will be about $500,000 per year and that Route S will cause an estimated reduction in retail sales and rents of $400,000 per year. What effect does this new information have on the B/C analysis?

SOLUTION These new "costs" should be considered as disbenefits. Since the disbenefits of Route S are $100,000 less than those of Route N, this difference is *added* to the benefits of Route S to give a total net benefit of $250,000 + $100,000 = $350,000 to the downtown alternative. Now we have

$$B/C = \frac{350,000}{345,250} = 1.01$$

and Route S is to be slightly favored. In this case the inclusion of disbenefits has reversed the earlier decision.   ////

Sec. 11.3

Example 11.7 Purchase of an asset with first cost, salvage, and cash flows as in Example 11.3 is anticipated. Compute the payback period.

SOLUTION Using Eq. (11.3), for $n' = 5$,

$$-18,000 + 5(3,000) + 3,000 = 0$$

Therefore, the asset must be retained at least five years to return the investment with *no interest*.

COMMENT Note the tremendous difference that interest makes. At a 15% required return, this asset would have to be retained 15.3 years, while with no interest only five years of ownership are required. This characteristic of longer required ownership is always present with $i > 0\%$ because of the time value of money. The pattern of cash flows is the primary determinant of the difference in required retention periods.

This result indicates that the only use of the payback-period method of analysis is if investment capital is in very short supply and management requests recovery of capital in a short period. The payback calculation will give the amount of time required to recover the invested *dollars*; but from an economic, rational, and time value of money point of view, the payback-period method of analysis is incorrect.

////

Sec. 11.6

Example 11.8   Consider the two machines detailed below.

Machine 1	Machine 2
$P = \$12{,}000$ Net income per year $=$ 3,000 Maximum $n = 7$ years	$P = \$8{,}000$ Net income per year $=$ 1,000 (years 1–5) $=$ 3,000 (years 6–15) Maximum $n = 15$ years

The increase in cash flow for Machine 2 is anticipated due to its versatility and future application in the manufacture of new products. Which machine should be purchased (*a*) if payback analysis is used and capital is to be recovered as rapidly as possible and (*b*) if service-life analysis is used and a 15% rate of return is required? Assume the machine will be required for 15 years.

SOLUTION

(*a*)   For Machine 1, using Eq. (11.4),

$$n' = 12{,}000/3{,}000 = 4 \text{ years}$$

For Machine 2, using Eq. (11.3) and assuming $n' = 6$ years,

$$0 = -8{,}000 + 5(1{,}000) + 3{,}000$$

Therefore, $n' = 6$ years and select Machine 1 to recover the investment in four years of ownership.

(*b*)   The service life is found by equating the EUAC expressions and solving for the n value. For $n = 7$,

$$\begin{aligned}
\text{EUAC}_1 &= -12{,}000(A/P, 15\%,7) + 3{,}000 = \$116 \\
\text{EUAC}_2 &= -8{,}000(A/P, 15\%,7) + [1{,}000(P/A,15\%,5) \\
&\quad + 3{,}000(P/A, 15\%,2)(P/F, 15\%,5)](A/P,15\%,7) \\
&= \$-534
\end{aligned}$$

Thus $\text{EUAC}_1 - \text{EUAC}_2 = \$650$. At $n = 10$, $\text{EUAC}_1 - \text{EUAC}_2 = \$-290 - 70 = \$-360$, assuming reinvestment in Machine 1 after 7 years and annualizing over 10 years. Interpolation yields a service life of 8.9 years. Since the difference in EUAC is decreasing as n increases, Machine 2 is favored above 8.9 years. Notice that this decision is contradictory to that of part (*a*), in which Machine 1 is selected due to its rapid recovery of invested capital.

COMMENT   By simple inspection of the machines' characteristics, you might select Machine 2, but payback favors the rapid recovery of investment in Machine 1. To give you an idea of how the respective machine rates of return develop, Table 11.2 presents the return for each asset for 4 to 15 years of ownership. The rates of return support the selection of Machine 2 for periods greater than 8.9 years at a MARR $\geqslant 15\%$, as shown by the service life analysis above. These results show that the conclusions of the payback-period analysis are incorrect because it neglects the time value of money.

Table 11.2 ACTUAL RATES OF RETURN FOR A
GIVEN NUMBER OF YEARS OF
OWNERSHIP FOR EXAMPLE 11.8

Years owned	Rate of return	
	Machine 1	Machine 2
4	0 %	< 0 %
6	13.02	< 0
7	16.45	7.08
10	10.69	16.00
12	14.41	18.51
15	12.42	20.11

If the two machines are compared using EUAC analysis at 15%,

$$EUAC_1 = -12,000(A/P,15\%,7) + 3,000 = \$116$$
$$EUAC_2 = -8,000(A/P, 15\%,15) + 1,000$$
$$+ 2,000(F/A, 15\%,10)(A/F, 15\%,15)$$
$$= \$485$$

This is a clear decision in favor of the profit generated by Machine 2 and is again contradictory with the payback period method of decision favoring Machine 1. If payback is used as an analysis tool it is important to realize (1) that it neglects the *time value of money*, since computations are made at $i = 0\%$, and (2) that it also neglects all cash flows after the computed payback (n') (and thereby neglects the actual economic life of the asset). As a result of the second "oversight" of the method, a short-lived asset is usually favored; that is, it will be selected for purchase rather than a longer-lived asset, which may bring a higher return, as shown in this example.

////

Sec. 11.7

REFERENCE

1. **L. D. James** and **R. L. Lee**, "Economics of Water Resources Planning," pp. 161-192, McGraw-Hill Book Company, New York, 1971.

BIBLIOGRAPHY

Since virtually all books discuss payback period ($i = 0\%$) and its pitfalls, rather than service life ($i > 0\%$), we give the payback-period references here.

Barish, pp. 179-181 (payback).
DeGarmo and **Canada**, pp. 448-457 (B/C); pp. 225-226, 290-291 (payback).
Fabrycky and **Thuesen**, pp. 193-198 (B/C); pp. 110-111 (payback).
Grant and **Ireson**, pp. 135-144 (B/C); pp. 347, 405, 528 (payback).

Ostwald, pp. 36–37, 378–382 (B/C).
Park, pp. 32–37 (payback).
Reisman, pp. 128, 149 (payback).
Riggs, pp. 181–184, 245–248 (B/C); pp. 243–245 (payback).
Smith, pp. 275–290 (B/C); pp. 116–117 (payback).
Taylor, pp. 425–435.
Thuesen, Fabrycky, and Thuesen, pp. 84, 87, 121–122, 224 (B/C); pp. 218, 222, 225, 229, 232 (payback).

PROBLEMS

P11.1   Why should disbenefits be subtracted from benefits rather than added to costs?

P11.2   Classify the following cash flows as either benefits, costs, or disbenefits:
    (*a*) Drive-in theatre had to be destroyed because of highway right-of-way
    (*b*) $10 million expenditure for new highway
    (*c*) Less disbursement by motorists because of new highway
    (*d*) Archeological sites inundated by new reservoir
    (*e*) $2 million paid for right-of-way for new highway

P11.3   The U.S. Army Corps of Engineers is considering the feasibility of constructing a small flood-control dam in an existing arroyo. The initial cost of the project will be $2.2 million, with inspection and upkeep costs of $10,000 per year. In addition, minor reconstruction will be required every 15 years at a cost of $65,000. If flood damage will be reduced from the present cost of $90,000 per year to $10,000 annually, use the benefit/cost method to determine if the dam should be constructed. Assume the dam will be permanent and the interest rate is 5% per year.

P11.4   A state highway department is considering the construction of a new highway through a scenic rural area. The road is expected to cost $6 million, with annual upkeep estimated at $20,000 per year. The improved accessibility is expected to result in additional income from tourists of $350,000 per year. If the road is expected to have a useful life of 25 years, use the (*a*) B − C method and (*b*) B/C method at an interest rate of 6% per year to determine if the road should be constructed.

P11.5   If the highway in problem P11.4 would result in agricultural-income losses of $15,000 the first year, $16,000 the second, and amounts increasing by $1,000 per year, by how much would the tourist income have to increase each year (starting in year 2) in order for the highway to become economically feasible?

P11.6   The U.S. Bureau of Reclamation is considering a project to extend irrigation canals into a desert area. The initial cost of the project is expected to be $1.5 million, with annual maintenance costs of $25,000 per year. If agricultural revenue is expected to be $175,000 per year, make a B/C analysis to determine whether the project should be undertaken, using a 20-year study period and an interest rate of 6% per year.

P11.7 Calculate the B/C ratio for problem P11.6 if the canal must be dredged every three years at a cost of $60,000 and there is a $15,000 per year disbenefit associated with the project.

P11.8 Two routes are under consideration for a new interstate highway. The long route would be 22 miles in length and would have an initial cost of $21 million. The transmountain route would be 10 miles long and would have an initial cost of $45 million. Upkeep costs are estimated at $40,000 per year for the long route and $65,000 per year for the transmountain route.

Regardless of which route is selected, the volume of traffic is expected to be 400,000 vehicles per year. If the vehicle operating expense is assumed to be $0.12 per mile, determine which route should be selected by (a) B/C analysis and (b) B − C analysis. Assume a 20-year life for each road and an interest rate of 6% per year.

P11.9 The U.S. Army Corps of Engineers is considering three sites for flood-control dams (designated as Sites A, B, and C). The construction costs are $10 million, $12 million, and $20 million, and maintenance costs are expected to be $15,000, $20,000, and $23,000, respectively, for Sites A, B, and C. In addition, a $75,000 expenditure will be required every ten years at each site. The present cost of flood damage is $2 million per year. If only the dam at Site A is constructed, the flood damage will be reduced to $1.6 million per year. If only the dam at Site B is constructed, the flood damage will be $1.2 million per year. Similarly, if the Site C dam is built, the damage will be reduced to $1.4 million per year.

Since the dams would be built on different branches of a large river, either one or all of the dams could be constructed and the decrease in flood damages would be additive. If the interest rate is 5% per year, determine which ones, if any, should be built on the basis of their B/C ratios. Assume the dams will be permanent.

P11.10 Highway Department officials are considering the economics of resurfacing an existing highway versus construction of a new one. The existing highway is 12 miles long and would cost $2 million to resurface. Annual upkeep cost is expected to be $5,000 the first year, $10,000 the second, and amounts increasing by $5,000 per year until year 10, at which time the road would have to be resurfaced again.

If a new road is constructed, the initial cost would be $15 million for a road 10 miles long. The maintenance is expected to cost $5,000 the first year, $7,000 the second, and amounts increasing by $2,000 per year until year 10, after which the cost will be $23,000 per year. If the new road is constructed, the cost of auto accidents is expected to decrease by $500,000 per year. If vehicle operating cost is assumed to be $0.10 per mile and 600,000 vehicles per year travel the road, use the benefit/cost method to determine which road should be constructed at an interest rate of 6%.

P11.11 The U.S. Forest Service is considering two locations for a new federal park. Location E would require an investment cost of $3 million and $50,000 per year in maintenance. Location W would cost $7 million to construct, but the forest service will receive an additional $25,000 per year in park-use fees. The operating cost of Location W will be $65,000 per year. The revenue to park concessionaires will be $500,000 per year at Location E and $700,000 per year at W. The disbenefits associated with each location are $30,000 per year for Location E and $40,000 per year for Location W. Use (*a*) the B/C method and (*b*) the B − C method to determine which location if either should be selected, using an interest rate of 5% per year. Assume the park will be maintained indefinitely.

P11.12 The Bureau of Reclamation is considering the lining of the main canals of its irrigation ditches. The initial cost of lining is expected to be $4 million, with $25,000 per year required for maintenance. If the canals are not lined, a weed-control and dredging operation will have to be instituted, which will have an initial cost of $700,000 and a cost of $50,000 the first year, $52,000 the second, and amounts increasing by $2,000 per year for 25 years. If the canals are lined, less water will be lost through infiltration so that additional land can be cultivated for agricultural use. The agricultural revenue associated with the extra land is expected to be $120,000 per year. Use (*a*) the B/C and (*b*) the B − C method to determine if the canals should be lined. Assume the project life is 25 years and the interest rate is 6% per year.

P11.13 Why must an incremental B/C analysis be conducted when only one proposal can be selected from three or more proposals?

P11.14 What form will Eq. (11.2) assume if the cash flow is the same in each year?

P11.15 Why is it wrong to use a payback-period analysis?

P11.16 Determine the number of years that the investor of problem P10.5 must keep the property to make a 15% return. Assume the sales price would be $14,000 if kept one to ten years and $20,000 thereafter.

P11.17 (*a*) Determine the 8% service life for an asset that initially costs $8,000, has a salvage of $500 when sold, and has a net profit of $900 per year.
(*b*) If the asset will be used for five years by the owner, should it be purchased?

P11.18 Determine the number of years that the two machines of Plan B of problem P9.18 must be kept to make a 10% return if the annual income of the plan is $4,000. Assume the estimated salvage values apply for all years.

P11.19 How many years must the owner of the artificial Christmas tree in problem P10.14 use the tree to make 20% on the investment?

P11.20 The R-U-There Detective Agency would like to purchase some new camera equipment. The following data are estimates

$$\text{First cost} = \$1,050$$
$$\text{Annual operating cost} = 70 + 5(k) \qquad k = 1,2,3,\ldots$$
$$\text{Annual income} = 200 + 50(k) \qquad k = 1,2,3,\ldots$$
$$\text{Salvage value} = \$600$$

(a) Compute the service life to make a return of 5% on the investment. (b) Should the camera be purchased if the expected useful period is seven years? (c) Answer (a) and (b) above for a return of 10%.

P11.21 Decide which alternative is more economic in problem P10.4 using the service-life method of analysis and a required return of 12%.

P11.22 El Ranchero Investment Company has been offered the two apartment-house resales detailed below.

	Apartment 1	Apartment 2
Actual purchase price	$−220,000	$−390,000
Annual operating cost	−8,000	−11,500
Annual rental income:		
Years 1–3	+72,000	+117,000
Years 4–6	+80,000	+117,000
After year 7	+82,000	+117,000
Expected sales price	+250,000	+450,000

The investment company never retains ownership of a property for more than five years and requires a minimum return of 25% on all invested capital. Answer the following questions:

(a) After determining the breakeven service life, plot the EUAC versus service life. Which apartment should be chosen for investment?

(b) How long could the investment be retained before the decision in (a) would change if the same MARR is required?

(c) Which alternative is more advantageous if you use present-worth analysis and the five-year retention period as the expected life?

P11.23 The lease cost of trucks is $275 per truck per month. As an alternative the purchase price per truck is $700 with a monthly payment of $300 for four years. A truck can be sold for an average of $1,200 regardless of the length of time of ownership. The prospective owner, Speede Drug Company, requires five trucks and a nominal return of 12% per year. If the drug company must pay operation, maintenance, and insurance for the leased or purchased trucks, this cost is equal and, therefore, neglected. (a) How many years must the lease or purchase plan be adhered to in order to just break even? (b) If a purchased truck has an expected life of six years, should the trucks be leased or purchased?

P11.24 Determine which alternative machine should be selected in Example 11.8 if a 10% return is required and this type of machine would be kept (a) two years and (b) ten years.

12
REPLACEMENT, RETIREMENT, AND BREAKEVEN ANALYSIS

The objective of this chapter is to aid you in the economic comparison of two assets: one which you presently own, the other which can be considered a replacement of the owned asset.

The replacement of an asset is usually contemplated prior to the time of its anticipated sale because of physical deterioration or obsolescence (planned or unplanned). Planned obsolescence is usually the result of "bigger and better" equipment, a fact that has become accepted in today's business world. Unplanned obsolescence usually results from an unanticipated decrease in product demand or the possibility of leasing, rather than owning, the equipment.

CRITERIA

When you are able to accomplish the following, you will have completed this chapter:

1 Describe the consultant's viewpoint for replacement analysis, determine the value of a sunk cost, and state the values of first cost, life, annual cost, and salvage value for a defender and challenger, given the appropriate data.

2 Select the better of a defender and challenger plan, given their initial costs, lives, operating costs, interest rate, market value of the defender, and the planning horizon.

3 Compute the replacement value of a defender given the selected challenger plan, defender's remaining life, salvage value, operating costs, and the MARR.

4 Determine the minimum cost life of an asset using the EUAC method, given the first cost or market value now, future salvage values, annual operating costs, and the required rate of return.

5 Calculate the breakeven point between two alternatives, given the necessary data for each alternative.

EXPLANATION OF MATERIAL

12.1 The Defender/Challenger Concept

Here, as in previous chapters, we are comparing two alternatives; however, we now own one of the assets, referred to as the *defender*, and are considering the purchase of its replacement or *challenger*. The defender/challenger terminology is borrowed from the publications by George Terbough of the Machinery and Allied Products Institute (MAPI).[1,2]

In the actual comparison we must take the *consultant's viewpoint*, as in Chaps. 8–11. Thus, we *pretend* we own neither asset. In order to "purchase" the defender, therefore, we would have to pay the going market value for this used asset. We then use the present, fair market value as the first cost (P) of the defender. Likewise, there will be an associated salvage value (SV), economic life (n), and annual operating cost (AOC) for the defender. Even though the values may all be different than the original data, it makes no difference to us, because we are using the consultant's viewpoint, thus making all previous data irrelevant to the present economic evaluation.

In replacement analysis it is important that the role of a *sunk cost* be understood and treated correctly. The sunk cost is defined as

$$\text{Sunk cost} = \text{present book value} - \text{present realizable salvage value}$$

If incorrect estimates have been made about the utility or market value of an asset (as is usual since no one can be perfect in their estimation of the future), there is a positive sunk cost, *which cannot be recovered*. However, some analysts try to "recover" the sunk cost of the defender by adding it to the first cost of the challenger. The sunk cost, rather, should be charged to an account entitled Unrecovered Capital, or the like, which will ultimately be reflected in the company's income statement for the year in which the sunk cost was incurred. From the tax viewpoint the value of this sunk cost will be important in that a capital gain or loss is involved (Chaps. 15 and 16). Therefore, *for replacement analysis the sunk cost should not be included in the economic comparison*. The following example illustrates the correct data to use in a defender/challenger situation in which a sunk cost is involved.

Example 12.1 A dump truck was purchased three years ago for $12,000 with an estimated life of eight years, salvage of $1,600, and annual operating cost of $3,000. Straight-line depreciation has been used for the truck.

A challenger is now offered for $11,000 and a trade-in value of $7,500 on the old truck. The company estimates challenger life at ten years, salvage at $2,000, and annual operating costs at $1,800 per year. New estimates for the old truck are made as follows: realizable salvage, $2,000; remaining life, three years; same operating costs.

What values should be used for P, n, SV, and AOC for each asset?

SOLUTION Using the consultant's viewpoint the most current information is applicable.

Defender	Challenger
$P = \$7,500$	$P = \$11,000$
AOC = 3,000	AOC = 1,800
SV = 2,000	SV = 2,000
$n = 3$	$n = 10$

COMMENT A sunk cost is incurred on the defender if it is replaced. The computation of this sunk cost is the only reason for specifying the depreciation method. Using straight-line depreciation, annual write-off by Eq. (7.1) is

$$D = \frac{P - SV}{n} = \frac{12,000 - 1,600}{8} = \$1,300$$

and present book value from Eq. (7.2) is

$$BV_3 = 12,000 - 3(1,300) = \$8,100$$

Therefore,

$$\text{Sunk cost} = 8,100 - 7,500 = \$600$$

The $600 is not added to the first cost of the challenger, since this action would (1) try to "cover up" past mistakes of estimation and (2) penalize the challenger because the capital to be recovered each year would be higher due to the increased first cost.   ////

Problems P12.1–P12.5

12.2 Replacement Analysis Using a Specified Planning Horizon

The planning horizon (also called a study period) is the number of years in the future which is to be used in comparing the defender and the challenger. Typically, one of two situations is present: (1) the anticipated remaining life of the defender equals the life of the challenger or (2) the life of the challenger is greater than that of the defender. We will discuss both possibilities in order.

If the defender and challenger have equal lives, any of the evaluation methods can be used with the *most current data*. The example below compares owning versus leasing to clearly depict this situation.

Example 12.2 The Pak-Mor Transport Company owns two vans but has always leased more vans on a yearly basis as needed. The two vans were purchased two years ago for $60,000 each. The company plans to keep the vans for ten more years. Fair market value for a two-year-old van is $42,000 and for a 12-year-old van, $8,000. Annual fuel, maintenance, tax, etc., costs are $12,000 per year. Lease cost is $9,000 per year with annual operating charges of $14,000. Should the company lease all of its vans if a 12% rate of return is required?

SOLUTION Consider a ten-year life of the owned van (defender) and the leased van (challenger).

Defender	Challenger
$P = \$42,000$	Lease cost = $ 9,000 per year
AOC = 12,000	AOC = 14,000
SV = 8,000	
$n = 10$	$n = 10$

The EUAC for the defender (EUAC_D) by Eq. (9.1) is

$$\begin{aligned}
\text{EUAC}_D &= P(A/P, i\%, n) - \text{SV}(A/F, i\%, n) + \text{AOC} \\
&= 42{,}000(A/P, 12\%, 10) - 8{,}000(A/F, 12\%, 10) + 12{,}000 \\
&= \$18{,}977
\end{aligned}$$

EUAC_C for the challenger is

$$\text{EUAC}_C = 9{,}000 + 14{,}000 = \$23{,}000$$

Clearly, the firm should retain ownership of the two vans.   ////

In many instances, an asset is to be replaced by another having an estimated life different than that of the defender's remaining life. For analysis, the length of the planning horizon must first be selected, usually coinciding with the life of the longer-lived asset. Selection of the planning horizon makes the assumption that the EUAC value of the shorter-lived asset is the same throughout the planning horizon. This implies that the service performed by the shorter-lived asset can be acquired at the same EUAC as presently computed for its expected service life.

Example 12.3 A company has owned a particular machine for three years. Based on current market value the asset has an EUAC of $5,210 per year and an anticipated remaining life of five years due to rapid technological growth. The possible replacement for the asset has a first cost of $25,000, salvage value of $3,800, life of 12 years, and an annual operating cost of $720 per year. If the company uses a minimum rate of return of 10% on asset investments and plans to retain the new machine for its full anticipated life, should the old asset be replaced?

SOLUTION A planning horizon of 12 years to correspond with the challenger's life is appropriate.

$$\text{EUAC}_\text{D} = \$5,210$$
$$\text{EUAC}_\text{C} = 25,000(A/P, 10\%,12) - 3,800(A/F, 10\%,12) + 720$$
$$= \$4,211$$

Thus, purchase of the new asset is less costly than retention of the presently owned machine.

COMMENT When using the planning horizon for different-lived assets and the present-worth value is desired, it is important to realize that a horizon of 12 years assumes you plan to purchase a new, similar asset if the shorter-lived asset is retained. In other words, in this problem a defender-similar asset would be purchased at the end of five years. Thus, present-worth computation would be as follows:

$$\text{PW}_\text{D} = 5,210(P/A, 10\%,12) = \$35,500$$
$$\text{PW}_\text{C} = 4,211(P/A, 10\%,12) = \$28,693$$

Of course, the decision to purchase the new machine is still made.   ////

Often management is skeptical of the future. Reflection of this skepticism is a management desire to use abbreviated periods of time for the planning horizon. In this situation it is assumed that only the time in the planning horizon is allowed for the recovery of invested capital and a specified return; that is, n values in the computations will reflect the shortened horizon. The following example indicates what occurs when a shortened horizon is specified.

Example 12.4 Consider the data of Example 12.3, except use a five-year planning horizon. Management specifies five years because it is leery of the technological progress being made in this area, progress that has already called into question retention of presently owned, operational equipment.

SOLUTION The approach is as in the preceding example, except a capital-recovery period of only five years is used for the challenger.

$$\text{EUAC}_\text{D} = \$5,210$$
$$\text{EUAC}_\text{C} = 25,000(A/P, 10\%,5) - 3,800(A/F, 10\%,5) + 720$$
$$= \$6,693$$

Now, retention of the defender is less costly, thus reversing the decision made with a 12-year horizon.

COMMENT By not allowing the full anticipated life of the challenger to be used, management has ruled out its use. However, the decision to not consider the use of this new asset past five years is one of management responsibility. The reason why the decision is reversed in this example is quite simple, actually. The challenger is given only five years to recover the same investment and a 10% return as in the previous example for 12 years. Reasonably, EUAC_C must increase drastically. It would be possible to recognize unused value in the challenger by increasing the salvage value from \$3,800 to the estimated fair market value after five years of service, if such a value can be predicted.   ////

Selection of the planning horizon is a difficult decision, one which must be based on sound judgment and data. The use of a short horizon may often bias the economic decision in that the capital-recovery period for the challenger may be abbreviated to much less than the anticipated life. This is the case in Example 12.4, where only five years were allowed for recovery of invested capital plus a 10% return. However, use of a large horizon is also often detrimental due to the uncertainty of the future and its estimate. In this case, the direction of bias is less certain than in the case of a too short horizon. A common practice is augmentation; that is, the defender is augmented with a newly purchased asset to make it comparable in ability (speed, volume, etc.) with the challenger. Since the analysis is similar to that covered here, a sample solution is included in Solved Examples.

Example 12.8
Problems P12.6–P12.14

12.3 Computation of Replacement Value for a Defender

Once a challenger is defined, the market value of the defender is of critical importance to the replacement analysis. Rather than initially obtaining market-value quotes, it may be advisable to compute a breakeven market value of the defender in terms of the contemplated challenger. The replacement value is computed most easily using the general form

$$0 = EUAC_D - EUAC_C \qquad (12.1)$$

where $EUAC_D$ is the EUAC for the defender and $EUAC_C$ is for the challenger. After calculating the replacement value, the company can then go into the field to see if the defender market value is greater or less than the replacement value. If greater, the asset should be replaced, since $EUAC_D$ will be greater than $EUAC_C$.

Example 12.5 A three-year-old asset has annual operating costs of \$9,500, salvage of \$3,500, and a useful life of seven more years. The selected challenger will cost \$28,000, have a life of 14 years, salvage of \$2,000, and annual operating costs of \$5,500. What is the minimum trade-in deal that the owner can accept and still purchase the new asset, if this class of asset has a 15% rate of return?

SOLUTION To compute the replacement value of the defender we need the EUAC values. If we let RV represent the replacement value of the defender, we have

$$
\begin{aligned}
EUAC_D &= RV(A/P,\ 15\%,7) - 3,500(A/F,15\%,7) + 9,500 \\
&= 0.24036(RV) + 9,184 \\
EUAC_C &= 28,000(A/P,\ 15\%,14) - 2,000(A/F,\ 15\%,14) + 5,500 \\
&= \$10,342
\end{aligned}
$$

Equation (12.1) results in a replacement value of RV = \$4,818. If a trade-in value greater than \$4,818 can be obtained, the challenger should be purchased.

COMMENT To verify the result, assume the trade-in offer is \$5,000. Then

$$\text{EUAC}_D = 5,000(A/P, 15\%, 7) - 3,500(A/F, 15\%, 7) + 9,500$$
$$= \$10,386$$

Thus, because EUAC_D is greater than EUAC_C, replacement is economically justified.

////

Problems P12.15–P12.19

12.4   Determination of Minimum Cost Life

There are several situations in which an analyst desires to know how long an asset should be used before it is removed from service. Determination of this time (an n value) is given several names. If the asset is a defender to be replaced by a potential challenger, the n value is the *remaining life of the defender*. If the function performed by the asset will be discontinued or assumed by some other facility in the company, the n value is often referred to as the *retirement* life. Finally, a similar analysis may be performed for an anticipated asset purchase, in which case the n value may be called *expected life*. Regardless of what it is called, this value should be found by determining the number of years (n) that will yield a minimum present-worth or EUAC value. This approach is often called a *minimum-cost-life analysis*. In this chapter we consistently use EUAC analysis.

To find the minimum cost life, n is increased from zero to the maximum expected life and the EUAC found for each value of n. Each EUAC is the equivalent annual cost of the asset if it were used for n years. The number of years corresponding to the minimum EUAC indicates the minimum cost life. The respective EUAC and n values are those which should be used in a replacement analysis or alternative evaluation. Since n varies, the annual cost values and salvage value must be estimated for each possible n. The EUAC approach for replacement analysis is illustrated in Example 12.6.

Example 12.6   An asset purchased three years ago is now challenged by a new piece of equipment. Market value of the defender is \$13,000. Anticipated salvage values and annual operating costs for the next five years are given in columns (2) and (3), respectively, of Table 12.1. What is the minimum cost life to be used when comparing this defender with a challenger if capital is worth 10%?

SOLUTION   Column (4) of Table 12.1 presents the equivalent uniform amount for capital recovery and a 10% return using the salvage sinking-fund method, Eq. (9.1). Computation for a particular n value ($n = 1, 2, \ldots, 5$) is obtained by computing

Table 12.1 COMPUTATION OF MINIMUM COST LIFE FOR A PRESENTLY OWNED ASSET

Year, n (1)	Salvage, SV, after n years (2)	Annual operating costs, AOC (3)	EUAC for n years		Total (6) = (4) + (5)
			Capital recovery and return (4)	Equivalent operating costs (5)	
1	$9,000	$2,500	$5,300	$2,500	$7,800
2	8,000	2,700	3,681	2,595	6,276
3	6,000	3,000	3,415	2,717	6,132*
4	2,000	3,500	3,670	2,886	6,556
5	0	4,500	3,429	3,150	6,579

*Total EUAC $= 13,000(A/P,10\%,3) - 6,000(A/F,10\%,3) + [2,500(P/F,10\%,1)$
$+ 2,700(P/F,10\%,2) + 3,000(P/F,10\%,3)](A/P,10\%,3)$
$= \$6,132.$

$13,000(A/P,10\%,n) - SV(A/F,10\%,n)$. Similarly, column (5) presents equivalent uniform annual operating costs for a particular n value, found by

$$\left[\sum_{k=1}^{n} AOC_k(P/F, 10\%, k) \right](A/P, 10\%, n)$$

Column (6) = (4) + (5), which represents total EUAC if the defender is retained for n years. A sample computation for $n = 3$ is shown in Table 12.1. The minimum EUAC of $6,132 per year for $n = 3$ indicates that three years should be the anticipated remaining life of this asset, when compared with a challenger.

COMMENT It should be realized that the approach presented in this example is general. It can be utilized to find the minimum cost life of any asset, whether it is owned presently and to be replaced or retired or if it is a contemplated purchase. A further example of minimum cost life is given in the Solved Examples. ////

Example 12.9
Problems P12.20–P12.24

12.5 Computation of Breakeven Points Between Alternatives

In some economic comparisons, one or more of the elements of cost are either very questionable or vary as a function of output or usage. Under such circumstances, it is sometimes more convenient to express the uncertain parameter as a variable function and find the value of the variable at which the two proposals break even. This concept is similar to that of service life wherein the minimum number of years of service is determined that justifies a specified expenditure. Breakeven analysis, however, usually involves a variable cost that is *common to both alternatives*, such as a variable

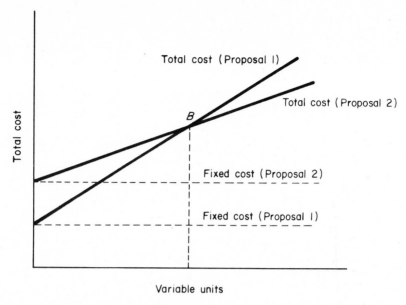

FIG. 12.1 Graphical illustration of breakeven.

operating cost or production cost. Figure 12.1 graphically illustrates the breakeven concept for two proposals (identified as Proposal 1 and Proposal 2). As shown in the figure, the fixed cost (which may be simply the initial investment cost) of Proposal 2 is greater than that of Proposal 1, but Proposal 2 has a lower variable cost (as shown by its smaller slope). The point of intersection (B) of the two lines represents the breakeven point between the two proposals. Thus, if the variable units (such as hours of operation or level of output) are expected to be greater than the breakeven amount, Proposal 2 would be selected, since the total cost of the operation would be lower with this alternative. Conversely, an anticipated level of operation below the breakeven number of variable units would favor Proposal 1.

Instead of plotting the total costs of each alternative and finding the breakeven point graphically, it is generally easier to calculate the breakeven point algebraically. Although the total cost can be expressed as either a present worth or equivalent uniform annual cost, the latter is generally preferable because the variable units are oftentimes expressed on a yearly basis. Additionally, EUAC calculations are simpler when the alternatives under consideration have different lives. In either case, however, the first step in calculating the breakeven point is to *express the total cost of each alternative as a function of the variable that is sought*. Example 12.7 illustrates breakeven calculations.

Example 12.7 A sheet metal company is considering the purchase of an automatic machine for a certain phase of the finishing process. The machine has an initial cost of $23,000, a salvage value of $4,000, and a life of ten years. If the machine is purchased, one operator will be required at a cost of $12 an hour. The output with this machine would be 8 tons per hour. Annual maintenance and operation cost of the machine is expected to be $3,500.

Alternatively, the company can purchase a less sophisticated machine for $8,000, which has no salvage value and a life of five years. However, with this alternative, three laborers will be required at a cost of $8 an hour and the machine will have an annual maintenance and operation cost of $1,500. Output is expected to be 6 tons per hour for this machine. All invested capital must return 10%. (*a*) How many tons of sheet metal must be finished per year in order to justify the purchase of the automatic machine? (*b*) If management anticipates a requirement to finish 2,000 tons per year, which machine should be purchased?

SOLUTION

(*a*) The first step is to express each of the variable costs in terms of the units sought, which is tons per year in this case. Thus, for the automatic machine, the annual cost per ton would be

$$\text{Annual cost per ton} = \left(\frac{\$12}{\text{hour}}\right)\left(\frac{1 \text{ hour}}{8 \text{ tons}}\right)\left(\frac{x \text{ tons}}{\text{year}}\right) = \frac{12}{8}x$$

where x = number of tons per year for break even. Note that the final units are in dollars per year, which is what we want since we are trying to obtain the EUAC. The total EUAC for the automatic machine is

$$\text{EUAC}_{\text{auto}} = 23,000(A/P, 10\%, 10) - 4,000(A/F, 10\%, 10)$$

$$+ 3,500 + \frac{12}{8}x$$

$$= \$6,992 + 1.5x$$

Similarly, the EUAC of the manual machine is

$$\text{EUAC}_{\text{manual}} = 8,000(A/P, 10\%, 5) + 1,500 + \frac{3(8)}{6}x$$

$$= \$3,610 + 4x$$

Equating the two costs and solving for x yields

$$\text{EUAC}_{\text{auto}} = \text{EUAC}_{\text{manual}}$$
$$6,992 + 1.5x = 3,610 + 4x$$
$$x = 1,352.8 \text{ tons per year}$$

Thus, at an output of 1,352.8 tons per year, the EUAC of each method is the same. If the output is expected to be greater than this figure, the automatic machine should be purchased; if the output is to be less, then the less sophisticated machine should be purchased.

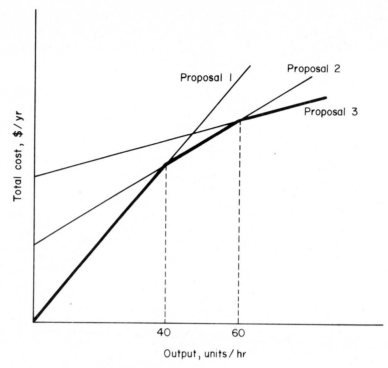

FIG. 12.2   Breakeven points for three proposals.

(b) Substituting the expected production level of 2,000 tons per year into the EUAC relations, we have $EUAC_{auto} = \$9,992$ and $EUAC_{manual} = \$11,610$. Therefore, purchase the automatic machine.

COMMENT Work the problem on a present-worth basis to satisfy yourself that either method results in the same breakeven point. A question that sometimes arises after the breakeven point is calculated is: How do you know which alternative should be selected when you are either above or below the breakeven point? As shown in Fig. 12.1, the alternative with the smaller slope (i.e., lower variable cost) should be selected when the variable units are above the breakeven point (and vice versa).   ////

While the preceding example dealt with only two alternatives, the same type of analysis can be made for three or more alternatives. In this case, it becomes necessary to compare the alternatives with each other in order to find their respective breakeven points. The results reveal the ranges through which each alternative would be the most economical one. For example, in Fig. 12.2, if the output is expected to be less than 40 units per hour, Proposal 1 should be selected. Between 40 and 60 units per hour Proposal 2 would be the most economical, and above 60 units per hour Proposal 3 would be favored.

If the variable-cost relations are nonlinear, analysis is more complicated. If the costs increase or decrease uniformly, mathematical expressions that allow direct determination of the breakeven point can be derived. For further discussion of breakeven analysis refer to the Bibliography.

Example 12.10
Problems P12.25–P12.36

SOLVED EXAMPLES

Example 12.8 Three years ago the city of Water purchased a new fire truck. Due to expanded growth in a certain portion of the city, new fire-fighting capacity is needed. An additional identical truck can be purchased now or a double-capacity truck can replace the presently owned asset. Data for each asset are presented in Table 12.2. Compare the assets at $i = 12\%$ using (a) a 12-year study period and (b) a nine-year period, which the city management believes to be more realistic due t) population growth.

SOLUTION Plan A is the retention of the previously owned truck and *augmentation* with the new identical-capacity vehicle; Plan B is purchase of the double-capacity truck. Details of each plan are below.

	Plan A		Plan B
	Presently owned	Augmentation	Double capacity
	$P = \$18,000$	$P = \$58,000$	$P = \$72,000$
	AOC = 1,500	AOC = 1,500	AOC = 2,500
	SV = 5,100	SV = 6,960	SV = 7,200
	$n = 9$	$n = 12$	$n = 12$

(a) For a full life 12-year horizon,

$$\text{EUAC}_A = (\text{EUAC of presently owned}) + (\text{EUAC of augmentation})$$
$$= [18,000(A/P, 12\%,9) - 5,100(A/F, 12\%,9) + 1,500]$$
$$+ [58,000(A/P, 12\%,12) - 6,960(A/F, 12\%, 12) + 1,500]$$
$$= 4,533 + 10,575$$
$$= \$15,108$$
$$\text{EUAC}_B = 72,000(A/P, 12\%,12) - 7,200(A/F, 12\%,12) + 2,500$$
$$= \$13,825$$

Table 12.2 DATA FOR FIRE TRUCK REPLACEMENT ANALYSIS

	Presently owned	New purchase	Double capacity
P	$51,000	$58,000	$72,000
AOC	1,500	1,500	2,500
Trade-in	18,000	–	–
SV	10% of P	12% of P	10% of P
n	12	12	12

Purchase the double-capacity truck (Plan B) with an advantage of $1,283 per year.

(b) The analysis for a truncated nine-year horizon is identical, except that $n = 9$ in each factor; that is, three fewer years are given to the augmentation and double-capacity truck to recover investment plus a 12% return. Here

$$\text{EUAC}_A = \$16,447 \qquad \text{EUAC}_B = \$15,526$$

and Plan B is again selected but now only by a margin of $921.

If the planning horizon were truncated more severely, at some point the decision would be reversed. ////

Sec. 12.2

Example 12.9 Assume an asset can be purchased for $5,000 and will have a negligible salvage value. The annual operating costs are expected to follow a gradient of $200 per year with a base amount of $300 in year 1. Find the number of years the asset should be kept if interest is not considered important.

SOLUTION The main effects of $i = 0\%$ will be to decrease the total annual cost values and make computations simpler; however, cost patterns will be similar to those for $i > 0\%$, except that the minimum cost life may change. Table 12.3 presents the entire solution to the problem. Column (3) gives cumulative AOC according to the gradient, while the average AOC is given in column (4) $(= (3)/n)$. A sample computation is given below the table, which indicates that $n = 7$ is the minimum cost life with a total annual cost of $1,614.

COMMENT Due to the tremendous regularity of this type of problem and the fact that $i = 0\%$, a quick formula can be derived to find the minimum cost life (n^*). We can write

Table 12.3 COMPUTATION OF MINIMUM COST LIFE FOR $i = 0\%$

Year, n (1)	Operating cost			Average first cost (5)	Total annual cost (6)
	Annual (2)	Cumulative (3)	Average (4)		
1	$ 300	$ 300	$ 300	$5,000	$5,300
2	500	800	400	2,500	2,900
3	700	1,500	500	1,667	2,167
4	900	2,400	600	1,250	1,850
5	1,100	3,500	700	1,000	1,700
6	1,300	4,800	800	833	1,633
7	1,500	6,300	900	714	1,614*
8	1,700	8,000	1,000	625	1,625
9	1,900	9,900	1,100	555	1,655

*Total annual cost is computed as $(6) = (4) + (5) = (3)/n + 5{,}000/n$.
For $n = 7$, $6{,}300/7 + 5{,}000/7 = \$1{,}614$.

Total annual cost (TAC) = average operating cost + average first cost

For year n this may be expressed as

$$TAC_n = \frac{\sum_{j=1}^{n} AOC_j}{n} + \frac{P}{n}$$

where TAC_n = total annual cost for n years of ownership
AOC_j = annual operating cost through year j ($j = 1, 2, \ldots, n$)
However, we can make the substitution

$$\frac{\sum_{j=1}^{n} AOC_j}{n} = B + \left(\frac{n-1}{2}\right)G$$

where B = base amount of the gradient
G = amount of gradient
Then

$$TAC_n = B + \left(\frac{n-1}{2}\right)G + \frac{P}{n} \qquad (12.2)$$

The general shape of the terms and TAC_n itself is shown in Fig. 12.3. If we take the derivative of Eq. (12.2) and solve for an optimum life value (n^*), we have

$$\frac{d\,TAC_n}{dn} = \frac{G}{2} - \frac{P}{n^2} = 0$$

$$n^* = \left(\frac{2P}{G}\right)^{\frac{1}{2}}$$

Substitution of $P = \$5,000$ and $G = \$200$ for this example yields

$$n^* = \left(\frac{10,000}{200}\right)^{\frac{1}{2}} = 7.07 \text{ years}$$

which is, for all practical purposes, the same as $n = 7$ obtained in Table 12.3.   ////

Sec. 12.4

Example 12.10 The Junk-O Toy Company currently purchases the metal parts which are required in the manufacture of certain toys, but there has been a proposal that the company make these parts themselves. Two machines will be required for the operation: Machine A will cost $18,000, have a life of six years, and a $2,000 salvage value; Machine B will cost $12,000, have a life of four years, and a $-500 salvage value. Machine A will require an overhaul after three years costing $3,000. The annual operating cost of Machine A is expected to be $6,000 per year and for Machine B $5,000 per year. A total of four laborers will be required for the two machines at a

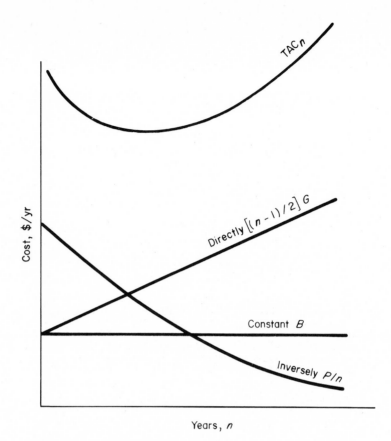

FIG. 12.3 General plot of TAC_n and component terms used to compute minimum cost life, Example 12.9.

cost of $2.50 per hour per worker. In a normal eight-hour day, the machines can produce parts sufficient to manufacture 1,000 toys. Using a **MARR** of 15% per year, (*a*) if the company's present price for these parts is $0.50 per toy, how many toys must be manufactured each year in order to justify the purchase of the machines, and (*b*) if the company expects to produce 75,000 toys per year, what maximum expenditure could be justified for the more expensive machine, assuming its salvage value and all other costs would be the same?

SOLUTION

(*a*) We have variable costs for the laborers and fixed costs for the machines. For the variable-cost component, the variable is the number of toys produced per year. This variable can be expressed as an annual cost:

$$\text{Variable cost/year} = (\text{cost/unit})(\text{units/year}) = \frac{4(2.50)(8)}{1{,}000} \, x = 0.08x$$

where $x =$ number of units per year. For Machine A, the fixed cost per year would be

$$\text{Fixed EUAC}_A = 18,000(A/P, 15\%,6) - 2,000(A/F, 15\%,6)$$
$$+ 6,000 + 3,000(P/F, 15\%,3)(A/P, 15\%,6)$$

For Machine B, the fixed cost per year is

$$\text{Fixed EUAC}_B = 12,000(A/P, 15\%,4) + 500(A/F, 15\%,4) + 5,000$$

Equating the annual costs of the purchase option $(0.50x)$ and the manufacture option yields

$$0.50x = \text{fixed EUAC}_A + \text{fixed EUAC}_B + \text{variable cost}$$
$$= 18,000(A/P, 15\%, 6) - 2,000(A/F, 15\%,6) + 6,000$$
$$+ 3,000(P/F, 15\%,3)(A/P, 15\%, 6) + 12,000(A/P, 15\%,4)$$
$$+ 500(A/F, 15\%,4) + 5,000 + 0.08x \qquad (12.3)$$

$$0.42x = 20,352.43$$
$$x = 48,458 \text{ units per year}$$

Therefore, at least 48,458 toys must be produced each year to justify the manufacture proposal.

(b) Substituting 75,000 for x and P for the $18,000 in Eq. (12.3), the amount that could be spent for the more expensive Machine A is found to be $60,187.

COMMENT You should work the problem on a present-worth basis. Keep in mind that it will be necessary to use the least common multiple of years between the two machines and the variable costs will have to be multiplied by the appropriate P/A factor.   ////

Sec. 12.5

REFERENCES

1. **G. D. Terbough,** "Dynamic Equipment Policy," McGraw-Hill Book Company, New York, 1949.
2. **G. D. Terbough,** "Business Investment Management," Machinery and Allied Products Institute, Washington, D.C., 1967.

BIBLIOGRAPHY

Barish, pp. 240–244 (replacement); pp. 634–670 (breakeven).
DeGarmo and **Canada,** pp. 381–406 (replacement); pp. 38–42 (breakeven).
Emerson and **Taylor,** pp. 15-3 –15-7 (replacement).
Fabrycky and **Thuesen,** pp. 128–140 (replacement); 278–291 (breakeven).

Grant and **Ireson,** pp. 394–410 (replacement); pp. 252–259 (breakeven).
Park, pp. 164–177 (breakeven).
Reisman, pp. 131–140 (breakeven).
Riggs, pp. 64–68, 207–215 (replacement); pp. 46–64 (breakeven).
Smith, pp. 445–456 (replacement); pp. 156–165 (breakeven).
Taylor, pp. 212–240, 250, 381.
Thuesen, Fabrycky, and **Thuesen,** pp. 166–200 (replacement); pp. 201–206 (breakeven).

PROBLEMS

P12.1   If the difference between present book value and realizable salvage value is negative, how would you treat this sum in a replacement analysis?

P12.2   Why is the original asset cost irrelevant in a replacement analysis? What does this fact have to do with the consultant's viewpoint in replacement analysis?

P12.3   A new meat display counter was purchased by a supermarket four years ago at a cost of $28,000. Book value is presently $10,000 with five years remaining before a salvage value of $1,000 is reached. Due to lagging sales, the owners wish to trade for a new, smaller counter, which costs $13,000 and has an installation charge of $500. As estimated by the owners, the old counter will last another eight years and has a trade-in value of $18,000 now. A review of the accounts shows annual repair costs averaging $150 for the old counter. (*a*) Determine the values of *P, n, SV,* and AOC for the existing counter to be used in a replacement analysis. (*b*) Is there a sunk cost involved here? If so, what is its amount?

P12.4   The owners of a downtown shoe shop are considering the possibility of moving to a rented shop in a suburban shopping center. They purchased their shop 15 years ago for $8,000 cash. They estimate their annual investment in property improvement at $500 and believe the shop should have a current book value that includes the purchase price and these improvements at 8% interest. The annual insurance, utility, etc., costs have averaged $1,080 per year. If they stay downtown, they hope to retire in ten years and will give the shop to their son-in-law. They will ask $25,000 for the shop if they sell at this time.

If the owners move to the shopping center, they must sign a ten-year lease for $6,600 per year with no additional yearly charges. They must pay a $750 deposit when they sign the lease, but this amount is returned at the time of lease expiration. Determine (*a*) the values of *P, n, SV,* and AOC for the two alternatives and (*b*) the amount of the sunk cost if one exists.

P12.5   An asset purchased two years ago can now be traded for an "improved" version for 40% of the first cost. If the asset was purchased for $18,000 and is being depreciated over eight years for tax purposes and twelve years for company income purposes by the double declining-balance method, compute the sunk cost (*a*) to be reported in the tax reports and (*b*) to be used by company economists.

P12.6   The Moon City Bus Lines has 20 buses purchased five years ago for $22,000 each. The company president plans to have a major overhaul done on these buses next year at a cost of $1,800 each. However, the vice president wants to trade these 20 buses in on 25 of a new, smaller model. The trade-in value is $4,000 and the new models cost $22,500 each. The president estimates a remaining life of seven years for the old buses once the overhaul is completed and further states that annual operational costs per bus are $3,000, and a $800 salvage is reasonable when sold to an individual as a "vacation van." The vice president interjects the comment that the smaller buses can maneuver in traffic more easily, will cost $1,000 per year less to operate, will last for eight years, and will have a salvage value of $500 when sold to day camps. With this wealth of knowledge, you, the "general everything" for the company, are requested to determine which officer's desire is economically correct at the firm's MARR of 10%.

P12.7   Rework problem P12.6 if the new buses have a life of 12 years, which happens to correspond with the planning horizon selected by the vice president.

P12.8   Rework problem P12.6 using a planning horizon of five years.

P12.9   Perform a replacement analysis for problem P12.3 if the new counter has an expected life of ten years, annual cost of $30, and a salvage value of $1,500. Use an interest rate of 15%.

P12.10   Determine which alternative is better in problem P12.4 at a 10% interest rate.

P12.11   A new earth mover was purchased by the Never Dri Cement Company three years ago and has been used to transport raw material from the quarry to the crushers. When purchased the mover possessed the following characteristics: $P = \$55,000$, $n = 10$ years, $SV = \$5,000$, and capacity $= 180,000$ tons per year. With increased construction in industrial parks around the city, an additional mover with a capacity of 240,000 tons per year is needed. Such a vehicle can be purchased. If bought, this new asset will have $P = \$70,000$, $n = 10$ years, $SV = \$8,000$.

However, the company could have constructed a conveyor system to move the material from the quarry. This system will cost $115,000, have a life of 15 years, no salvage value, and carry 400,000 tons of material per year. The company will need to have some way to move material to the conveyor in the quarry. If the presently owned mover is used, it will more than suffice. However, a new smaller-capacity mover can be purchased. A $15,000 trade-in on the old mover will be given on any new mover. This one will have $P = \$40,000$, $n = 12$ years, $SV = \$3,500$, capacity $= 400,000$ tons per year over this short distance.

Monthly operating, maintenance, and insurance costs average $0.01 per ton-mile for the movers; similar costs for the conveyor are expected to be $0.0075 per ton. The company wants to make 12% on this investment. Records show that the mover must travel an average of 1.5 miles from the quarry to the crusher pad. The conveyor will be placed to reduce this distance to 0.2 mile. Should the old mover be augmented by a new mover or should the conveyor be considered as a replacement and, if so, which method of moving the material in the quarry should be used?

P12.12 Solve problem P12.11 under this condition: only a four-year planning period is possible because management feels that this spurt in business is very short-lived.

P12.13 Machine A, purchased two years ago, is wearing out more rapidly than expected. It has a remaining life of two years, annual operating costs of $3,000, and no salvage value. To continue the function of this asset, Machine B can be purchased and a trade-in value of $9,000 will be allowed for Machine A. Machine B has $P = \$25,000$, $n = 12$ years, AOC $= \$4,000$, and SV $= \$1,000$.

As an alternative, Machine C can be bought to replace A. No trade-in will be allowed for A, but it can be sold for $7,000. This new asset will have $P = \$38,000$, $n = 20$ years, AOC $= \$2,500$, and SV $= \$1,000$. If the retention of A is called Plan A, Plan B is the acquisition of B, and Plan C is the purchase of C, use a 20-year period and a MARR $= 8\%$ to determine which plan is more economic.

P12.14 A family presently own a ten-year-old built-in range, which is not working properly. Repairs would cost $50, but the service shop estimates a remaining life of only five years with no salvage. The utility company advertises that a range costs an average of $24 per year to operate. If replaced, the range will be sold at a garage sale for $15 in its present condition.

The family would like to have the range fixed and buy a microwave oven for $450. This oven will cost $50 to build in and has a life of ten years and a salvage of $25. Its annual operating cost is $6; the operating cost of the range is $9 per month.

The electric company promotion department personnel suggest a combination electric-microwave range for all cooking. This range costs $650, will last ten years, has a salvage value of $25, and costs $12 per year to operate.

If money is currently worth 7% in the bank and the family doesn't think they will stay in their present house for more than five years, should they buy the combination range or should they repair the old range and buy the microwave oven?

P12.15 What is the replacement value of the old display counter described in problem P12.3 if, as in problem P12.4, the new counter has $n = 10$, AOC $= \$30$, SV $= \$1,500$, and $i = 15\%$?

P12.16 A construction company bought a 180,000-ton-per-year-capacity earth mover three years ago at a cost of $55,000; the expected life at time of purchase was 10 years with a $5,000 salvage value and an annual operating cost of $2,700. A 480,000-ton-per-year replacement mover is under consideration. This mover will cost $40,000, have a life of 12 years, a salvage of $3,500, and an annual operating cost of $7,200. Compute the trade-in value of the presently owned mover if the replacement mover is bought.

P12.17 (*a*) Solve problem P12.16 using a planning horizon of four years.
   (*b*) How does this truncation of the horizon affect the replacement value of the presently owned mover?

P12.18 Assume that for the situation of problem P12.14 a neighbor offers to buy the old range. What price must be asked to alter the decision made in the previous case?

P12.19 An asset presently owned can last for six more years with costs of $24,000 this year and increasing by 10% per year. A desirable challenger would cost $70,000, last for six years, have an annual cost of $12,000 and a salvage value of $4,000. What is a trade-in value of the old asset that will make replacement economic if a 5% return is desired?

P12.20 Ms. Adams just bought a new car for $5,800; she financed the purchase at 5% compounded per year for three years and put $400 down. The resale values for the next six years are $2,200 after the first year and decrease by $400 per year to year 5, after which the resale value remains at $600. Annual costs of repairs, insurance, gas, etc., are expected to be $1,000 the first year and increase by 10% each year. If money is worth 7%, how many years should the car be retained? Assume the owner will pay off the entire loan with interest if she sells the car before she has owned it three years.

P12.21 Rework problem P12.20 at $i = 0\%$ rate of return and find the difference between the two answers.

P12.22 Machine H was purchased five years ago for $40,000 and had an expected life of ten years. The past and estimated future maintenance and operating costs and salvage values are given below. At a value of $i = 10\%$, determine the number of years the asset should be kept in service before replacement.

Year	Operating cost	Maintenance cost	Salvage value
1	$1,500	$2,000	$25,000
2	1,600	2,000	25,000
3	1,700	2,000	22,000
4	1,800	2,000	22,000
5	1,900	2,000	15,000
6	2,000	2,100	5,000
7	2,100	2,700	5,000
8	2,200	3,300	0
9	2,300	3,900	0
10	2,400	4,500	0

P12.23 One year ago the Bullwinkle Pool Company purchased a machine to blow concrete onto the walls of a new swimming pool, thereby greatly reducing the time necessary to construct a pool. The machine cost $8,000 and is expected to last 14 more years. The owner has already seen newer, improved versions. To compare these challengers to the defender, knowing the most economic life of the old version would be of benefit. If costs were $500 for the first year and are expected to increase by $100 per year, compute the most economic life for $i = 0\%$. Assume salvage value is zero for all years.

P12.24 Rework problem P12.23 at $i = 5\%$ and compare the answers.

P12.25 Two pumps can be used for pumping a corrosive liquid. A pump with a brass impeller costs $800 and is expected to last three years. A pump with a stainless steel impeller will cost $1,900 and last five years. An overhaul costing $300 will be required after 2,000 operating hours for the brass impeller pump while an overhaul costing $700 will be required for the stainless pump after 9,000 hours. If the operating cost of each pump is $0.50 per hour, how many hours per year must the pump be required to justify the purchase of the more expensive pump? Use an interest rate of 10% per year.

P12.26 The Sli-Dog Company is considering two proposals for improving the employees' parking area. Proposal A would involve filling, grading, and blacktopping at an initial cost of $5,000. The life of the parking lot constructed in this manner is expected to be four years with annual maintenance costs of $1,000. Alternatively, the parking area would be paved under Proposal B, in which case the life would be extended to 16 years. The annual maintenance cost will be negligible for the paved parking area, but the markings will have to be repainted every two years at a cost of $500. If the company's minimum attractive rate of return is 12% per year, how much could it afford to spend for paving the parking area so that the proposals would break even?

P12.27 The Redi-Bilt Construction Company is considering the purchase of a dirt scraper. A new scraper will have an initial cost of $75,000, a life of 15 years, a $5,000 salvage value, and an operating cost of $30 a day. Annual maintenance cost is expected to be $6,000 per year.

Alternatively, the company can lease a scraper and driver as needed for $210 per day. If the company's minimum attractive rate of return is 12% per year, how many days per year must the scraper be required in order to justify the purchase?

P12.28 The Slo-Stitch Pant Company is considering the purchase of an automatic cutting machine. The machine will have a first cost of $22,000, a life of ten years, and $500 salvage value. The annual maintenance cost of the machine is expected to be $2,000 per year for the range of usage anticipated. The machine will require one operator at a cost of $24 a day. Approximately 1,500 yards of material can be cut each hour with the machine.

Alternatively, if manual labor is used, five workers each earning $18 a day can cut 1,000 yards per hour. If the company's minimum attractive rate of return is 8% per year, how many yards of material must be cut each year in order to justify the purchase of the automatic machine?

P12.29 A couple have an opportunity to buy a fire-damaged house for what they believe to be a bargain price of $28,000. They estimate that remodeling the house now will cost $12,000 and annual taxes will be approximately $800 per year. They estimate that utilities will cost $500 per year and that the house must be repainted every three years at a cost of $400. At the present time, resale houses are selling for $16 per square foot, but they expect this price to increase by $1.50 per square foot per year. They will lease the house for $2,500 per year until it is sold. If the house has 2,500 square feet and they want to make 8% on their investment, (*a*) how long must they keep the house before they can break even, and (*b*) what would the selling price be at the time of the sale?

P12.30 The Rawhide Tanning Company is considering the economics of furnishing an in-house water-testing laboratory instead of sending samples to independent laboratories for analysis. If the lab is completely furnished so that all tests could be conducted in-house, the initial cost would be $25,000. A technician will be required at a cost of $13,000 per year. The cost of power, chemicals, etc., will be $5 per sample. If the lab is only partially furnished, the initial cost will be $10,000. A parttime technician will be required at an annual salary of $5,000. The cost of in-house sample analysis will be $3 per sample, but since all tests cannot be conducted by the company, outside testing will be required at a cost of $20 per sample. If the company elects to continue the present condition of complete outside testing, the cost will be $55 per sample. If the laboratory equipment will have a useful life of 12 years and the company's MARR is 10% per year, how many samples must be tested each year in order to justify (*a*) the complete laboratory and (*b*) the partial laboratory? (*c*) If the company expects to test 175 samples per year, which of the three alternatives should be selected?

P12.31 The Quik-Kil slaughterhouse and packing plant currently pays the city $4,000 a month in wastewater discharge fees. The company is considering constructing a treatment plant of its own. The treatment plant will require a parttime operator at $300 per month. In addition, the operating cost is expected to be $500 per month. The treatment plant will require minor repairs every four years costing $1,500 and is expected to last for 20 years. If the company's MARR is a nominal 12% per year, how much could the company afford to spend on a treatment plant and not exceed its present costs?

P12.32 A city engineer is considering two methods for lining water-holding tanks. A bituminous coating can be applied at a cost of $2,000. If the coating is touched up after four years at a cost of $600, its life can be extended two more years.

Alternatively, a plastic lining can be installed, which will have a life of 15 years. If the city uses an interest rate of 5% per year, how much money could be spent for the plastic lining so that the two methods just break even?

P12.33 A family who are planning to build a new house are trying to decide between purchasing a lot in the city or in the suburbs. A $\frac{1}{4}$-acre lot in the city will cost $10,000 in the area in which they want to buy. If they purchase a lot outside the city limits, a similar parcel will cost only $2,000. For the size of house they plan to build, they expect annual taxes to amount to $1,200 per year if they build in the city and only $150 per year in the suburbs. If they purchase the lot outside the city limits, they will have to drill a well for $4,000. With their own well, they will save $150 per year in water charges, but they expect the city to provide water to their area in five years, after which time they will purchase the city water. They estimate that the increased travel distance will cost $325 the first year, $335 the second year, and amounts increasing by $10 per year. Using a 25-year analysis period and an interest rate of 6% per year, how much extra could the family afford to spend on the house outside the city limits and still have the same total investment? Assume the land can be sold for the same price as its initial cost.

P12.34   The I. M. Rite family are considering insulating their attic to prevent heat loss. They are considering R-11 and R-19 insulation. They can install R-11 for $160 and R-19 for $240. They expect to save $35 per year in heating and cooling with R-11. If the interest rate is 6%, how much money must they be able to save per year in order to justify the R-19 insulation if they want to recover their investment in seven years?

P12.35   A waste-holding lagoon situated near the main plant receives sludge on a daily basis. When the lagoon is full, it is necessary to remove the sludge to a site located three miles from the main plant. At the present time, whenever the lagoon is full, the sludge is removed by pumping it into a tank truck and hauling it away. This requires a portable pump that costs $800 and has an eight-year life. The company supplies the labor to operate the pump at a cost of $25 per day, but the truck and driver must be rented at a cost of $110 per day.

Alternatively, the company can install a pump and pipeline to the remote site. The pump would have an initial cost of $600, a life of ten years, and would cost $3 per day to operate. The company's MARR is 15%. (*a*) If the pipeline would cost $1.10 per foot to construct, how many days per year must the lagoon require pumping in order to justify construction of the pipeline? (*b*) If the company expects to pump the lagoon one time per week, how much money could it afford to spend on the pipeline in order to just break even? Assume pipeline life is ten years.

P12.36   A building contractor is considering two alternatives for improving the exterior appearance of a commercial building that is being renovated. The building can be completely painted at a cost of $2,800. The paint is expected to remain attractive for four years, at which time the job would have to be redone. Every time the building is painted, the cost will increase by 20% over the previous time. Alternatively, the building can be sandblasted now and every ten years at a cost 40% greater than the previous time. The remaining life of the building is expected to be 38 years. If the company's MARR is 10%, what is the maximum amount that could be spent now on the sandblasting alternative so that the two alternatives will just break even?

13
BONDS

The objective of this chapter is to teach you how to make an economic analysis of transactions involving bonds.

CRITERIA

To complete this chapter, you must be able to do the following:

1 Define *mortgage bond, collateral bond, equipment trust bond, debenture bond, convertible debenture, municipal bond, general obligation bond, revenue bond,* and *bond rating.*
2 Calculate the interest payable (receivable) per period from the sale (purchase) of a bond, given the face value of the bond, the bond interest rate, and the interest payment period.
3 Calculate the present worth of a bond, given the face value, bond interest rate, interest payment period, date the bond matures, and the desired rate of return.
4 Calculate the nominal and effective rates of return that would be received from the purchase of a bond, given the face value, purchase price, bond interest rate, compounding period, and date the bond matures.

EXPLANATION OF MATERIAL

13.1 Bond Classifications

A *bond* is a long-term note issued by a corporation or governmental entity for the purpose of financing major projects. In essence, the borrower receives money now in return for a promise to pay later, with interest paid between the time the money was borrowed and the time it was repaid. In general, bonds may be classified as mortgage bonds, debenture bonds, and municipal bonds. These types of bonds can be further subdivided.

A *mortgage bond* is one which is backed by a mortgage on specified assets of the company issuing the bonds. If the company is unable to repay the bondholders at the time the bonds mature, the bondholders have the option of foreclosing on the mortgaged property. Mortgage bonds can be subdivided into first-mortgage and second-mortgage bonds. As their names imply, in the event of foreclosure by the bondholders, the first-mortgage bonds take precedence during liquidation. The first-mortgage bonds, therefore, generally provide the lowest rate of return. Second-mortgage bonds, when backed by collateral of a subsidiary corporation, are referred to as *collateral bonds*. An *equipment trust bond* is one in which the equipment purchased through the bond serves as collateral. These types of bonds are generally issued by railroads for purchasing new locomotives and cars.

Debenture bonds are not backed by any form of collateral. The reputation of the company is important for attracting investors to this type of bond. As further incentive for investors, debenture bonds are often *convertible* to common stock at a fixed rate as long as the bonds are outstanding. For example, a $1,000 convertible debenture bond issued by the Get Rich Quik (GRQ) Company may have a conversion option to 50 shares of GRQ common stock. If the value of 50 shares of GRQ common stock exceeds the value of the bond at any time prior to bond maturity, the bondholder has the option of converting his bond to common stock. Debenture bonds generally provide the highest rate of interest because of the increased risk associated with them.

The third general type of bonds are *municipal bonds*. Their attractiveness to investors lies in their income tax-free status. As such, the interest rate paid by the governmental entity is usually quite low. Municipal bonds can be either *general obligation bonds* or *revenue bonds*. General obligation bonds are issued against the taxes received by the governmental entity (i.e., city, county, or state) that issued the bonds and are backed by the full taxing power of the issuer. School bonds are an example of general obligation bonds. Revenue bonds are issued against the revenue generated by the project financed, as a water treatment plant or a bridge. Taxes cannot be levied for repayment of revenue bonds.

In order to assist prospective investors, all *bonds* are rated by various companies according to the amount of risk associated with their purchase. One such rating is Standard and Poor's, which rates bonds from AAA (highest quality) to DDD (bond in default). In general, first-mortgage bonds carry the highest rating, but it is not

Table 13.1 CLASSIFICATION AND CHARACTERISTICS OF BONDS

Classification	Characteristics	Types
Mortgage	Bonds backed by mortgage or specified assets	First mortgage Second mortgage Equipment trust
Debenture	No lien to creditors	Convertible Nonconvertible
Municipal	Income tax free	General obligation Revenue

uncommon for debenture bonds of large corporations to carry a AAA rating, or ratings higher than first-mortgage bonds of smaller, less reputable companies. The concepts presented in this section are summarized in Table 13.1.

Problem P13.1

13.2 Bond Terminology and Interest

As stated in the preceding section, a bond is a long-term note issued by a corporation or governmental entity for the purpose of obtaining needed capital for financing major projects. The conditions for repayment of the money obtained by the borrower are specified at the time the bonds are issued. These conditions include the bond face value, the bond interest rate, the bond interest-payment period, and the bond maturity date.

The bond *face value*, which refers to the denomination of the bond, is usually an even denomination starting at \$100, with the most common being the \$1,000 bond. The face value is important for two reasons:

1 It represents the lump-sum amount that will be paid to the bondholder on the bond maturity date.
2 The amount of interest paid per period prior to the bond maturity date is determined by multiplying the face value of the bond by the bond interest rate per period.

Bond interest per period (I) is computed as

$$I = \frac{\text{(face value)(bond interest rate)}}{\text{compounding periods per year}} = \frac{Fb}{c} \qquad (13.1)$$

Often a bond is purchased at a discount (less than face value) or a premium (greater than face value), but only face value, not purchase price, is used to compute bond interest (I). Examples 13.1 and 13.2 illustrate the computation of bond interest.

Example 13.1 A shirt-manufacturing company planning an expansion issued 4%, $1,000 bonds for financing the project. The bonds will mature in 20 years with interest paid semiannually. Mr. John Doe purchased one of the bonds through his stockbroker for $800. What payments is Mr. Doe entitled to receive?

SOLUTION In this example, the face value of the bond is $1,000. Therefore, Mr. Doe will receive $1,000 on the date the bond matures, 20 years from now. In addition, Mr. Doe will receive the semiannual interest the company promised to pay when the bonds were issued. The interest every six months will be computed using $F = \$1,000$, $b = 0.04$, and $c = 2$ in Eq. (13.1):

$$I = \frac{1,000(0.04)}{2} = \$20 \text{ every 6 months} \qquad ////$$

Example 13.2 Determine the amount of interest you would receive per period if you purchased a 6%, $5,000 bond which matures in ten years with interest payable quarterly.

SOLUTION Since interest is payable quarterly, you would receive the interest payment every three months. The amount you would receive would be

$$I = \frac{5,000(0.06)}{4} = \$75$$

Therefore, you would receive $75 interest every three months in addition to the $5,000 lump sum after ten years. $////$

Problems P13.2–P13.4

13.3 Bond Present-Worth Calculations

When a company or governmental agency offers bonds for financing major projects, investors must determine how much they are willing to pay for a bond of a given denomination. The amount they pay for the bond will determine the rate of return on the investment. Therefore, investors must determine the present worth of the bond that will yield a specified rate of return. These calculations are shown in Example 13.3.

Example 13.3 Ms. Jones wants to make 8% nominal interest compounded semiannually on a bond investment. How much should she be willing to pay now for a 6%, $10,000 bond that will mature in 15 years and pays interest semiannually?

SOLUTION Since the interest is payable semiannually, Ms. Jones will receive the following payment:

$$I = \frac{10,000(0.06)}{2} = \$300 \text{ every 6 months}$$

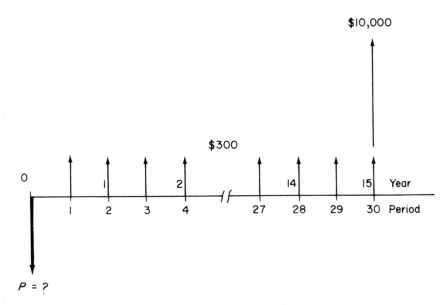

FIG. 13.1   Cash flow for a bond investment, Example 13.3.

The cash-flow diagram (Fig. 13.1) for this investment allows us to write a present-worth relation to compute the value of the bond now, using an interest rate of 4% per six-month period, the same as the interest payment period of the bond. Note in the following equation that I is simply an A value.

$$P = 300(P/A,4\%,30) + 10,000(P/F,4\%,30) = \$8,270.60$$

Thus, if Ms. Jones is able to buy the bond for $8,270.60, she will receive a nominal 8% rate of return on her investment. If she were to pay more than $8,270.60 for the bond, the rate of return would be less than 8%, and vice versa.

COMMENT It is important to note that the interest rate used in the present-worth calculation is the interest rate per period that Ms. Jones *wants to receive*, not the bond interest rate. Since she wants to receive a nominal 8% per year compounded semiannually, the interest rate per six-month period is 8%/2 = 4%. The bond interest rate is used *only* to determine the amount of the interest payment. If you desire to review nominal and effective interest rates, refer to Secs. 4.1–4.3.

////

When the investor's compounding period is either more often or less often than the interest payment period of the bond, it becomes necessary to use the techniques learned in Chap. 4. Example 13.4 illustrates the calculations when the investor's compounding period is less than the interest period of the bond.

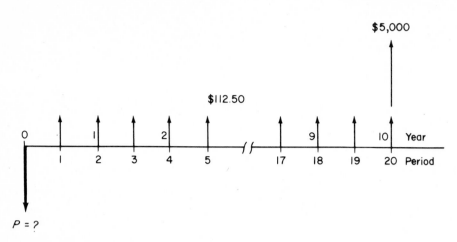

FIG. 13.2 Cash flow, Example 13.4.

Example 13.4 Calculate the present worth of a 4.5%, $5,000 bond with interest paid semiannually. The bond matures in ten years, and the investor desires to make 8% compounded quarterly on the investment.

SOLUTION The interest the investor would receive is

$$I = \frac{5,000(0.045)}{2} = \$112.50 \text{ every 6 months}$$

The present worth of the payments shown in Fig. 13.2 can be found in either of two ways:

1 Take each interest payment ($112.50) back to year zero separately and add to the present worth of $5,000. In this case, using Eq. (4.4), the interest rate would be 8%/4 = 2% per quarter and the number of periods would be double those shown in Fig. 13.2, since the interest payments are made semiannually while the desired rate of return is compounded quarterly. Thus,

$P = 112.50(P/F, 2\%,2) + 112.50(P/F, 2\%,4) + 112.50(P/F, 2\%,6)$
$\quad + \cdots + 112.50(P/F, 2\%,40) + 5,000(P/F, 2\%,40)$
$\quad = \$3,788$

2 Determine the effective interest rate compounded *semiannually* (the bond interest-payment period) that would be equivalent to the nominal 8% compounded quarterly (as stated in the problem), then use the *P/A* factor to compute the present worth of interest and add to the present worth of $5,000. The semiannual rate is 8%/2 = 4%. Since there are two quarters per six-month period, Table B-1 indicates that the effective semiannual rate is *i* = 4.04%. Alternatively, the effective semiannual rate can be computed from Eq. (4.3):

$$i = \left(1 + \frac{0.04}{2}\right)^2 - 1 = 0.0404$$

The present worth of the bond can now be determined with calculations similar to those in Example 13.3:

$$P = 112.50(P/A,4.04\%,20) + 5,000(P/F,4.04\%,20) = \$3,790 \qquad ////$$

In summary, the steps that should be followed in calculating the present worth of a bond investment are the following:

1 Calculate the interest payment (I) per period, using the face value (F), the bond interest rate (b), and the number of interest periods (c) per year, by $I = Fb/c$.
2 Draw the cash-flow diagram of the bond receipts to include interest and face value.
3 Determine the investor's desired rate of return per period. When the bond interest period and the investor's compounding period are not the same, it is necessary to use the effective-interest-rate formula to find the proper interest rate per period (Example 13.4).

Example 13.6
Problems P13.5–P13.13

13.4 Rate of Return on Bond Investment

To calculate the rate of return received on a bond investment, the procedures learned in this chapter and Chap. 10 should be followed. That is, the procedures of Secs. 13.1 and 13.2 should be used to establish the timing and the magnitude of the income associated with a bond investment; then, the rate of return on the investment can be determined by setting up and solving the rate-of-return equation, Eq. (10.4). The following example illustrates the general procedure for calculating the rate of return on a bond investment.

Example 13.5 In Example 13.1 it was stated that Mr. John Doe paid $800 for a 4%, $1,000 bond that would mature in 20 years with interest payable semiannually. What nominal and effective interest rates would Mr. Doe receive on his investment for semiannual compounding?

SOLUTION The income Mr. Doe will receive from the bond purchase is the bond interest every six months plus the face value in 20 years. The equation for calculating the rate of return using the cash flow of Fig. 13.3 would be

$$0 = -800 + 20(P/A,i\%,40) + 1,000(P/F,i\%,40)$$

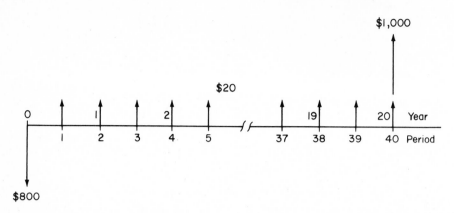

FIG. 13.3   Cash flow, Example 13.5.

which can be solved to obtain $i = 2.87\%$ compounded semiannually. The nominal interest rate is computed as interest rate per period times number of periods, which is

$$\text{Nominal } i = 2.87\%(2) = 5.74\%$$

From Table B-1 or Eq. (4.3), the effective rate is 5.81% per year.   ////

Example 13.7
Problems P13.14–P13.22

SOLVED EXAMPLES

Example 13.6   Ms. Big wants to invest in some 4%, $10,000, 20-year mortgage bonds with interest paid semiannually. If she requires a rate of return of 10% per year compounded semiannually and can purchase the bonds through a broker at a discount price of $8,375, (*a*) should she make the purchase, and (*b*) if she does purchase the bond, what will be her total gain in dollars?

SOLUTION

(*a*)   The interest each six months is

$$I = \frac{10,000(0.04)}{2} = \$200$$

For the nominal rate of 10%/2 = 5% per six months,

$$P = 200(P/A, 5\%, 40) + 10,000(P/F, 5\%, 40)$$
$$= \$4,852$$

If Ms. Big must pay $8,375 per bond, she cannot even come close to 10% compounded semiannually so, she should not buy these bonds.

(*b*) If she buys at $8,375 per bond, we can find dollars gained by computing the future worth assuming Ms. Big reinvests all interest at 10% compounded semiannually:

$$F = 200(F/A,5\%,40) + 10,000 = \$34,159$$

Thus, she stands to gain a total of $34,159 − $8,375 = $25,784. However, as stated in part (*a*), her rate of return would be much less than 10% per year.

||||

Sec. 13.3

Example 13.7 In the preceding example, Ms. Big would naturally be saddened by her inability to make 10% compounded semiannually if she pays $8,375 for a 4%, $10,000, 20-year bond. Compute the (*a*) actual nominal and effective return of the bond and (*b*) actual dollar gain if this rate is used for reinvestment.

SOLUTION

(*a*) The rate of return equation is

$$0 = -8,375 + 200(P/A,i\%,40) + 10,000(P/F,i\%,40)$$

Solution and interpolation show that $i = 5.40\%$ per year nominally (2.70% semiannually) and $i = 5.47\%$ effectively. You should definitely verify these values to see if you can work a bond rate-of-return problem.

(*b*) Using a nominal rate of 5.40% per year compounded semiannually, the future worth of the bond is

$$F = 200(F/A,2.7\%,40) + 10,000 = \$24,180$$

which represents a gain of $15,805.

||||

Secs. 13.4 and 13.3

BIBLIOGRAPHY

Barish, pp. 144, 152, 213.
DeGarmo and **Canada**, pp. 191–198.
Fabrycky and **Thuesen**, pp. 72–74.
Grant and **Ireson**, pp. 101, 116–117.
Reisman, pp. 284–286.
Riggs, pp. 225–226.
Smith, pp. 108–109.
Taylor, p. 116.
Thuesen, Fabrycky, and **Thuesen**, pp. 94–95.

PROBLEMS

P13.1 What is the difference between (*a*) mortgage bonds and debenture bonds and (*b*) general obligation bonds and revenue bonds?

P13.2 What would be the interest payment and payment period on a 5%, $5,000 bond that is payable semiannually?

P13.3 What is the frequency and amount of the interest payments on a 6%, $10,000 bond for which interest is payable quarterly?

P13.4 What are the interest payments and their frequency on a $4\frac{1}{2}$%, $5,000 bond that pays monthly interest?

P13.5 How much should you be willing to pay for a 4%, $10,000 bond that is due ten years from now if you want to make a nominal 8% rate of return compounded semiannually? Assume the bond interest is payable semiannually.

P13.6 What is the present worth of a 5%, $50,000 bond that has interest payable semiannually? Assume the bond is due in 25 years and the desired rate of return is 6% compounded semiannually.

P13.7 A $5\frac{1}{2}$%, $15,000 bond with interest payable quarterly is due 20 years from now. What is the present value of the bond if the purchaser desires to make a nominal 6% rate of return compounded quarterly?

✓ *P13.8* You have been offered a 6%, $20,000 bond at a 4% discount. If interest is paid quarterly and the bond is due in 15 years, can you make a nominal 8% rate of return compounded quarterly?

P13.9 How much should you be willing to pay for the bond in problem P13.5 if you want to make an effective 8% rate of return on your investment?

P13.10 How much should you be willing to pay for the bond in problem P13.5 if you desire to make a nominal 8% rate of return compounded quarterly? Compounded monthly?

P13.11 What would the discount price of the bond in problem P13.7 have to be if the purchaser desired to make an effective 7% rate of return on the investment?

P13.12 If a $1\frac{1}{2}$%, $10,000 bond that has interest payable semiannually is for sale for $8,500, when would the bond have to be due for the purchaser to make a nominal 8% rate of return on his investment?

P13.13 When would the bond in problem P13.12 have to be due if the purchaser wanted to earn an effective 8% rate of return on the investment?

P13.14 A 4%, $10,000 bond is offered for sale for $8,000. If the bond interest is payable semiannually and the bond becomes due in 13 years, what nominal rate of return will the purchaser make on the investment?

P13.15 A 5%, $50,000 bond is offered for sale for $43,500. If the bond interest is payable quarterly and the bond becomes due in 15 years, what nominal rate of return would the purchaser make on the investment?

P13.16 An investor purchased a 5%, $1,000 bond for $825. The interest was payable semiannually, and the bond was to become due in 20 years. The bond was kept for only eight years and sold for $800 immediately after the 16th interest payment. What nominal rate of return was made on the investment?

P13.17   What effective rate of return would the purchaser of the bond in problem P13.14 receive?

P13.18   What effective rate of return would the purchaser of the bond in problem P13.15 receive?

P13.19   At what bond interest rate would a $10,000 bond yield a nominal 6% per year compounded semiannually, if the purchaser pays $9,000 and the bond becomes due in 15 years? Assume the bond interest is payable semiannually.

P13.20   At what bond interest rate would a $20,000 bond that has interest payable semiannually yield an effective 8% rate of return if the price of the bond is $18,000 and the bond becomes due in 20 years?

P13.21   What would the bond interest rate have to be in problem P13.19 if the purchaser wanted to make a nominal 8% rate of return compounded quarterly?

P13.22   The Sli-Dog Company plans to sell 200 4%, $1,000 bonds. Interest will be paid quarterly, and the bonds will be retired after 15 years. If the management wants to set up a fund that will be specifically used to pay interest and retire the bonds at maturity date, what equivalent annual amount must be placed in the fund? Assume the fund will earn 8% compounded quarterly.

LEVEL IV

Even though this level of material may be considered optional to a basic understanding of economic analysis, it is vital when conducting a thorough alternative evaluation. If you master all the chapters in this level you will have a good understanding of the tax factors that should be considered when comparing alternatives. You will also learn how to select one project from many rather than selecting one of two plans, as we have done previously. A brief introduction to capital budgeting shows you how to select several projects from many while not violating given restrictions. The setting of the MARR (minimum attractive rate of return), which in preceding chapters has been assumed, is also investigated. The order in which you tackle the chapters is immaterial, except that Chap. 16 should be studied after Chap. 15.

Chapter	Subject
14	Accounting
15	Corporate taxes
16	After-tax analysis
17	Multiple alternatives
18	Establishing MARR
19	Sensitivity, risk analysis

14

ACCOUNTING FOR ENGINEERING ECONOMISTS

The overall objective of this chapter is to give the engineering economist an idea of the basics of financial statements and cost-accounting records, the information they contain, and how that information may be used in an economics study.

By no means is this chapter thorough enough to permit the engineering economist who has mastered it to practice accounting. However, some of the basic approaches of cost accounting discussed here will assist the engineering economist in gathering data for the study to be performed.

CRITERIA

To complete this chapter you must be able to do the following:

1 For a balance sheet, name and explain the major accounting categories and state the basic equations used in its preparation. Given the financial data, prepare a balance sheet.
2 For an income statement and a cost-of-goods-sold statement, name and explain the major categories and state the basic equations used in their preparation. Given the financial data, prepare each of these statements.

3 Compute and give a meaning for each of the following accounting ratios, given a balance sheet and income statement: current ratio, acid-test ratio, equity ratio, operating ratio, and income ratio.

4 Compute overhead rates, given the allocation basis, total overhead expenses, and estimated activity level.

5 Compute the actual cost of overhead and the variance, given the allocation basis, overhead rate, and observed activity level.

EXPLANATION OF MATERIAL

14.1 The Balance Sheet

At the end of each fiscal year a company publishes a balance sheet. A sample balance sheet for the Wimble Corporation is presented in Table 14.1. This is a yearly presentation of the state of the firm at a stated time, for example, December 31, 19XX; however, in addition a balance sheet is usually prepared quarterly and monthly. Note that three main categories are used. These divisions and their meanings are the following:

Assets A summary of all resources owned by or owed to the firm. Two classes of assets are distinguishable. *Current* assets represent the short-lived, working capital of the company (cash, accounts receivable, etc.), which can be converted to cash in approximately one year. Long-lived assets are referred to as *fixed*.

Table 14.1 SAMPLE BALANCE SHEET

	Wimble Corporation				
	Balance Sheet				
	December 31, 19XX				
Assets			**Liabilities**		
Current					
Cash	$ 10,500		Accounts payable	$ 19,700	
Accounts receivable	18,700		Dividends payable	7,000	
Interest accrued			Long-term notes		
receivable	500		payable	16,000	
Inventories	52,000		Bonds payable	20,000	
Total current assets		$ 81,700	Total liabilities		$ 62,700
Fixed			**Net worth**		
Land	$ 25,000				
Building and			Common stock	$275,000	
equipment $438,000			Preferred stock	100,000	
Less:					
Depreciation					
allowance 82,000	356,000		Retained earnings	25,000	
Total fixed					
assets		381,000	Total equity		400,000
			Total liabilities		
Total assets		$462,700	and equity		$462,700

Table 14.2 SAMPLE INCOME STATEMENT

Wimble Corporation
Income Statement
Year Ended December 31, 19XX

Revenues		
Sales	$505,000	
Interest revenue	3,500	
Total revenues		$508,500
Expenses		
Cost of goods sold (Table 14.3)	$290,000	
Selling	28,000	
Administrative	35,000	
Other	12,000	
Total expenses		365,000
Income before taxes		143,500
Taxes for year		71,750
Net profit for year		$ 71,750

Conversion of these holdings (land, equipment, etc.) to cash in a short period of time would require a major company reorientation.

Liabilities A summary of financial obligations of the company.

Net worth (equity) A summary of the worth of ownership, including outstanding stock issues and earnings retained for expansion.

The balance sheet is constructed using the basic equation

$$\text{Assets} = \text{liabilities} + \text{net worth}$$

You will notice in Table 14.1 that each major category is subdivided into particular titles. For example, "Accounts receivable" represents all money owed to the firm by its customers. Further utilization of some of these subcategories will be made when accounting ratios are discussed.

Problem P14.1

14.2 The Income Statement and the Cost-of-Goods-Sold Statement

A second important financial statement utilized to present the relationships between the major categories in the balance sheet is the *income statement* (Table 14.2). The income statement shows this relation by summarizing the profit or loss situation of the firm for a preceding, stated amount of time. The major categories of an income statement are the following:

Revenues Includes all sales and interest revenue that the company has received in the past accounting period.

Expenses A summary of all expenses for the accounting period. Some expense values, for example, income taxes and cost of goods sold, are itemized in other statements.

Table 14.3 SAMPLE COST–OF–GOODS–SOLD STATEMENT

Wimble Corporation
Statement of Cost of Goods Sold
Year Ended December 31, 19XX

Materials		
Inventory, January 1, 19XX	$ 5,800	
Purchases during year	180,000	
Total	$185,800	
Less: Inventory, December 31, 19XX	7,300	
Cost of materials		$178,500
Direct labor		110,000
Prime cost		288,500
Factory expense		7,000
Factory cost		295,500
Less: Increase in finished goods inventory during year		5,500
Cost of goods sold (Table 14.2)		$290,000

The income statement is published at the same time as the balance sheet. The income statement uses the equation

$$\text{Revenues} - \text{expenses} = \text{profit (or loss)}$$

The *cost of goods sold* is an important accounting term to the engineering economist. It represents the net cost of producing the product marketed by the firm. The economist will find a statement of the cost of goods sold, such as that shown in Table 14.3, useful in determining exactly how much it cost to make a particular product over a stated time period, usually one year. Note that the total of the cost-of-goods-sold statement is entered as an expense item on the income statement. This total is determined using the relations

$$\text{Direct materials} + \text{direct labor} = \text{prime cost}$$
$$\text{Prime cost} + \text{factory expense} = \text{cost of goods sold} \qquad (14.1)$$

The item "Factory expense" includes all overhead charged to produce a product. Overhead accumulation systems are discussed later in this chapter. Cost of goods sold is also called *factory cost*.

Problems P14.2, P14.3

14.3 Accounting Ratios

Accountants frequently utilize ratio analysis in discussing the state of the company over time and in relation to industry norms. Because the engineering economist must communicate with the accountant, the economist should have a general idea of the types of ratios commonly employed by accountants. Information for these ratios is extracted from the balance sheet and income statement. Tables 14.1 and 14.2 are used for the numerical examples of this section.

14.3.1 Current ratio This ratio, utilized to analyze the company's working-capital conditions, is defined as

$$\text{Current ratio} = \frac{\text{current assets}}{\text{current liabilities}}$$

Current liabilities include all short-term debts, such as accounts and dividends payable. Note that only balance-sheet data are utilized in the current ratio; thus, no association with revenues or expenses is made. For the balance sheet of Table 14.1, current liabilities amount to $19,700 + $7,000 = $26,700$ and

$$\text{Current ratio} = \frac{81,700}{26,700} = 3.06$$

Since current liabilities are those debts payable in the next year, the current-ratio value of 3.06 means that the current assets would "cover" short-term debts approximately three times. Current-ratio values of 2 to 3 are common. For comparison purposes, it is necessary to compute the current ratio, and all other ratios, for several companies in the same industry. Industry-wide median ratio values are periodically published by firms such as Dunn and Bradstreet. You should realize that the current ratio assumes that the working capital invested in inventory can be converted to cash quite rapidly. Often, however, a better idea of a company's *immediate* financial position can be obtained by using the acid-test ratio.

14.3.2 Acid-test ratio (quick ratio) This ratio, defined by the relation

$$\text{Acid-test ratio} = \frac{\text{quick assets}}{\text{current liabilities}} = \frac{\text{current assets} - \text{inventories}}{\text{current liabilities}}$$

is meaningful for the emergency situation when the firm must cover short-term debts using its readily convertible assets. For the Wimble Corporation,

$$\text{Acid-test ratio} = \frac{81,700 - 52,000}{26,700} = 1.11$$

Comparison of this and the current ratio shows that approximately two times the current debts of the company are invested in inventories. However, an acid-test ratio of approximately 1.0 is generally regarded as a strong *current* position, regardless of the amount of assets in inventories.

14.3.3 Equity ratio This ratio has historically been a measure of financial strength since it is defined as

$$\text{Equity ratio} = \frac{\text{stockholder's equity}}{\text{total assets}}$$

For the Wimble Corporation,

$$\text{Equity ratio} = \frac{400,000}{462,700} = 0.865$$

that is, 86.5% of Wimble is stockholder owned. The $25,000 retained earnings is also called *equity* since it is actually owned by the stockholders, not the corporation. An equity ratio in the range 0.80 to 1.0 usually indicates sound financial condition, with little fear of forced reorganization because of unpaid liabilities. However, a company with virtually no debts, that is, one with a very high equity ratio, may not have a promising future, due to its inexperience in dealing with short- and long-term debt financing.

14.3.4 Operating ratio A measure of how much of the revenue from customers is required to cover expenses is determined by computing the operating ratio.

$$\text{Operating ratio} = \frac{\text{total revenues}}{\text{total expenses}}$$

Since this ratio is a measure of the effectiveness of the firm's attempts to raise revenue, it is closely watched by management. Referring to the income statement for the Wimble Corporation (Table 14.2), we see that

$$\text{Operating ratio} = \frac{508,500}{365,000} = 1.39$$

Note that taxes are not included in the expense figure, since taxes are an unavoidable expense imposed by company-exterior sources, namely federal, state, or city governments. However, since the firm's revenue-making ability must also cover taxes, some managements do include taxes when computing the operating ratio.

A figure consistently somewhat above 1.0 is desirable, basically because a value of 1.0 implies the absence of gross profit for the firm. It is also necessary to realize that this ratio would be computed for specific product lines to determine their effectiveness with respect to the rest of the firm. For example, assume the Tebb Company has an overall operating ratio of 1.6 and the four product lines, A through D, have ratio values of 1.9, 2.1, 0.9, and 1.7, respectively. The third line is being "carried" by the other three, since the firm can well meet its expenses, but this line has an operating ratio less than 1.0.

14.3.5 Income ratio This often quoted, and often misused ratio, is defined as

$$\text{Income ratio} = \frac{\text{net profit}}{\text{total revenue}} (100\%)$$

Net profit is the after-tax value from the income statement. For the Wimble Corporation,

$$\text{Income ratio} = \frac{71,750}{508,500} (100\%) = 14.1\%$$

Such a high percentage as this for the Wimble Corporation may, at first glance, look highly favorable. However, often a large income ratio reflects an inability to earn a high rate of return on resources owned and employed by the firm. In recent years, corporations have pointed to small income ratios, say 2.5% to 4.0%, as indications of sagging economic conditions. In truth, for a relatively large-volume, high-turnover business, an income ratio of 3% is quite healthy. Of course, a steadily decreasing ratio indicates rising company expenses, which absorb the potentially higher net profit after taxes.

Naturally, there are many other ratios a company's personnel can use in various circumstances; however, the ones presented here are commonly used by both accountants and engineering economists.

Problems P14.4–P14.7

14.4 Allocation of Factory Expense

All costs incurred in the production of an item are accounted for by the cost-accounting system. It can be generally stated that the statement of cost of goods sold (Table 14.3) is the end product of this system. Cost accounting accumulates the material, direct labor, and factory-expense costs by using *cost centers*. For example, a department or a machine may be a cost center. Thus all costs incurred in the department or in utilizing the machine are collected under one cost-center title, such as Machine 300. Since direct materials and labor are assignable to a cost center, the accountant has only to keep track of these costs. Of course, this is no easy chore and often the cost of the cost-accounting system prohibits collection of direct costing data in the detail that the accountant or the engineering economist might desire.

By far the most burdensome chore of cost accounting is the allocation of factory expense, or overhead. The costs associated with property taxes, service and maintenance departments, personnel, supervision, utilities, etc., must be allocated to the using cost center. Detailed collection of these data is cost prohibitive and often impossible; thus, allocation schemes are utilized to distribute the expenses on a reasonable basis. A description of some of the bases is given in Table 14.4. The most common bases used are direct labor cost and direct labor hours.

Most allocation is accomplished utilizing a predetermined factory expense rate, computed using the relation

Table 14.4 FACTORY EXPENSE ALLOCATION BASES

Overhead category	Allocation basis
Taxes	Space occupied
Heat, light	Space, usage, number of outlets
Power	Space, direct labor hours, direct labor cost, machine hours
Receiving, purchasing	Cost of materials, number of orders, number of items
Personnel, machine shop	Direct labor hours, direct labor cost
Building maintenance	Space occupied, direct labor cost

$$\text{Overhead rate} = \frac{\text{estimated overhead costs}}{\text{estimated basis level}} \qquad (14.2)$$

The estimated overhead cost is the amount allocated to the particular cost center. For example, if a company has two producing departments, the overhead allocated to *each department* is used in Eq. (14.2) when determining the departmental overhead rate. Example 14.1 illustrates the use of Eq. (14.2) when the cost center is a machine.

Example 14.1 An engineering economist is attempting to compute overhead rates for the production of glass bookends. The following information is obtained from last year's budget for the three machines used to produce this product.

Cost source	Allocation basis	Estimated activity level
Machine 1	Labor cost	$10,000
Machine 2	Labor hours	2,000 hours
Machine 3	Material cost	$12,000

Determine overhead rates for each machine if the estimated factory expense budget for producing the bookends is $5,000 per machine.

SOLUTION Applying Eq. (14.2) for each machine, rates for the year are

$$\text{Machine 1 rate} = \frac{\text{overhead budget}}{\text{direct labor cost}} = \frac{5,000}{10,000}$$

$$= \$0.50 \text{ per dollar labor}$$

$$\text{Machine 2 rate} = \frac{\text{overhead budget}}{\text{direct labor hours}} = \frac{5,000}{2,000}$$

$$= \$2.50 \text{ per labor hours}$$

$$\text{Machine 3 rate} = \frac{\text{overhead budget}}{\text{material cost}} = \frac{5,000}{12,000}$$

$$= \$0.42 \text{ per material dollar}$$

COMMENT Once the product has been manufactured and actual direct labor costs, hours, and material costs computed, each dollar of direct labor cost spent on Machine 1 implies that $0.50 in factory expense will be added to the cost of the product. Similar expense values will be added for Machines 2 and 3.   ////

The amount of overhead allocated to an asset is a component of the annual operating cost (AOC) used in computations in previous chapters. Other elements of AOC are directly assignable expenses such as insurance and direct labor cost.

Example 14.3
Problems P14.8–P14.10

Table 14.5 ACTUAL DATA USED FOR OVERHEAD ALLOCATION FOR EXAMPLE 14.2

Cost source	Machine number	Actual cost	Actual hours
Material	1	$ 3,800	–
	3	19,550	–
Labor	1	2,500	650
	2	3,200	750
	3	2,800	720

14.5 Factory-Expense Computation and Variance

Once the overhead rates are computed and data collected, it is possible to actually determine the cost of production. This cost may be to operate a department or a machine or to manufacture a product, depending upon the manner in which the cost centers are set up. A *cost center* is simply a term used for the specific overhead account; for example, Department A may be one cost center.

Assuming the estimated overhead budget was correct, total allocated overhead should equal estimated overhead. However, since some error in budgeting always exists, there will be some over- or under-allocation of overhead, termed *variance*. Experience in overhead estimation assists in reducing the variance at the end of the accounting period. Example 14.2 illustrates overhead allocation and variance computation.

Example 14.2 The economist who collected the data and determined rates in Example 14.1 is now ready to compute the actual cost of producing the product. Assume you are the economist and do the computation, if the actual cost data of Table 14.5 are available from cost-accounting records for the year.

SOLUTION To determine actual production costs, the economist should use the factory cost relation given in Eq. (14.1). That is,

$$\text{Factory cost} = \text{materials} + \text{labor} + \text{factory expense}$$

To determine factory expense, which is the overhead expense, the rates from Example 14.1 are utilized as follows:

$$\text{Machine 1 overhead} = (\text{labor cost})(\text{rate}) = 2{,}500(0.50)$$
$$= \$1{,}250$$
$$\text{Machine 2 overhead} = (\text{labor hours})(\text{rate}) = 750(2.50)$$
$$= \$1{,}875$$
$$\text{Machine 3 overhead} = (\text{material cost})(\text{rate}) = 19{,}550(0.42)$$
$$= \$8{,}211$$

$$\text{Total factory overhead expense} = \$11{,}336$$

Thus, factory cost is the sum of material, labor, and overhead charges, or a total $43,186 for the year.

COMMENT The variance for factory expense in this example is 3($5,000) − $11,336 = $3,664 under the budget. The $15,000 budget for the three machines represents a 32.3% allocation over actual overhead. This analysis should allow a more realistic overhead budget for the product in the future years.

Actually, once estimates of overhead costs are determined, it is possible to make an economic analysis of the present operation versus a proposed or anticipated operation. Such a study is explained in the Solved Examples.   ////

Examples 14.4, 14.5
Problems P14.11–P14.13

SOLVED EXAMPLES

Example 14.3 The ABC Company produces many products, two of which are stereo cabinets and Formica-topped dinette sets. A total of $33,000 is allocated to overhead for next year. Management wants to determine overhead rates on the basis of direct labor hours for the two processing and one finishing departments (a) individually and (b) using an overall (blanket) basis. Table 14.6 presents departmental overhead allocation and processing time. Develop the rates for management.

SOLUTION

(a) The rate for each department is computed using Eq. (14.2):

$$\text{Stereo processing} = \frac{145,000}{20,000} = \$7.25 \text{ per hour}$$

$$\text{Dinette processing} = \frac{145,000}{10,000} = \$14.50 \text{ per hour}$$

$$\text{Finishing} = \frac{10,000}{6,000} = \$1.67 \text{ per hour}$$

Table 14.6 OVERHEAD ALLOCATION FOR ONE YEAR FOR EXAMPLE 14.3

Allocation	Stereo processing	Dinette processing	Finishing
Overhead dollars	$145,000	$145,000	$10,000
Direct labor hours	20,000	10,000	6,000

(*b*) A blanket rate is found by computing

$$\text{Overhead rate} = \frac{\text{total overhead allocation}}{\text{total labor hours}}$$

$$= \frac{300{,}000}{36{,}000}$$

$$= \$8.33 \text{ per hour}$$

COMMENT Of course, an overall rate is easier to compute and use; however, it cannot account for differences in the nature of the work of departments "blanketed" under the rate. See Example 14.4.　　　　////

Sec. 14.4

Example 14.4 Use the two sets of rates determined in Example 14.3 to compute factory expense if the following processing times per unit are correct:

$$\text{Stereo: processing} = 2 \text{ hours}$$
$$\text{finishing} = 8 \text{ hours}$$
$$\text{Dinette: processing} = 7 \text{ hours}$$
$$\text{finishing} = 3 \text{ hours}$$

SOLUTION

(*a*) For the stereo, total overhead per unit is computed as $2(\$7.25) + 8(\$1.67) = \$27.86$; for the dinette sets $7(\$14.50) + 3(\$1.67) = \$106.51$. The larger expense is allocated to the dinette due to its large processing time at high overhead cost compared to low costs in finishing. Factory expense is the sum or $\$134.37$ per unit.

(*b*) For the blanket rate we use a total of ten hours to manufacture either item, so overhead for both is $10(\$8.33) = \$83.33$. Factory expense is $\$166.66$ per unit.

COMMENT Factory expense doesn't change much but, by using the blanket rate, the stereo is overcharged and the dinette undercharged in overhead. For this reason, blanket rates can be inaccurate and misleading.　　　　////

Sec. 14.5

Example 14.5 For several years a certain company has purchased the motor and frame assembly of its major product line for an annual cost of $\$1.5$ million. The thought now is to make the components in-house, using existing departmental facilities. For the three departments involved the overhead rates, estimated material,

Table 14.7 PRODUCTION COST ESTIMATES FOR EXAMPLE 14.5

Department	Overhead Basis (hours)	Rate ($/hour)	Allocated hours	Material cost	Direct labor cost
A	Labor	$10	25,000	$200,000	$200,000
B	Machine	5	25,000	50,000	200,000
C	Labor	15	10,000	50,000	100,000
				$300,000	$500,000

labor, and hours are quoted in Table 14.7. The "Allocated hours" column is the time necessary to produce the motor and frame only.

To produce the products, equipment would have to be purchased. The machinery would have a first cost of $2 million, a salvage value of $50,000, and a life of ten years. Perform an economic analysis of the suggestion to make the components, assuming a 15% return is required.

SOLUTION For making the components in-house, the annual operating costs (AOC) are composed of labor, material, and overhead costs. Using the data of Table 14.7, overhead is

Department A:	25,000(10) =	$250,000
Department B:	25,000(5) =	125,000
Department C:	10,000(15) =	150,000
		$525,000

Thus,

$$AOC = 500,000 + 300,000 + 525,000 = \$1,325,000$$

The EUAC, using Eq. (9.1), is

$$
\begin{aligned}
\text{EUAC}_{make} &= P(A/P,i\%,n) - SV(A/F,i\%,n) + AOC \\
&= 2,000,000\,(A/P,15\%,10) - 50,000\,(A/F,15\%,10) \\
&\quad + 1,325,000 \\
&= \$1,721,037
\end{aligned}
$$

Currently,

$$\text{EUAC}_{buy} = \$1,500,000$$

Therefore, it is cheaper to continue to buy the motor and frames assembly.　////

Sec. 14.5

BIBLIOGRAPHY

Barish, pp. 31–42, 191–208.
DeGarmo and Canada, pp. 9–20.
Emerson and Taylor, pp. 11-1 - 11-15.

Fabrycky and Thuesen, pp. 152–157.
Grant and Ireson, pp. 155–167, 311–315.
Ostwald, pp. 53–67.
Park, pp. 178–186.
Smith, pp. 14–19.
Taylor, pp. 23, 138.
Thuesen, Fabrycky, and Thuesen, pp. 237–253.

PROBLEMS

The following financial data have been gathered for the month of July 19XX for the Stop-Gap Tool and Die Company. You will use this information in problems P14.1–P14.5.

PRESENT SITUATION, JULY 31, 19XX

Account	Balance
Accounts payable	$ 35,000
Accounts receivable	29,000
Bonds payable (20 year)	110,000
Buildings (net value)	605,000
Cash on hand	17,000
Dividends payable	8,000
Inventory value (all inventories)	31,000
Land value	450,000
Long-term mortgage payable	450,000
Retained earnings	154,000
Stock value outstanding	375,000

TRANSACTION FOR JULY, 19XX

Category		Amount
Direct labor		$ 50,000
Expenses		
Insurance	$ 20,000	
Selling	62,000	
Rent and lease	40,000	
Salaries	110,000	
Other	62,000	
Total		294,000
Income taxes		20,000
Increase in finished goods inventory		25,000
Materials inventory, July 1, 19XX		46,000
Materials inventory, July 31, 19XX		25,000
Materials purchases		20,000
Overhead charges		75,000
Revenue from sales		500,000

P14.1 Using the data summary above, (*a*) construct a balance sheet for the Stop-Gap Company as of July 31, 19XX, and (*b*) determine the value of each term in the basic equation of the balance sheet.

P14.2 (*a*) Prepare the cost-of-goods-sold statement for July, 19XX. (*b*) What was the net change in materials inventory value during the month?

P14.3 Use the account summary above to develop (*a*) an income statement for July 19XX and (*b*) the basic equation of the income statement. (*c*) What percent of revenue is reported as after-tax income?

P14.4 (*a*) Compute the value of each accounting ratio that uses only balance-sheet information from the statement you constructed in problem P14.1. (*b*) What percent of the company's current debt is "tied up" in inventory?

P14.5 (*a*) Compute the operating ratio and income ratio for the Stop-Gap Company. (*b*) What percent of each sales dollar can the company rely upon as profit?

P14.6 The Horseradish Supreme Company reported the following income-statement values at the end of the year.

$$Sales = \$800,000$$
$$Expenses \text{ (without taxes)} = 675,000$$
$$Net \text{ profit after taxes} = 50,000$$

The president of the company has learned, in a conversation with a competitor, that the latter sold 150,000 units last year at an average of $10 per unit. If the published financial records of the competitor state that the operating ratio was 1.75 and the income ratio was 18.0%, compute the total expenses and net profit of the competitor. Compare the two companies' ratios and comment on the income picture of each as depicted by these ratios.

P14.7 The ratio of total revenue to net worth is a measure of the turnover rate of the capital invested by company ownership. If the capital invested in the two companies of problem P14.6 has turned over 2.8 times in the year, (*a*) compute the percentage return on the net worth of each company and (*b*) compare this value with the percentage return on sales for the year.

P14.8 A company has a processing department with 25 machines. Due to the nature and use of three of these machines, each is considered a separate account for factory overhead. The remaining 22 machines are grouped under one account number, M104. Machine hours are used as an overhead allocation basis for all machines. A total of $25,000 in overhead is allocated to the department for next year. Using the data below, determine (*a*) the overhead rate for each account and (*b*) the blanket rate for the department.

Account number	Overhead allocated	Estimated machine hours
M101	$ 5,000	600
M102	5,000	200
M103	5,000	800
M104	10,000	1,600

P14.9 A machining department's overhead is allocated by accounting. The department manager has just obtained records of allocation rates and actual charges for the prior three months and estimates for this month (May) and next month. However, the basis of allocation is not indicated and accounting has no record of the basis used. You, an engineer with the company, are asked to investigate the allocation for each month. You are given the following information by the department manager as it came from accounting.

		Overhead	
Month	Rate	Allocated	Charged
February	$1.40	$2,800	$2,600
March	1.33	3,400	3,800
April	1.37	3,500	3,500
May	1.03	3,600	–
June	0.92	6,000	–

You collect the following data from departmental and accounting records.

	Direct labor		Material	Space
Month	Hours	Costs	costs	(sq ft)
February	640	$2,560	$5,400	2,000
March	640	2,560	4,600	2,000
April	640	2,560	5,700	3,500
May	640	2,720	6,300	3,500
June	800	3,320	6,500	3,500

(a) Determine the actual allocation basis used each month and (b) comment on the decreasing overhead rate published each month by accounting.

P14.10 An electronics manufacturing company has five separate departments. Departmental charges for a certain month are given below. Also detailed are the space allocations, direct labor hours and direct labor costs for each producing department.

	Overhead	Space	Direct labor	
Department	costs	(sq ft)	Hours	Costs
Preparation	$20,000	10,000	480	$1,680
Subassemblies	15,000	18,000	1,000	3,250
Final assembly	10,000	6,000	600	2,460
Quality control	5,000	1,200	–	–
Engineering	9,000	2,000	–	–

Determine the departmental overhead rates to be used in making a redistribution of the overhead cost for quality control and engineering ($14,000) to the other three departments on the basis of (a) space, (b) direct labor hours, and (c) direct labor costs.

P14.11 Compute the actual overhead allocation and department variance for problem P14.8 using (a) the individual account rates and (b) the blanket rate. The actual hours credited to each account are as follows: M101, 700 hours; M102, 350 hours; M103, 650 hours; M104, 1,300 hours.

P14.12 Overhead allocation rates and the allocation bases for the six producing departments of the E-Z-Duz-It Calculator Manufacturers are listed below. (a) Use the reported data to distribute overhead to the departments. (b) Compute the variance of the allocation if a total of $750,000 had been allocated.

Department	Allocation* basis	Rate	Direct labor hours	Direct labor cost	Machine hours
1	DLH	$2.50	5,000	$20,000	3,500
2	MH	0.95	5,000	35,000	25,000
3	DLH	1.25	10,500	44,100	5,000
4	DLC	5.75	12,000	84,000	40,000
5	DLC	3.45	10,200	54,700	10,200
6	DLH	1.00	29,000	89,000	60,500

*DLH = direct labor hours; MH = machine hours; DLC = direct labor costs.

P14.13 Perform the overhead allocation of problem P14.10 using the rates you determined. Use a basis of direct labor hours for the preparation and subassembly departments and direct labor cost for final assembly.

P14.14 (a) The electronics firm in P14.10 presently makes all the components required by the preparation department. The firm is considering the possibility of purchasing, rather than making, all these components. The firm has a quote of $67,500 a month from an outside contractor to make the items. If the costs for the month stated in P14.10 are representative and if $41,000 worth of materials were charged to preparation, make an economic comparison of the present versus proposed situation. Assume that the preparation department's share of the quality control and engineering costs is a total of $3,230 per month. (b) Another alternative for the company is to purchase new equipment in the preparation department. The machinery will cost $375,000, have a five-year life, no salvage value, and a monthly operating cost of $475. This purchase is expected to reduce the quality control and engineering costs by $2,000 and $3,000, respectively, and reduce direct labor hours to 200 and direct labor cost to $850 for the preparation department. The redistribution of the overhead costs in quality control and engineering to the three production departments is on the basis of direct labor hours. If other costs remain the same, compare the present cost of making the components with the cost of the equipment purchase alternative if a return of a nominal 12% per year compounded monthly is required on all capital investments.

15

CORPORATE TAX STRUCTURE

There are two primary objectives of this chapter. The first is to give you a basic knowledge of tax definitions and tax rates as applied to corporations. The second is to compare the depreciation methods of Chap. 7 from the tax viewpoint.

This chapter is only an introduction to the effect of taxes on corporate income and engineering-economy studies. Further and deeper analysis of tax considerations are studied in the next chapter. To perform an economy study without somehow accounting for tax effects is foolish and misleading because taxes can reverse the before-tax decision. It is not, however, necessary to know all the details of corporation taxing in order to accomplish a realistic study with taxes considered. Of course, the introduction of taxes increases the complexity of the study. Also important is that taxes are another disbursement that *must* be included in the economy study.

In this text we study corporate taxes only[1,2] and, in fact, a much simplified corporate tax structure for the engineering economist. Personal tax problems should be resolved by consulting the appropriate Internal Revenue Service bulletins.

CRITERIA

To complete the material in this chapter, you must be able to do the following:

1 Define *gross income, taxable income, capital gain, capital loss, investment tax credit,* and *operating loss.*
2 Compute the income tax using normal and surtax rates or an effective tax rate, given taxable income and applicable tax rates.
3 Compute the resulting net capital gains or losses and the income tax, given the gain or loss values, federal tax laws for gains and losses, and applicable tax rates.
4 Compute the investment tax credit allowed on a section 38 property asset, given expected life, initial cost, and tax credit laws.
5 State the tax law as it concerns operating losses occurring in a particular year.
6 Show the advantage of one depreciation method over another by computing the present worth of the taxes involved, given the depreciation methods, asset data, tax rate, and an after-tax rate of return.
7 Compute an inflated before-tax rate of return, given the effective tax rate and after-tax rate of return.

EXPLANATION OF MATERIAL

15.1 Some Tax Definitions

To better understand the tax rates and formulas discussed below some basic definitions are presented here.

> *Gross income* Total of all incomes from revenue-producing sources. These include all items listed in the revenue section of an income statement (Table 14.2).
>
> *Expenses* All costs incurred while transacting business (Table 14.2).
>
> *Taxable income* The dollar value remaining upon which taxes are to be paid. It is computed as follows:

$$\text{Taxable income} = \text{gross income} - \text{expenses} - \text{depreciation} \qquad (15.1)$$

> *Capital gain* When the selling price of depreciable property (assets) or real property (land) exceeds the book value, a capital gain is incurred. Thus, at sale time the computation would be

$$\text{Gain} = \text{selling price} - \text{book value}$$

> where the capital gain > 0. If the sales date occurs within six months of purchase date, the capital gain is referred to as *short-term gain* (STG); if the ownership period is longer than six months, the gain is a *long-term gain* (LTG). An STG and an LTG are taxed differently.
>
> *Capital loss* Here the capital loss is computed as

$$\text{Loss} = \text{book value} - \text{selling price}$$

The terms *short-term loss* (STL) and *long-term loss* (LTL) are defined in a fashion similar to capital gains, that is, a six-month break point. The concept of sunk cost, briefly discussed in Sec. 12.1, results in a capital loss.

Investment tax credit This is a tax credit given to the purchaser of new or used equipment that qualifies as section 38 property. The tax credit is given to encourage the purchase and use of modern equipment.

Operating loss When a corporation experiences a year of net loss rather than net profit, it has an operating loss. Special tax considerations are made in an attempt to balance the lean and fat years. The anticipation of operating losses, and thus, the ability to take them into account in an economy study is, of course, virtually impossible.

Problem P15.1

15.2 Basic Tax Formulas and Computations

Taxes are computed using the relation

$$\text{Taxes} = (\text{gross income} - \text{expenses} - \text{depreciation})T \qquad (15.2)$$

where T = tax rate. Since Eq. (15.2) uses the definition of taxable income (TI) from Eq. (15.1), we have

$$\text{Taxes} = (TI)T \qquad (15.3)$$

However, to give the small businessman a slight assist, corporate taxes are actually computed as

$$\text{Corporate taxes} = (TI)(\text{normal tax rate}) + (TI - \$25,000)(\text{surtax rate}) \qquad (15.4)$$

Presently, the federal normal tax rate is 22% and surtax rate is 26%. Thus, for all TI over \$25,000 the applicable (effective) tax rate is 22% + 26% = 48%.

Example 15.1 For a particular year, the Muche Company has a gross income of \$2,750,000 with expenses and depreciation totaling \$1,950,000. What is the amount of taxes to be paid for the year?

SOLUTION

$$TI = 2,750,000 - 1,950,000 = \$800,000$$
$$\text{Taxes} = 800,000(0.22) + (800,000 - 25,000)(0.26) = \$377,500$$

COMMENT Note that the \$25,000 deduction produces a \$6,500 decrease in taxes, since a straight 48% tax rate requires a total of \$800,000(0.48) = \$384,000 in taxes to be paid. ////

It is important to realize that the total quantity which is classed as *depreciation* is deducted from gross income in computing TI. This is an advantage sought after by corporation-management tax people. More will be said about taxes and depreciation later in this chapter.

For the sake of simplicity the tax rate used in an economy study is often a "one-figure" effective tax rate, which serves to account for federal, state, and city taxes. Commonly used effective tax rates are 50% or 52%, but the applicable rate in a particular case is easily approximated. One advantage in applying an effective rate is that state taxes are deductible from federal taxes. Thus, you can use the following relation to compute the effective tax rate as a decimal fraction

$$\text{Incremental effective tax rate} = \text{state rate} + (1 - \text{state rate})(\text{federal rate}) \qquad (15.5)$$

where the federal rate of 48% is usually applicable, because TI is already above $25,000; that is, the incremental federal rate is 48%.

Example 15.2 Compute the income tax for the data presented in Example 15.1 using an effective tax rate, if the state rate is 8% and the company uses a 48% federal rate.

SOLUTION First the effective rate is computed using Eq. (15.5):

$$\text{Effective rate} = 0.08 + (1 - 0.08)(0.48) = 0.5216$$

Now,

$$\begin{aligned}
\text{Taxes} &= \text{TI(effective tax rate)} \\
&= 800,000(0.5216) \\
&= \$417,280
\end{aligned}$$

COMMENT Do not compare this $417,280 with results of Example 15.1, since the latter do not include a state tax.

Example 15.7
Problems P15.2–P15.7

15.3 Tax Laws for Capital Gains and Losses

Capital gains and losses are individually treated according to the following schedule.

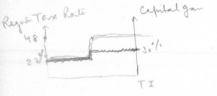

Category	Effect
LTG (TI ≤ 25,000)	Normal rate (22%)
LTG (TI > 25,000)	30%
STG	Taxed as ordinary income
LTL	Offset LTG
STL	Offset STG

The exact procedure used to compute taxes in which capital gains and losses are involved is now described. An LTG is taxed at a rate of 30% of gain incurred for the year, if taxable income exceeds $25,000. For a small firm, the normal tax rate of 22% of gain is applicable. An LTL can only be used to offset an LTG. Thus, if in a particular year LTG = $5,500 and LTL = $3,200, there is a net LTG = $2,300 to be taxed. The treatment of STG and STL is similar, except that the net gain is taxed as regular income. Once the losses have been used to offset gains, the net result is obtained. This value is then used for tax computation, not the individual gain and loss values. In any event, the net losses claimed in any one year cannot exceed the gains for the year. Examples 15.3 and 15.4 will clarify the actual procedure.

Example 15.3 A company has the following income, gain, and loss values for one year.

$$\text{Taxable income} = \$50,000$$
$$\text{STG} = \quad 4,000$$
$$\text{LTG} = \quad 7,500$$
$$\text{STL} = \quad (500)$$
$$\text{LTL} = \quad (9,000)$$

(Parentheses are used to indicate losses.) Compute the income taxes.

SOLUTION Using losses to offset gains, the following results are obtained:

LTG	$ 7,500	STG	$4,000
LTL	(9,000)	STL	(500)
Net LTL	$(1,500)	Net STG	$3,500

Result: Net STG $2,000

Therefore, $52,000 is actually taxed as ordinary income, as follows:

Taxable income	$50,000
STG	2,000
Ordinary income	$52,000

Income tax computation is

$$\text{Tax} = 52,000(0.22) + 27,000(0.26) = \$18,460 \qquad ////$$

Example 15.4 Assume the following losses and gain occur in one year.

$$\text{STL} = \$(3,000)$$
$$\text{LTL} = \quad (500)$$
$$\text{STG} = \quad 1,000$$

How would the resulting net gain or loss be treated in tax computation?

SOLUTION The result is a net loss of $2,500. However, the law provides that losses only up to the amount of gains can be deducted from taxable income. Therefore, of the $2,500 loss only $1,000, the amount of the gain, can be taken from TI to reduce taxes.

COMMENT Tax laws, however, do allow the undeducted difference of losses and gains to be carried back three years as a STL, or forward five years, if not completely absorbed in carry-back. Therefore, the remaining $1,500 loss may be used to offset capital gains until exhausted over a maximum nine-year period.   ////

Problems P15.8–P15.11

15.4 Tax Laws for Investment Tax Credit

Corporations are encouraged to invest in modern equipment because of a tax advantage applicable in the purchase year of the asset. The advantage is termed the *investment tax credit,* first referred to in Sec. 7.2 of this book. As its name implies, it is a tax credit and in no way affects TI. Table 15.1 presents the percent tax credit for an asset with initial cost P and expected life of n years. The actual tax credit is 7% of P; however, the entire 7% is creditable only for assets with $n \geq 8$ years. If $n = 6$ years, for example, the tax credit is $\frac{2}{3}(7\%) = 4\frac{2}{3}\%$ of P. The $2\frac{1}{3}\%$ in Table 15.1 is computed as $\frac{1}{3}(7\%)$.

Section 38 property, for which a tax credit may be claimed, refers to an asset which is depreciable, an integral member of the manufacturing process, and has an expected life of at least three years. To avoid confusion, we will assume that newly acquired assets are *not* section 38 property unless specifically stated.

Example 15.5 Assume the purchase of the following section 38 property in one year.

Asset number	Life, n years	Cost, P
309	12	$75,000
318	5	8,000

Compute the actual cash outflow for the year if the appropriate tax credit is claimed.

SOLUTION Neglecting all other factors, the cash outflow due to the asset purchases without the tax credit is $83,000. Tax credits are

Asset number	Tax credit	
	% of P	Amount
309	7	$5,250
318	$4\frac{2}{3}$	373

Therefore, cash flow for the year is actually $-83,000 + $5,250 + $373 = $-77,377.   ////

Table 15.1 INVESTMENT TAX CREDIT
PERCENTS

Expected life, n years	Tax credit, % of P
0-2	None
3 or 4	$2\frac{1}{3}$
5, 6, or 7	$4\frac{2}{3}$
8 or more	7

It is assumed that the investment tax credit is creditable at the time of asset purchase, that is, year zero. The tax credit that can be claimed in a year is limited by the income tax for the year. If the credit exceeds this amount, the unclaimed portion can be carried back for three and forward for seven succeeding tax years. In addition, this tax credit does not reduce the total depreciable value of an asset. Thus in Example 15.5, the full first cost to the extent of the salvage value can still be written off.

Problems P15.12–P15.14

15.5 Tax Laws on Operating Losses

We have discussed situations in which capital losses and unclaimable tax credits can be carried backward and forward for several tax years. Another important tax advantage is the tax law that allows an operating loss to be carried backward for three and forward for five years until the loss is completely exhausted. The amount of operating loss claimed in any one year cannot exceed taxable income. Since only the amount of the loss is recoverable, this and all carry back–carry forward laws present a question of strategy, that is, *when* to utilize the tax advantage.

Problem P15.15

15.6 Tax Effects of Depreciation Models

This section is included in this chapter so that the reader who does not proceed to Chap. 16 will have an idea of how the choice of a particular depreciation model affects the taxes that must be paid. Here we assume that some after-tax rate of return is to be used for the analysis. Example 15.6 compares two depreciation methods. The results of this example show that for a constant tax rate, annual gross income greater than or equal to annual depreciation, and write-off down to the salvage value, the following are always true:

1 The total taxes paid are equal.
2 The present worth of taxes are less for faster write-off depreciation methods.

Table 15.2 TAX COMPUTATION FOR ASSET USING
STRAIGHT-LINE DEPRECIATION

Year	CFBT	Depreciation*	TI	Taxes
0	$-100,000	–	–	–
1-9	+20,000	$10,000	$10,000	$5,000
9	+10,000	–	–	–

*From Eq. (7.1), depreciation = $(100,000 - 10,000)/9 = \$10,000$ per year.

Example 15.6 Assume an asset with the following characteristics is purchased.

$$P = \$100,000$$
$$SV = 10,000$$
$$CFBT = 20,000 \text{ per year}$$
$$n = 9 \text{ years}$$

The value CFBT = $20,000 is the *cash flow before taxes*, that is, the net value of gross income minus expenses before any taxes are computed. (You will see this term often hereafter.) If the effective tax rate is 50% and an after-tax rate of return of 8% is used, compare (*a*) straight-line and (*b*) sum-of-year-digits depreciation from the tax viewpoint.

SOLUTION

(*a*) We first study the tax situation for straight-line depreciation. Annual taxable income (TI) is cash flow before taxes (CFBT) minus depreciation, or

$$TI = CFBT - depreciation$$

In Eq. (15.1), CFBT is the difference of gross income and expenses. Table 15.2 presents tax computation for the entire asset's life and Fig. 15.1 is a *tax*

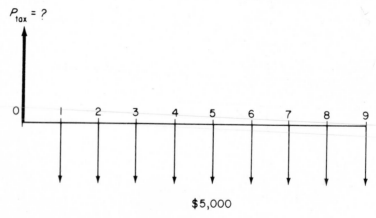

$$P_{tax} = 5,000 \ (P/A, 8\%, 9) = \$31,235$$

FIG. 15.1 Tax cash flow for asset using straight-line depreciation, Example 15.6.

**Table 15.3 TAX COMPUTATION FOR ASSET USING
SUM-OF-YEAR-DIGITS DEPRECIATION**

Year	CFBT	Depreciation*	TI	Taxes
0	$-100,000	–	–	–
1	+20,000	$18,000	$ 2,000	$ 1,000
2	+20,000	16,000	4,000	2,000
3	+20,000	14,000	6,000	3,000
4	+20,000	12,000	8,000	4,000
5	+20,000	10,000	10,000	5,000
6	+20,000	8,000	12,000	6,000
7	+20,000	6,000	14,000	7,000
8	+20,000	4,000	16,000	8,000
9	+20,000	2,000	18,000	9,000
9	+10,000	–	–	–
				$45,000

*From Eq. (7.3), depreciation $= [(9 - m + 1)/45] \, (100,000 - 10,000)$,
$(m = 1, 2, \ldots, 9)$.

cash-flow diagram. Total taxes are $5,000(9) = $45,000$. Present worth of the
taxes, (P_{tax}), as calculated below the cash-flow diagram, is $31,235.
(*b*) Table 15.3 and Fig. 15.2 present tax computation and the tax cash flow,
respectively, using sum-of-year-digits. Total taxes are again $45,000, but the

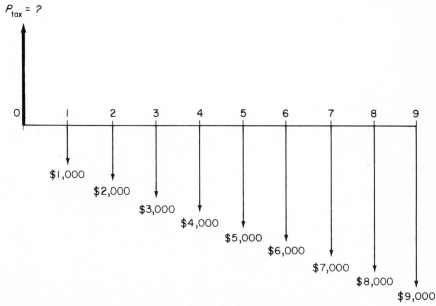

$$P_{tax} = 1,000 \, (P/A, 8\%, 9) + 1,000 \, (P/G, 8\%, 9) = \$28,055$$

FIG. 15.2 Tax cash flow using sum-of-year-digits depreciation.

present worth of the taxes is $28,055, or 10.2% less than the corresponding value for straight-line depreciation.

COMMENT From this illustration it is clear that any depreciation that writes off faster than straight-line gives a tax advantage, since the present-worth value will be greater for straight-line depreciation. This is true, since with rapid write-off the taxes are "moved" to the later years of ownership, which decreases their present worth when compared to the constant taxes applicable for straight-line depreciation. Examples with another depreciation method and introduction to the effect of an asset's life on taxes are presented in Solved Examples.   ////

Examples 15.8, 15.9
Problems P15.16–P15.22

15.7 Using an Inflated Before-Tax Rate of Return

If an engineering economist does not wish to be bothered with tax considerations, he may choose to increase the rate of return used in the before-tax study to include an approximation of the tax effect. A simple relation to use is

$$\text{Before-tax rate of return} = \frac{\text{after-tax rate of return}}{1 - \text{tax rate}} \qquad (15.6)$$

Thus, if the effective tax rate is 52% and a 6% after-tax return is required, the study could be performed using a before-tax rate of return of $0.06/(1 - 0.52) = 0.125$, or 12.5%. Use of a before-tax rate of return is the reason why most of the rates employed prior to this chapter are in the 12% to 20% range.

Problems P15.23–P15.25

SOLVED EXAMPLES

Example 15.7 The Wa-Out Fruit Company is considering the purchase of one of two new lines of automatic sorter, cleaner, and packer machines. Details of each machine are as follows:

	Wonderama	Spee-de Fix
	$P = \$325,000$	$P = \$270,000$
	$AOC = 30,000$	$AOC = 10,000$
	$SV = 150,000$	$SV = 0$
	$n = 6$	$n = 6$

If the effective tax rate is $T = 0.52$, straight-line depreciation is used, and gross income is expected to be the same from either machine, compute the tax benefit for one of the assets.

SOLUTION   This problem can be solved in a fashion similar to that described in Sec. 7.8; however, here we first compute the taxes for the individual assets, then the tax difference between the two. The effect on taxes is exactly the same. The annual depreciation for Wonderama (D_w) and Spee-de Fix (D_s) are

$$D_w = \frac{325,000 - 150,000}{6} = \$29,167$$

$$D_s = \frac{270,000}{6} = \$45,000$$

We can set up a tax relation for each asset from Eqs. (15.1) and (15.3):

$$\text{Taxes}_w = (\text{GI} - \text{AOC} - D_w)\text{T} = (\text{GI} - 30,000 - 29,167)(0.52)$$
$$\text{Taxes}_s = (\text{GI} - 10,000 - 45,000)(0.52)$$

If we use the Wonderama as a basis for *tax difference* between the two assets, we have

$$\text{Taxes}_w - \text{taxes}_s = (\text{GI} - 30,000 - 29,167 - \text{GI} + 10,000$$
$$+ 45,000)(0.52)$$
$$= (-4,167)(0.52)$$
$$= -\$2,167$$

The negative sign indicates that the asset which is *not* the base pays more taxes. So, the Wonderama has a tax benefit of $2,167 per year.

COMMENT   Logically, the asset with more deductions in computing TI will have the tax advantage. Since Wonderama has $59,167 in deductions and Spee-de Fix has $55,000, the former gets the tax advantage.   ////

Sec. 15.2

Example 15.8   As another illustration of the tax advantage for a fast write-off depreciation model, compute the present worth of taxes for Example 15.6 using the double declining-balance (DDB) method (Sec. 7.6).

SOLUTION   The depreciation rate for DDB is $d = 2/n = 2/9 = 0.22222$. Table 15.4 details the taxes for the life of the asset. The operating losses in years 1 and 9 are not carried back or forward (as is permitted by the tax laws explained in Sec. 15.5) in this example for two reasons: first, there are many different strategies allowed by the tax laws, the best of which would be dependent upon the overall corporate tax picture; second, the purpose of this example is to illustrate the tax advantage of rapid write-off depreciation, rather than income tax manipulations. From Table 15.4, the present worth of the taxes is

$$P_{\text{tax}} = (-1,111)(P/F,8\%,1) + 1,358(P/F,8\%,2)$$
$$+ \cdots (-208)(P/F,8\%,10)$$
$$= \$27,516$$

Table 15.4 TAX COMPUTATION FOR ASSET USING DOUBLE
DECLINING–BALANCE DEPRECIATION

Year	CFBT	Depreciation*	TI	Taxes
0	$-100,000	–	–	–
1	+20,000	$22,222	$– 2,222	$–1,111
2	+20,000	17,284	2,716	1,358
3	+20,000	13,443	6,557	3,279
4	+20,000	10,456	9,544	4,772
5	+20,000	8,132	11,868	5,934
6	+20,000	6,325	13,675	6,837
7	+20,000	4,920	15,080	7,540
8	+20,000	3,826	16,174	8,087
9	+20,000	2,976	17,024	8,512
9	+10,000	–	–416	–208
				$45,000

*Depreciation $D_m = BV_{m-1}$ (0.22222) by Eq. (7.7).

which is less than the comparable value for straight-line ($31,235) or sum-of-year-digits ($28,055) methods. Since DDB results in the lowest present-worth value, it is the most advantageous for tax purposes.

COMMENT The negative TI and tax values are computed as follows. In year 1, depreciation exceeds CFBT; thus 0.5($2,222) = $1,111 is a *reduction* of taxes. In year 9 the book value is $10,416, and $416 in depreciation is unclaimed; thus, there is a tax *reduction* of 0.5($416) = $208.

Finally, we should comment that the above computations imply a long-term loss (LTL) of $416 in year 9, when the asset is sold for $10,000. This would be present since the asset is disposed of for less than book value.   ////

Sec. 15.6

Example 15.9 The Know-It-All Manufacturing Company uses the common practice of keeping two sets of books on depreciable assets, one for its own internal use and one to show to IRS for tax computation purposes. The company owns an asset with $P = \$10,000$, SV = $1,000, and actual $n = 9$ years. However, IRS will allow a life of six years for tax purposes on this class of equipment. Show the tax advantage afforded the company by the shorter life if CFBT = $3,000 per year, a tax rate of 50% applies, invested money is returning 10% after taxes, and straight-line depreciation is used.

SOLUTION What method would you use to compare the two life values? Since this is supposed to be a tax advantage and since taxes will be the same for each life value, you should use the present worth of taxes. For $n = 9$, the following are true *for each year.*

$$\text{Depreciation} = \$1,000$$
$$\text{TI} = 3,000 - 1,000 = \$2,000$$
$$\text{Taxes} = 2,000(0.50) = \$1,000$$

Table 15.5   TAX COMPUTATION FOR SHORTENED LIFE
FOR EXAMPLE 15.9

Year	CFBT	Depreciation*	TI	Tax
0	$-10,000	–	–	–
1-6	+3,000	$1,500	$1,500	$ 750
7-9	+3,000	0	3,000	1,500
9	+1,000	–	–	–
				$9,000

*Depreciation = 9,000/6 = $1,500.

Then

$$P_{tax} = 1,000(P/A,10\%,9) = \$5,759$$

For $n = 6$, Table 15.5 presents the tax computation. Then

$$P_{tax} = 750(P/A,10\%,6) + 1,500(P/A,10\%,3)(P/F,10\%,6) = \$5,372$$

Note that a total of $9,000 in taxes is paid by both the nine- and six-year life. However, the more rapid write-off allowed for $n = 6$ results in a present-worth tax saving of $5,759 - $5,372 = $387.

Sec. 15.6

REFERENCES

1. *Corporations and the Federal Income Tax,* U.S. Internal Revenue Service Publication 542, published annually.
2. *Sales and Other Dispositions of Assets,* U.S. Internal Revenue Service Publication 544, published annually.

BIBLIOGRAPHY

Barish, pp. 91-92, 104-114, 133.
DeGarmo and Canada, pp. 351-363, 172-173.
Emerson and Taylor, pp. 10-1 - 10-13, 12-7 - 12-16, 14-1 - 14-11.
Fabrycky and Thuesen, pp. 162-173.
Grant and Ireson, pp. 348-372.
Smith, pp. 210, 230-235.
Taylor, pp. 318-336.
Thuesen, Fabrycky, and Thuesen, pp. 279-288, 375-384.

PROBLEMS

P15.1   The situations listed below apply to a dairy, the Pure Milk Company, in the past year. For each situation state which of the following is involved: gross income, taxable income, capital gain, capital loss, investment tax credit, or operating loss.

(a) An asset with a book value of $8,000 was retired and sold for $8,450.

(b) An artificial milk-making machine was purchased and had a first-year depreciation of $9,600 by the straight-line method.

(c) The company estimates that it will report a $−75,000 net profit to IRS on the tax return.

(d) The asset in part (b) will have a $4,200-per-year interest cost.

(e) An asset that had a life of eight years has been owned for 14 years and has a book value of $800. It was sold this year for $275.

(f) The cost of goods sold in the last year was $468,290.

P15.2   Two small businesses have the following data on their tax returns:

	Dough Company	Broke Company
Sales	$1,500,000	$820,000
Interest revenue	31,000	25,000
Expenses	754,000	591,000
Depreciation	48,000	54,000

If both concerns do business in the state of No-Taxes, compute the federal income tax for the year, using the federal normal and surtax rates.

P15.3   Compute the taxes for the situation of problem P15.2 without the $25,000 deduction in taxable income (TI). What percent reduction in taxes is allowed by the decrease in TI?

P15.4   A-to-Z Car Dealers will have a $250,000 taxable income (TI) this year. If an advertising campaign is initiated, the TI is estimated to increase to $290,000 for the year. Neglecting any state and local taxes, using the federal normal and surtax rates, compute the

(a) effective federal tax rate on TI = $250,000

(b) effective federal tax rate on only the additional taxable income

(c) effective federal tax rate on the entire $290,000 taxable income

(d) after-tax profit on the additional taxable income

P15.5   A company has a gross income of $3.9 million for the year. Depreciation and all expenses amount to $2.45 million. If the combined state and local tax rate amounts to 6.5% and an effective federal rate of 48% is applicable, compute the income taxes using the effective tax rate formula.

P15.6   Wallbanger Contractors reported a taxable income of $80,000 last year. If the state tax rate is 8%, compute the following:

(a) federal effective tax rate

(b) overall effective tax rate

(c) total taxes to be paid by the company

P15.7 The taxable income for a small partnership business is $150,000 this year. An effective tax rate of 46% is used by the owners. Investment in some new machinery was considered in the first quarter of the year. The equipment would have cost $35,000, have a life of five years, be salvaged for an estimated $5,000, and be written off by the sum-of-year-digits method. The purchase would have increased taxable income by $10,000 and expenses $1,000 for the year. Compute the change in income taxes for the year if the purchase had been made.

P15.8 The following capital gains and losses are present this year for a small clothing manufacturer.

$$LTG = \$2{,}800$$
$$LTL = 500$$
$$STL = 2{,}000$$

If TI = $80,000, compute the income taxes the company must pay if state taxes are 5% and regular federal rates apply.

P15.9 Compute the capital gains and losses for all the asset transactions below and use them to compute the annual income taxes. Total sales for the year were $80,000, while expenses and accumulated depreciation amounted to $39,400.

(*a*) Sold a three-year-old straight-line-depreciated asset, which had a first cost (P) of $50,000, a salvage value (SV) of $0.1P$, and a life of nine years, for $0.68P$.

(*b*) A machine that was only five months old was replaced due to extreme technological obsolescence. The asset had $P = \$10,000$, $SV = \$1,000$, $n = 4$ years, and was depreciated by the sum-of-year-digits method. The trade-in deal allowed the company $8,000 on a new machine. (Use 50% of annual depreciation for the five-month period.)

(*c*) Land purchased four months ago for $8,000 was sold for a 10% profit.

(*d*) A 23-year-old asset was sold for $500. When purchased the asset was entered on the books with $P = \$18,000$, $SV = \$200$, $n = 20$ years. Straight-line depreciation was used for the life of the machine.

(*e*) A capital loss of $22,000 incurred four years ago is not completely exhausted. A total of $3,500 remains on the books. The company's tax specialists want to remove one-half of this loss from the accounts this year.

P15.10 Referring back to problem P15.7, assume that the asset purchase had been made, but in December, due to lagging sales, the equipment had to be sacrificed for $28,000. Compute the change in income taxes for the year as a result to this asset sale.

P15.11 Four similar assets were purchased seven years ago (1972) and written off by the group method (Sec. 7.7) using straight-line depreciation. Details of the asset purchases are given below:

Asset number	First cost	Expected life, years	Salvage value
108	$25,000	5	$3,000
109	25,000	5	3,000
110	25,000	5	3,000
111	25,000	5	3,000

Details of asset disposal and yearly taxable income *before depreciation* are as follows:

Year	Asset sold	Actual salvage	Taxable income
1972	–	–	$170,000
1973	–	–	190,000
1974	109	$15,000	175,000
1975	108	16,500	210,000
1976	–	–	130,000
1977	110	1,300	90,000
1978	111	3,500	150,000

If the company uses an effective tax rate of 50%, compare the total taxes for the years 1972 through 1978 under these conditions: (*a*) assets were group depreciated as explained; (*b*) each asset was individually straight line depreciated and gains or losses accounted for in the applicable year.

P15.12 Rework problem P15.7 assuming the purchase is classified as section 38 property.

P15.13 Use the assets described in Example 15.5 (p. 260) to determine the first year's income taxes, if taxable income before removal of depreciation is $250,000 and an effective federal tax rate of 48% and a state tax rate of 10% are used in economy studies. Assume double declining-balance depreciation is used for both assets.

P15.14 Determine the investment tax credit for the following assets:

	First cost	Expected life, years	Salvage value
(*a*)	$10,000	10	$ 750
(*b*)	5,500	4	0
(*c*)	7,000	2	1,000
(*d*)	9,500	3	3,000
(*e*)	1,800	6	200

P15.15 An operating loss of $25,000 was incurred by the Bozo Health Spa in 1976. The effective tax rate is 50% and taxable income from 1973 to 1975 has been $110,000, $90,000, and $50,000, respectively. If the operating loss will be carried back only, compute the present worth in the year 1973 of the resulting taxes paid in 1973, 1974, and 1975 for the seven plans specified below ($i = 10\%$):

(a–c) Recover the entire loss in 1973, 1974, or 1975.

(d–f) Recover one-half the loss in each of two of the years 1973, 1974, or 1975.

(g) Recover one-third the loss in each of the years 1973, 1974, and 1975.

P15.16 Assume that the Wa-Out Fruit Company of Example 15.7 (p. 264) purchased the Wonderama sorter, cleaner, and packer machine. If annual income is expected to be $100,000 from this new asset and the effective tax rate is 48%, determine the percentage difference in taxes obtained by using (a) sum-of-year-digits and straight-line depreciation and (b) declining balance with $d = 0.15$ and sum-of-year-digits methods. Take the time value of money into account by using a rate of return of 10%. Constructing the tax cash flow diagram for each method will be helpful.

P15.17 Reconsider Example 7.11 (p. 123). Table 7.2 indicates that a switch from the declining balance ($d = 0.188$) to straight line is advised in year 5. Compare the present worth of taxes of the declining balance and straight-line methods, assuming the switch is made. Use an interest rate of 8%, a CFBT of $20,000 per year, and $T = 0.50$.

P15.18 What is the difference in the present worth of total tax paid for the following situation? Asset A can be purchased to produce a CFBT of $65,000 or five of Asset B can be purchased to produce the same CFBT. The interest rate is 12% and sum-of-year-digits (SYD) or double declining-balance (DDB) depreciation is used, as indicated below. Neglect any capital gain or loss at sale time.

	Asset A	Five of asset B
Total first cost	$250,000	$260,000
Total salvage value	25,000	25,000
Total annual CFBT	65,000	65,000
Depreciation method	DDB	SYD
Tax rate	50%	50%
Life, years	8	8

P15.19 Determine the salvage value that must be used for straight-line depreciation in Example 15.6 (p. 262) to make the present worth of taxes equal to the present worth of taxes for the sum-of-year-digits method, where SV = $10,000.

P15.20 Once again, review the situation discussed in Example 15.6. Compute the n value for the straight-line method that would make its present worth of taxes equal to that for the sum-of-year-digits methods, where $n = 9$ years.

P15.21 The Whiz Bang Computer Software Company has just bought an asset that has $P = $88,000$, SV = $8,000$, and $n = 8$ years. For tax purposes, the company is allowed to (a) use a life of six or seven years and depreciate by the straight-line or (b) use a life of seven years and depreciate by the double declining-balance method. If the interest rate is 10%, expected CFBT is $25,000 per year, and effective tax rate is 52%, determine what n value and depreciation method should be used to minimize the time value of taxes under the following condition: all operating losses and capital losses are treated as a tax advantage (Example 15.8) and capital gains are taxed as prescribed by law.

P15.22   Review Example 15.9 (p. 266). If the IRS will allow the use of a life of five through ten years for tax purposes, determine what n value is the most advantageous for tax purposes. Assume that if the asset is retired after ten years of service, it will be done by realizing a $1,000 capital loss, which can be used as a tax advantage in the sale year. Further assume that CFBT in year 10 would be $3,000, as in previous years.

P15.23   Compute the before-tax rate of return for problem P15.5 if an after-tax rate of return of 8% is required.

P15.24   Compare the plans in problem P15.18 by the EUAC method with the (a) after-tax return and (b) computed inflated before-tax return. (c) Compare the answers using the after-tax and before-tax returns. (Hint: CFBT − taxes = cash flow after taxes.)

P15.25   If a company uses a before-tax rate of return of 19% and an after-tax return of 8%, what percent of income is assumed to be absorbed in taxes?

16

AFTER–TAX ECONOMIC ANALYSIS

The objective of this chapter is to study in more detail the effects of taxes on an engineering-economy study. Rather than attempt to utilize an inflated before-tax rate of return in this chapter, sufficient detail of the income tax effect is considered so that the rate of return reflects the situation after taxes.

CRITERIA

To complete this chapter you must be able to do the following:

1 Compute and tabulate cash flow after taxes, given the cash flow before taxes, effective tax rate, and depreciation method.

2 Select the better of two plans using present-worth or EUAC analysis, given details of the plans, the tax rate, and the after-tax rate of return.

3 Compute the after-tax rate of return for one asset or the breakeven rate of return for two competing assets, given details of the asset(s), the depreciation method, and effective tax rate.

4 Select between a challenger and defender using after-tax replacement analysis, given the challenger and defender plans, the defender market value, depreciation methods, before- and after-tax rates of return, and the effective tax rate.

5 Compute the tax effect on economy studies of the tax laws dealing with depletion, given the applicable depletion laws and details of the plans.

EXPLANATION OF MATERIAL

16.1 Tabulation of Cash Flow After Taxes

It is vital that you be able to correctly tabulate cash flow after tax (CFAT) so that the present-worth, EUAC, or rate-of-return computations reflect the correct after-tax situation. The formulas to be used are no different than those used previously; however, they are reviewed here for easy reference. We will consistently use the abbreviations CFBT and CFAT for "cash flow before taxes" and "cash flow after taxes," respectively, to conserve space. Continued use of the abbreviations TI for "taxable income" and T for "tax rate" is made.

$$CFBT = \text{gross income} - \text{disbursements} \qquad (16.1)$$

$$TI = CFBT - \text{depreciation} \qquad (16.2)$$

$$\text{Taxes} = TI(T) \qquad (16.3)$$

$$CFAT = CFBT - \text{taxes} \qquad (16.4)$$

If the numerical value of Eq. (16.3) is negative, it will be assumed that this negative tax will affect other taxes for the same year attributable to other income-producing assets owned by the company. This simplifying procedure is used in lieu of the carry back–carry forward tax laws (Sec. 15.5), which may be more advantageous. This assumption is illustrated in Example 16.2.

Example 16.1 A proposal has been made that a new piece of equipment be purchased this year. Characteristics of the purchase plan are given at the top of the following page.

Table 16.1 TABULATION OF CASH FLOW AFTER TAXES
(CFAT) FOR EXAMPLE 16.1

Year	Income (1)	Disbursements (2)	CFBT (3) = (1) − (2)	Depreciation* (4)
0	–	$50,000	$−50,000	–
1	$27,000	10,000	+17,000	$10,000
2	26,000	10,500	+15,500	10,000
3	25,000	11,000	+14,000	10,000
4	24,000	11,500	+12,500	10,000
5	23,000	12,000	+11,000	10,000

*Depreciation = 50,000/5 = $10,000 per year.

$$P = \$50,000$$
$$SV = 0$$
$$n = 5 \text{ years}$$
$$\text{Expected income} = 28,000 - 1,000k \quad (k = 1,2,3,4,5)$$
$$\text{Expected disbursements} = 9,500 + 500k$$

If the effective tax rate is 40% on this type of asset and if straight-line depreciation is used, tabulate the cash flow after taxes (CFAT).

SOLUTION Table 16.1 details all tax information and CFAT for the asset, using Eqs. (16.1)–(16.4).

COMMENT If some portion of the $50,000 investment were borrowed from company-external sources (debt financing), a tax credit is offered in that the interest is tax deductible. See Solved Examples for an illustration. ////

Example 16.2 A capital asset qualified as section 38 property is purchased. When the purchase is made, the following data apply:

$$P = \$32,000$$
$$SV = 3,100$$
$$CFBT = 8,000 \text{ per year}$$
$$n = 6 \text{ years}$$
$$\text{Tax rate} = 52\%$$

However, the IRS requires that this type of asset be straight-line depreciated over ten years with a maximum salvage value of $2,000. Therefore, even though the company anticipates the asset will have a useful life of six years, it must be depreciated over ten years. The company will utilize a CFBT = $1,000 for years 7–10, when the asset will be in a standby condition. If the company anticipates a salvage value of $3,100 after ten years of use, tabulate the CFAT for the ten years.

Year	TI $(5) = (3) - (4)$	Taxes $(6) = 0.4(5)$	CFAT $(7) = (3) - (6)$
0	–	–	$–50,000
1	$7,000	$2,800	+14,200
2	5,500	2,200	+13,300
3	4,000	1,600	+12,400
4	2,500	1,000	+11,500
5	1,000	400	+10,600

Table 16.2 TABULATION OF CFAT FOR SECTION 38 PROPERTY WITH AN LTG

Year	CFBT	Depreciation*	TI	Taxes	CFAT
0	$-32,000	–	–	$-2,240†	$-29,760
1-6	+8,000	$3,000	$5,000	+2,600	+5,400
7-10	+1,000	3,000	-2,000	-1,040	+2,040
10	+3,100	–	–	+330‡	+2,770

*Depreciation = (32,000 − 2,000)/10 = $3,000 per year.
†Investment tax credit for ten-year life asset = 0.07(32,000) = $2,240.
‡Income taxes = LTG tax = 0.30(3,100 − 2,000) = $330.

SOLUTION Table 16.2 details the CFAT. Note the 7% investment tax credit (Sec. 15.4) and $1,100 long-term gain (Sec. 15.1), tax computations for which are detailed below in the table footnotes.   ////

Example 16.8
Problems P16.1–P16.6

16.2 After-Tax Analysis Using Present-Worth or EUAC Analysis

If the minimum after-tax rate of return is stated or known, present-worth or EUAC analysis can be used to select the most economic plan, in a fashion similar to that of Chaps. 8 and 9.

Example 16.3 Using a 7% after-tax return, select the more economic plan of those detailed in Examples 16.1 and 16.2, using (a) EUAC and (b) present-worth analysis. (The plans are summarized below for convenience.)

Plan A (Example 16.1)	Plan B (Example 16.2)
$P = \$50,000$	$P = \$32,000$
$SV = \quad 0$	$SV = \quad 2,000$ (IRS)
$n = 5$ years	$n = 10$ years (depreciable)
CFAT in Table 16.1	CFAT in Table 16.2

SOLUTION

(a) EUAC equations can be set up and solved at $i = 7\%$ as follows:

$$\begin{aligned}
\text{EUAC}_A = & [-50{,}000 + 14{,}200(P/F,7\%,1) \\
& + 13{,}300(P/F,7\%,2) + 12{,}400(P/F,7\%,3) \\
& + 11{,}500(P/F,7\%,4) \\
& + 10{,}600(P/F,7\%,5)](A/P,7\%,5) \\
= & \ \$327
\end{aligned}$$

(16.5)

$$\begin{aligned}
\text{EUAC}_B &= [-29{,}760 + 5{,}400\,(P/A,7\%,6) \\
&\quad + 2{,}040\,(P/A,7\%,4)(P/F,7\%,6) \\
&\quad + 2{,}770\,(P/F,7\%,10)]\,(A/P,7\%,10) \\
&= \$284 \qquad\qquad\qquad\qquad\qquad\qquad\qquad (16.6)
\end{aligned}$$

Plan A is selected, since EUAC values are positive (profit) and EUAC_A is larger. (b) Present-worth analysis would be based on a 30-year horizon to equalize the anticipated lives. Using the EUAC values above,

$$\text{PW}_A = \text{EUAC}_A(P/A,7\%,30) = \$4{,}058$$
$$\text{PW}_B = \text{EUAC}_B(P/A,7\%,30) = \$3{,}524$$

Again, Plan A is selected, because PW_A is larger.

COMMENT If only disbursement values before taxes, such as annual operating costs, are known, that is, CFBT < 0, the related taxes are a tax advantage to be applied against other interests of the company. ////

Problems P16.7–P16.14

16.3 Computation of After-Tax Rate of Return

The after-tax rate of return is usually about one-half the before-tax return. To determine the after-tax return, one of the two methods applied in Chap. 10 may be used:

1 Find the rate at which the *present worth* of CFAT equals zero.

$$0 = \sum_{j=1}^{n} \text{CFAT}_j(P/F,i\%,j)$$

2 Find the rate at which *EUAC* of CFAT equals zero.

$$0 = \left[\sum_{j=1}^{n} \text{CFAT}_j(P/F,i\%,j) \right] (A/P,i\%,n)$$

If two assets, A and B, are involved, the return is found using one of these methods; however, the equations take on these respective forms (as in Chap. 10):

$$0 = P_B - B_A \quad \text{and} \quad 0 = \text{EUAC}_B - \text{EUAC}_A$$

The second method (EUAC) will be used exclusively if two assets are involved, because it is compatible with the conventions used in earlier chapters and is usually computationally simpler for unequal-life assets. Unfortunately, as you will remember, this and all methods of rate-of-return calculation involve the use of trial and error. If similar computations are frequently made, a computer program is most advantageous. You will find such a program easy to write in FORTRAN, or you can refer to one of

Table 16.3 CASH FLOW AFTER TAXES USING STRAIGHT–LINE DEPRECIATION FOR EXAMPLE 16.4

Year	CFBT	Depreciation	TI	Taxes	CFAT
0	$-100,000	–	–	–	$-100,000
1-9	+20,000	$10,000	$10,000	$5,000	+15,000
9	+10,000	–	–	–	+10,000

the routines already published.[1] For illustration, the following two examples on after-tax return are presented. A more advanced illustration is included in Solved Examples.

Example 16.4 Using the asset purchase described in Example 15.6 and straight-line depreciation, compute the after-tax rate of return. (Summary: $P = \$100,000$, SV $= \$10,000$, $n = 9$, CFBT $= \$20,000$ per year, tax rate $= 50\%$.)

SOLUTION Table 16.3 presents the CFAT for the asset. The present-worth equation for the after-tax return is as follows:

$$0 = -100,000 + 15,000(P/A,i\%,9) + 10,000(P/F,i\%,9)$$

By trial and error, $i = 7.70\%$.

COMMENT If you, as an economist, want to use an inflated before-tax rate to approximate the tax effect on this type of asset, you could use Eq. (15.6) to obtain $i/(1 - T) = 0.0770/(1 - 0.50) = 0.1540$, or 15.4%. The actual before-tax return computed using the CFBT figures at Table 16.3 can be found from the equation

$$0 = -100,000 + 20,000(P/A,i\%,9) + 10,000(P/F,i\%,9)$$

which gives a value of $i = 14.56\%$. Comparison of 14.56% with the inflated before-tax rate of 15.4% shows that the tax effect is slightly overestimated by using a 15.4% before-tax return. ////

Example 16.5 If the minimum after-tax return for any plan is 7%, select one of the alternatives detailed in Example 16.3 by finding the breakeven rate of return.

SOLUTION Using EUAC analysis find, by trial and error (or via a computer program), the interest rate at which $0 = \text{EUAC}_B - \text{EUAC}_A$. Equations (16.5) and (16.6) can be used here, except that the breakeven i value is sought. Solution indicates that EUAC values are equal at $i = 7.31\%$. Since 7.31% > 7%, either project is justified. Figure 16.1 shows the EUAC values at 5% through 8% using Eqs. (16.5) and (16.6). Since at 7%, $\text{EUAC}_B < \text{EUAC}_A$, Plan B is selected as more economic in an after-tax consideration. ////

Example 16.9
Problems P16.15–P16.27

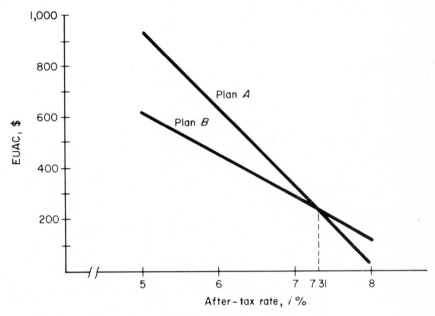

FIG. 16.1 Breakeven chart for after-tax analysis, Example 16.5.

16.4 After-Tax Replacement Analysis

When a defending asset is challenged by a new asset, the effects of income taxes may be considerable. To account for all the tax details in *replacement analysis* is sometimes neither time- nor cost-effective; however, it is worthwhile to account for a capital gain or loss, which would occur if the defender were replaced. Also important is the future tax advantage stemming from deductible operating and depreciation expenses. The following example will give you an idea of the impact of taxes on replacement analysis.

Example 16.6 Three years ago Thurston Mining purchased certain mining equipment. Due to the inadequacy of the machinery, a new piece of equipment is being considered. The defender and challenger characteristics are

	Defender	Challenger
$P =$	$6,000	$12,000
SV =	0	2,000
AOC =	700	150
	Original $n = 8$ years	$n = 5$ years

If a $4,000 trade-in is offered for the defender, perform (*a*) the before-tax and (*b*) the after-tax analysis, using a 15% return befores taxes and a 7% after-tax return. Tax rates are 52% on income and 30% on gains (LTG), and straight-line depreciation is applied to both assets.

SOLUTION

(a) The before-tax analysis will use $P = \$4,000$ and a five-year remaining life for the defender. The EUAC analysis appears as follows (EUAC$_D$ for the defender), Eq. (9.1):

$$EUAC_D = 4,000(A/P,15\%,5) + 700 = \$1,893$$
$$EUAC_C = 12,000(A/P,15\%,5) - 2,000(A/F,15\%,5) + 150 = \$3,433$$

The defender is retained by the large EUAC margin of $1,540. It actually seems pointless to consider the new asset, even after taxes, due to this large defender advantage; nevertheless, consider the analysis below.

(b) After-tax analysis for the *defender* proceeds as follows:

$$\text{Depreciation} = \frac{6,000}{8} = \$750 \text{ per year}$$
$$\text{Present book value} = 6,000 - 3(750) = \$3,750$$
$$\text{LTG on trade-in} = 4,000 - 3,750 = \$250$$
$$\text{LTG tax} = 0.3(250) = \$75$$
$$\text{Actual } P \text{ value} = 4,000 - 75 = \$3,925$$
$$\text{Tax saving} = (AOC + \text{depreciation})(0.52)$$
$$= (700 + 750)(0.52)$$
$$= \$754 \text{ per year}$$
$$EUAC_D = 3,925\,(A/P,7\%,5) + 700 - 754$$
$$= \$903$$

For the *challenger*, after-tax analysis is

$$\text{Depreciation} = \frac{12,000 - 2,000}{5} = \$2,000 \text{ per year}$$
$$\text{Tax saving} = (150 + 2,000)(0.52) = \$1,118 \text{ per year}$$
$$EUAC_C = 12,000\,(A/P,7\%,5) - 2,000(A/F,7\%,5)$$
$$+ 150 - 1,118$$
$$= \$1,611$$

Select the defender with an advantage of $708 per year. Even though before- and after-tax decisions are the same, the defender advantage after taxes is only 46% of the before-tax advantage. Since the adequacy of the defender is in question, this intangible factor may be enough to cause management to replace.

////

Problems P16.28–P16.31

16.5 Tax Effect of Depletion Laws

As was shown in Sec. 7.9, depletion is a vehicle similar to depreciation; that is, it is used to write off the initial investment of natural-resource recovery methods. Recall that there are two methods of depletion:

1 Factor depletion method

$$\text{Annual depletion} = \frac{\text{initial investment}}{\text{resource capacity}} \times \text{annual volume}$$

2 Depletion allowance, where a percentage of GI (gross income) is tax deductible, provided the result does not exceed 50% of TI (taxable income) prior to deduction of this depletion allowance.

The question of favoritism toward oil, gas, and many metal producers (those in the 22% allowance bracket) is always present, since they are allowed to deplete the property much beyond the original investment. Either of the two methods reviewed above may be used, but naturally the larger deduction is always used in tax computations. To illustrate the impact of the 22% depletion allowance, Example 16.7 compares identical initial investments and corresponding write-off using depreciation and depletion.

Example 16.7 A $240,000 investment has been made. The following income and expenses are realized for the 12-year life:

Year	Gross income	Expenses
1–10	$150,000	$50,000
11	75,000	20,000
12	25,000	5,000

The effective tax rate is 50%. Compute and compare the after-tax rate of return for the two situations below.

(*a*) The $240,000 is used to purchase an asset which is to be straight-line depreciation and has no salvage value.
(*b*) The $240,000 is used to develop a mine subject to a 22% depletion allowance. The ore is sold for $10 a ton, and 15,000 tons are sold for each of the years 1–10; 7,500 tons are sold in year 11, and 2,500 tons in year 12. Total expected tonnage in the mine was 160,000 tons at the time of purchase.

Table 16.4 EFFECT OF *DEPRECIATION* ON AFTER-TAX RATE OF RETURN* FOR EXAMPLE 16.7(*a*)

Year	GI	Expenses	CFBT	Depreciation†	TI	Taxes	CFAT
0	–	$240,000	$–240,000	–	–	–	$–240,000
1–10	$150,000	50,000	+100,000	$20,000	$80,000	$40,000	+60,000
11	75,000	20,000	+55,000	20,000	35,000	17,500	+37,500
12	25,000	5,000	+20,000	20,000	0	0	+20,000

*Rate-of-return computation: $-240,000 + 60,000(P/A,i\%,10) + 37,500(P/F,i\%,11) + 20,000$
$(P/F,i\%,12) = 0$ $i = 22.35\%$.
†Depreciation $= 240,000/12 = \$20,000$ per year.

SOLUTION

(*a*) For the asset, Table 16.4 details the after-tax situation. The after-tax return is 22.35%.

(*b*) Table 16.5 presents the tax situation using the depletion allowance or factor depletion, as allowed. The depletion factor is $240,000/160,000 = \$1.50$ per ton. Note that the depletion allowance is allowable for all 12 years, since it never exceeds 0.5(TI). To find the after-tax return the present worth of CFAT is equated to zero; but notice that a total depletion of $352,000 (Table 16.5), or 1.47 times the first cost, is claimed. The rate of return is 25.21%, or approximately 3% higher than the return for a depreciable asset with the same gross income, expenses, and life. Thus, the advantage of depletion is obvious—a larger after-tax return for natural-resource concerns.   ////

Example 16.10
Problems P16.32, P16.33

Table 16.5 EFFECT OF *DEPLETION* ON AFTER-TAX RATE OF RETURN* FOR EXAMPLE 16.7(*b*)

Year	Tons sold (1)	GI (2)	Expenses (3)	TI before depletion† (4) = (2) – (3)	0.5TI (5) = 0.5(4)
0	–	–	$240,000	$–240,000	–
1–10	15,000	$150,000	50,000	+100,000	$50,000
11	7,500	75,000	20,000	+55,000	27,500
12	2,500	25,000	5,000	+20,000	10,000

*Rate-of-return computation: $-240,000 + 66,500(P/A,i\%,10) + 35,750(P/F,i\%,11) + 12,750(P/F,i\%,12) = 0$ $i = 25.21\%$.
†Also the CFBT.

SOLVED EXAMPLES

Example 16.8 The Gutsy Cleaning Company plans to invest in a new dry cleaner. Details of the investment are:

$$P = \$15{,}000$$
$$SV = \quad\quad 0$$
$$\text{Income} = \quad 7{,}000 \text{ per year}$$
$$\text{Expenses} = \quad 1{,}000 \text{ per year}$$
$$\text{Tax rate} = 50\%$$
$$n = 5 \text{ years}$$

If straight-line depreciation is used, tabulate CFAT for the following conditions: (*a*) all $15,000 is from company funds (100% equity financing), and (*b*) one-half the investment is borrowed from a bank (50% equity − 50% debt financing) at 10% interest. Assume the 10% is simple interest on the total amount borrowed and repayment will be in five equal payments.

SOLUTION

(*a*) For 100% equity financing, Eq. (16.1) is used to obtain CFBT = $6,000 per year. Annual depreciation is $15,000/5 = $3,000. Table 16.6 details CFAT.

(*b*) The 50% debt financing requires that $7,500 be borrowed from outside the company or its stockholders. As in part (*a*), CFBT = $6,000. The loan repayment scheme will be as follows:

$$\text{Principal: } \frac{7{,}500}{5} = \$1{,}500 \text{ per year}$$

$$\text{Interest: } 7{,}500(0.10) = \$750 \text{ per year}$$

The $750 interest is tax deductible; however, the principal is *not deductible*. Therefore, the formulas for TI and CFAT, Eqs. (16.2) and (16.4), will be restated as

Year	Factor depletion (6) = 1.5(1)	Depletion allowance (7) = 0.22(2)	TI‡ (8)	Taxes (9) = 0.5(8)	CFAT (10) = (4) − (9)
0	—	—	—	—	$−240,000
1–10	$22,500	$ 33,000 §	$67,000	$33,500	+66,500
11	11,250	16,500 §	38,500	19,250	+35,750
12	3,750	5,500 §	14,500	7,250	+12,750
		$352,000			

‡TI = TI before depletion − depletion.
§Depletion claimed (note that the $33,000 is claimed for each of the first ten years).

Table 16.6 CFAT TABULATION FOR EXAMPLE 16.8(a)

Year	CFBT	Depreciation	TI	Taxes	CFAT
0	$-15,000	–	–	–	$-15,000
1-5	6,000	$3,000	$3,000	$1,500	4,500
	$ 15,000			$7,500	$ 7,500

$$\text{TI} = \text{CFBT} - \text{depreciation} - \text{interest} \qquad (16.7)$$

$$\text{CFAT} = \text{CFBT} - \text{taxes} - \text{interest} - \text{principal} \qquad (16.8)$$

Table 16.7 presents CFAT computation for 50% debt − 50% equity financing. You can see that the annual CFAT has decreased from $4,500 to $2,625 because of 50% debt financing. The $7,500 equity cash outflow in year zero is used because only 50% of the first cost comes from company funds while the remaining is from the lending bank.

COMMENT If only equity financing is involved, as in part (a), we can use the relation

$$\text{CFAT} = \text{depreciation} + \text{TI}(1 - \text{tax rate}) \qquad (16.9)$$

where $\text{TI}(1 - \text{tax rate})$ is the *profit*, that is, the portion of TI not absorbed by taxes. Thus, in Table 16.6,

$$\text{CFAT} = 3,000 + 3,000(1 - 0.50) = \$4,500$$

could be used in lieu of the method presented. But you must be careful because, if there is any debt financing whatsoever, as in part (b), this simple approach will not work. Let's try it on Table 16.7:

$$\text{CFAT} = 3,000 + 2,250(1 - 0.50) = \$4,125 \neq \$2,625$$

Table 16.7 CFAT TABULATION FOR EXAMPLE 16.8(b)

Year	CFBT (1)	Depreciation (2)	Interest (3)	Principal (4)	TI* (5)	Taxes (6)	CFAT† (7)
0	$-7,500	–	–	–	–	–	$-7,500
1-5	6,000	$3,000	$ 750	$1,500	$2,250	$1,125	+2,625
	$ 22,500		$3,750	$7,500		$5,625	$ 5,625

*In column notation, $(5) = (1) - (2) - (3)$, Eq. (16.7).
†$(7) = (1) - (6) - (3) - (4)$, Eq. (16.8).

Why doesn't it work? Simple! This method neglects the fact that interest is tax deductible *and* that CFAT is computed in a completely different way in the two cases. You might compare Eqs. (16.9) and (16.8) to verify this. Thus, we advise you to use the general formulas for CFAT tabulation, that is, Eqs. (16.1)–(16.4).   ////

Sec. 16.1

Example 16.9 Compute the after-tax rate of return for the two situations of Example 16.8: (*a*) 100% equity financing and (*b*) 50% equity − 50% debt financing. Compare the results!

SOLUTION

(*a*) For 100% equity financing, the CFAT values of Table 16.6 can be used to find *i* in the equation

$$0 = -15,000 + 4,500(P/A,i\%,5)$$
$$(P/A,i\%,5) = 3.3333$$

Hence, $i = 15.25\%$.

(*b*) For the 50% − 50% split on financing, Table 16.7 helps us set up the equation

$$0 = -7,500 + 2,625(P/A,i\%,5)$$
$$(P/A,i\%,5) = 2.8571$$
$$i = 22.22\%$$

Comparison of the two return values shows that debt financing increases the rate of return on company investments. Why? Because less of the firm's capital is tied up in investment.

COMMENT Why not use close to 100% debt financing and maximize the return? We will let you speculate about this dilemma until Sec. 18.7, where we answer this question. Please, venture a guess now! What if every time you wanted to make a purchase you had to borrow? How stable would your business be?   ////

Sec. 16.3

Example 16.10 Assume that the mine purchase discussed in Example 16.7 was 60% debt financed by a 12-year loan with interest computed at 3% per year on the original principal. Now, compute the after-tax rate of return.

SOLUTION For the 60% debt financing, the following data apply:

$$\text{Amount of loan} = 240,000(0.60) = \$144,000$$
$$\text{Principal payment} = \frac{144,000}{12} = \$12,000 \text{ per year}$$
$$\text{Interest} = 144,000(0.03) = \$4,320 \text{ per year}$$

Table 16.8   AFTER-TAX RETURN* WITH DEBT FINANCING FOR
EXAMPLE 16.10

Year	Principal payment	Interest payment	TI before depletion[†]	0.5TI	Factor depletion
0	–	–	$–96,000	–	–
1–10	$12,000	$4,320	95,680	$47,840	$22,500
11	12,000	4,320	50,680	25,340	11,250
12	12,000	4,320	15,680	7,840	3,750

*Rate-of-return computation: $-96,000 + 48,020(P/A,i\%,10) + 17,270$
$(P/F,i\%,11) - 5,730(P/F,i\%,12) = 0$ $i = 49.25\%$.
[†]TI before depletion = GI − expenses − interest(see Table 16.5).

Table 16.8 details the CFAT values. Note the outflow of $96,000 in year zero, not $240,000, since only 40% of the mine's first cost is due to equity funds. The factor and allowance values for depletion are taken from Table 16.5. The present-worth computation under Table 16.8 indicates that $i = 49.25\%$, which is 1.95 times the return for depletion with 100% equity funds. Again, we see the dramatic advantage of borrowed funds.   ////

Secs. 16.1, 16.5

REFERENCE

1. E. P. DeGarmo and J. R. Canada, "Engineering Economy," 5th ed., pp. 509–511, Macmillan Publishing Company, Inc., New York, 1973.

BIBLIOGRAPHY

DeGarmo and Canada, pp. 363–372.
Fabrycky and Thuesen, pp. 169, 178–180.
Grant and Ireson, pp. 337–348, 373, 392–394.
Smith, pp. 236–257.
Taylor, pp. 347–362.
Thuesen, Fabrycky, and Thuesen, pp. 288–294.

PROBLEMS

P16.1   An investment company plans to purchase an apartment complex for $350,000. Annual income before taxes of $28,000 is expected for the next eight years, after which the property will be sold for an estimated $420,000. The applicable tax rate is 52%, the estimated annual operating cost is $3,000,

Year	Depletion allowance	TI	Taxes	CFAT[‡]
0	–	–	–	$–96,000
1–10	$33,000 [§]	$62,680	$31,340	48,020
11	16,500 [§]	34,180	17,090	17,270
12	5,500 [§]	10,180	5,090	–5,730

[‡]CFAT = TI before depletion − taxes − principal − interest.
[§]Depletion claimed.

and the gain on property sale is taxed at 30%. Tabulate the cash flow after taxes for the years of ownership, if the property will be straight-line depreciated over a 20-year life with a 40% salvage value.

P16.2   Tabulate the after-tax cash flows for the Wonderama machine described in Example 15.7 (p. 264) if it is considered a section 38 property asset, will produce a CFBT of $50,000 annually, and is actually salvaged for $175,000 after six years.

P16.3   Rework problem P16.2 assuming that 40% of the first cost of Wonderama is borrowed at 12% "simple" interest on the declining balance to be repaid in six equal installments of $30,767 per year of which $9,100 is annual interest. (Note: The repayment scheme utilized here is one commonly used in practice, but it has not been discussed in this book.)

P16.4   An asset costing $10,000 has been owned by the Renuzit Plumbers for three years. Sum-of-year-digits depreciation with a salvage of $2,000 and life of four years has been used thus far. The company would like to switch to straight-line depreciation for the last year. Besides, management expects to retain the machine for three more years with the same annual CFBT ($5,000) as the last three years. (a) Tabulate the cash flow after taxes for an effective tax rate of 48%. (b) What net difference in total cash flow would have been realized if Renuzit had switched to straight line as soon as the law allowed?

P16.5   (a) Rework problem P16.2 using double declining-balance depreciation. (b) Which of the two depreciation methods should be used to obtain a larger net total CFAT value, if no switch will be made? Assume any operating loss is simply a tax savings for the company in the year of occurrence.

P16.6   Rework problem P16.4 assuming the purchase was 50% financed by a $5,000 loan with yearly interest at 3% on the original principal and repayment in five equal annual payments of $1,150 each.

P16.7   Revise problem P8.1 (p. 142) as follows: Both machines will be depreciated by the straight-line method, the tax rate is 50%, and an after-tax return of 7% is required. Select the more economic of the machines using after-tax analysis. Assume that annual maintenance and operating costs are a tax advantage.

P16.8   Rework problem P9.7 (p. 158) if the new machine were depreciated by the straight-line method and the used machine by the sum-of-year-digits method. Assume both machines are section 38 property purchases, that the effective tax rate is 50%, and an after-tax return of 5% is required.

P16.9   Update problem P9.6 (p. 158) as follows: Regardless of which machine is selected, a $10,000 loan will be necessary for the purchase. Repayment of this loan will be in five equal annual installments of $2,700 each ($2,000 principal, $700 interest). If the food-processing company is in the 52% tax bracket, uses straight-line depreciation on all assets, and an after-tax MARR of 6% is required, determine which labeling machine is more economic.

P16.10   The Siko-so-matic Drug Company, a subsidiary of Pure-Not Nature Food Manufacturers, has to decide between the two pill-forming machines detailed below.

	Roundee	Scored
First cost	$24,000	$15,000
Salvage value	6,000	3,000
Annual CFBT	4,000	2,000
Life, years	12	12

The machines have an anticipated useful life of 12 years as detailed above, but the IRS will allow straight-line depreciation over ten years with a zero salvage value. If an effective tax of 50% applies and an after-tax return of 10% is desired, compare the two machines using (*a*) present-worth analysis and (*b*) EUAC analysis.

P16.11   If in problem P16.10, Siko-so-matic had used the anticipated life and salvage values, would the decision have been the same?

P16.12   Select the more economic of the two alternatives detailed in Example 7.6 (p. 118) if the after-tax MARR is 8%.

P16.13   Rework problem P16.12 if one-half the first cost of each asset will be debt-financed with ten equal payments of $616.45 for A and $924.68 for B. For each loan, one-tenth of the principal amount is removed each year with the remainder of the payment necessary for interest.

P16.14   Compare the two plans below using an after-tax MARR of 10% and present-worth analysis.

	Plan A		Plan B
	Machine 1	Machine 2	
First cost	$5,000	$25,000	$40,000
Salvage value	0	1,000	5,000
Annual savings			
Years 1–4	500	10,000	15,000

	Plan A		Plan B
	Machine 1	Machine 2	
Annual savings			
Years 5–8	$ 500	$10,000	$20,000
Years 9–12	500	5,000	25,000
Tax rate	48%	48%	48%
Tax life, years	4	12	12
Expected life, years	6	12	12

Assume straight-line depreciation for all assets and use the minimum allowed life for each asset. Further, assume any operating loss will simply be a tax savings applicable to other taxable income for the year incurred.

P16.15 Compute the after-tax rate of return for problem P16.1.

P16.16 Compute the after-tax rate of return for problems (a) P16.2 and (b) P16.3. (c) Explain the difference in your answers.

P16.17 Determine the difference in the after-tax rate of return made by the two situations presented in problem P16.4.

P16.18 What is the after-tax rate of return for problem P16.6 where partial debt financing is involved? What difference does this financing make in the after-tax return? (See problem P16.17.)

P16.19 (a) Compute the breakeven after-tax rate of return for problem P16.7. (b) Plot the present-worth values and select the better plan if the after-tax MARR is 4%, (c) 6%, (d) 7%, and (e) 10%.

P16.20 Compute the after-tax return for each machine in problem P16.7, if gross income is $5,000 per year.

P16.21 Determine the after-tax return at which the machines of problem P16.9 are economically indifferent.

P16.22 At what annual value of depreciation will the after-tax return be (a) 5% and (b) 10% for the Wonderama machine of problem P16.2?

P16.23 (a) Set up an after-tax equation of the form $PW_B - PW_A = 0$ and (b) determine the breakeven after-tax return for problem P16.14.

P16.24 Determine the after-tax return for each alternative of problem P16.14.

P16.25 Consider the problem solved in Example 16.4 (p. 278). Assume the asset's owner is interested in having an after-tax return of 10%. If the tax rate remains at 50%, compute the value of the (a) first cost, (b) salvage value, and (c) annual depreciation at which this will occur. When determining any one of the values above, assume the remaining parameters retain the value detailed in the example.

P16.26 Compute the after-tax return for the following situation. The owners of Slightly Tipsy Underwater Diving Services (STUDS) wish to make an investment. They can purchase special diving equipment to handle a job. The equipment will cost $2,500, have a life of five years, and have no salvage value. They will receive $1,500 in year one, but believe they can make $300 each year using the equipment in the future. The diving equipment will be straight-line depreciated and the tax rate is 45%.

P16.27 If the equipment discussed in problem P16.26 is not purchased, the STUDS can pay another firm—DROWN—to do the job. If the STUDS can make 5% after taxes on its money, what percent of the $1,500 fee must they have to make the same return possible as if they purchase the equipment and retain it for the full five-year life?

P16.28 Rework Example 16.6 under the assumptions that the trade-in value of the defender is only $2,000 and that new remaining life and salvage value estimates are ten years and $750, respectively.

P16.29 Perform an after-tax analysis for problem P12.13 (p. 220) using a tax rate of 50% and an after-tax return of 4%. Use straight-line depreciation for all assets and assume that Asset A cost $20,000 when purchased and had an expected life of eight years.

P16.30 If the tax rate is 52%, what is (*a*) the breakeven rate of return between plans and (*b*) the replacement value of Asset A ($i = 4\%$) for problem P12.13 (p. 220)?

P16.31 (*a*) Compare the two plans detailed below using a tax rate of 48% and an after-tax return of 8%. (*b*) Is the decision different than the before-tax result? Use $i = 15\%$.

	Defender	Challenger
First cost	$28,000	$15,000
AOC when purchased	–	1,500
Actual AOC	1,200	–
Expected salvage when purchased	2,000	3,000
Trade-in value	18,000	–
Depreciation	SL	SL
Life, years	10	8
Years owned	2	–

P16.32 Rework problem P7.27 (p. 127) assuming that one-half the first cost of the mine is debt financed at 4% to be repaid in ten years with payments of $245,000 per year of which interest accounts for $70,000. Assume a tax rate of 50%.

P16.33 Compute the rates of return for the investments detailed below. Assume that Alternative A is depreciable by the straight-line method and that B is depletable using the larger of a 10% allowance, or factor of $1 per unit extracted.

	A	B
Investment	$ 50,000	$ 50,000
CFBT	15,000	15,000
Gross income	100,000	100,000
Tax rate	50%	50%
Life, years	10	10
Annual units	–	5,000

17

EVALUATION OF MULTIPLE ALTERNATIVES

This chapter has two objectives, both dealing with the management of capital. First, the chapter is designed to assist you in the selection of one alternative from several mutually exclusive alternatives. Second, a fundamental approach to the selection of one or more independent (non–mutually exclusive) alternatives, while not violating a stated budget, is studied; this is referred to as capital budgeting.

You should bear in mind throughout this chapter that the material is quite basic; it is not intended to make you an expert in capital management. While there are several wrong ways of approaching these problems, this chapter studies the generally accepted methods.

CRITERIA

To complete this chapter, you must be able to do the following:

1 Define the term *mutually exclusive alternative* and state the criteria used in selection of the best alternative when using the rate-of-return method of evaluation.
2 Select one alternative from several mutually exclusive alternatives using the incremental rate-of-return method, given initial cost, salvage value, life, and cash flows for each alternative, and the minimum attractive rate of return.

3 Same as *2*, except use incremental benefit/cost ratio analysis.
4 Define the capital-budgeting problem by stating three prominent characteristics.
5 Solve a capital-budgeting problem using the rate-of-return method, given the project investment, lives, cash flows, budget constraint, and a minimum attractive rate of return.
6 Same as *5*, except using a present-worth method.

EXPLANATION OF MATERIAL

17.1 Selection from Mutually Exclusive Alternatives

The primary difference between the material in this and that in preceding chapters is that here more than two alternatives are considered. This is referred to as *multiple-alternative evaluation*. Furthermore, when selection of one alternative precludes the acceptance of other alternatives, the alternatives are termed *mutually exclusive*. Thus, if a particular part is to be purchased and four vendors are available, only one vendor can be selected to supply the part.

As in any selection problem in engineering economics, there are several correct solution techniques. The present-worth and EUAC methods are the simplest and most straightforward techniques. Using a specified MARR, the total present-worth or EUAC value is computed for each mutually exclusive alternative. The alternative that has the most favorable present-worth or EUAC value is the one selected. The EUAC method is illustrated in the Solved Examples section (Example 17.6). Two other methods frequently used are rate-of-return and benefit/cost ratio. When the rate-of-return method is applied, the firm will require that the investment return at least the minimum attractive rate of return (MARR). When several alternatives return at least the MARR, at least the alternative requiring the lowest investment is justified. However, since more capital can be invested in other acceptable alternatives, the incremental investment required must also be justified. If the return on the extra investment exceeds the MARR, the entire investment should be made in order to maximize the return. Thus, for rate-of-return analysis, the following criteria are used to select one mutually exclusive project: select the one alternative that (1) requires the *largest investment* and (2) indicates that the *incremental investment over another acceptable alternative is justified*, in that the return is at least the MARR.

When benefit/cost analysis is used, incremental-investment justification is also necessary. The criteria are the same as that used for rate-of-return evaluation, except that the incremental B/C value must be greater than 1. These latter two methods are discussed in detail in the following two sections because they are more difficult to apply than the present-worth or EUAC methods.

Problems P17.1, P17.2

17.2 Selection Using Incremental Rate of Return

You will recall from Secs. 10.5 and 10.6 that the incremental-analysis procedure determines rate of return on the *extra investment* that is required by the plan having the higher investment cost. As discussed there, if the rate of return on the extra investment is greater than the MARR, the plan requiring the extra investment should be selected. This same procedure is followed when analyzing mutually exclusive alternatives, but now it becomes important to determine *which* alternatives must be compared with each other (and therefore, *which* increments will be involved). In this regard, the most important rule that must be remembered when evaluating alternatives by the incremental-investment rate-of-return method is that *an alternative can never be compared with one for which the incremental investment has not been justified.* The procedure to be used when evaluating multiple, mutually exclusive alternatives can conveniently be summarized as follows:

1 Rank the alternatives in terms of increasing *initial* investment.
2 Considering the "do nothing" alternative as a defender, compute the overall rate of return i for the alternative with the lowest initial investment.
3 If $i <$ MARR, remove the lowest investment alternative from further consideration and compute the overall rate of return for the next higher investment alternative. Repeat this step until $i \geqslant$ MARR for one of the alternatives. When $i \geqslant$ MARR, the lowest investment alternative becomes the defender and the next higher investment alternative is the challenger.
4 Determine the incremental costs and incomes between the challenger and the defender.
5 Calculate the rate of return on the incremental investment required in the challenger.
6 If the rate of return calculated (on the increment of investment) in step *5* is greater than the MARR, the challenger becomes the defender and the previous defender is removed from further consideration. Conversely, if the rate of return in step *5* is less than the MARR, the challenger is removed from further consideration and the defender remains as the defender against a new challenger.
7 Repeat steps *4-6* until only one alternative remains.

Note that in the incremental analysis (steps *4-6*), only *two* alternatives are compared at any one time. It is very important, therefore, that the correct alternatives be compared. Unless the procedure is followed as presented above, the wrong alternative can be selected from the incremental analysis. The procedure detailed above is illustrated in Examples 17.1 and 17.2.

Example 17.1 Four different building locations have been suggested, of which only one will be selected. Data for each site are detailed in Table 17.1. Annual CFAT varies due to different tax structures, labor costs, and transportation charges resulting in different annual receipts and disbursements. If the MARR is 10% after taxes, use incremental rate-of-return analysis to select a building location.

Table 17.1   FOUR ALTERNATE BUILDING LOCATIONS

	Location			
	A	B	C	D
Building cost	$—200,000	$—275,000	$—190,000	$—350,000
Annual CFAT	+22,000	+35,000	+19,500	+42,000
Life, years	30	30	30	30

SOLUTION   The steps outlined above result in the following procedure:

1 Order the alternatives according to increasing initial investment. This is done in the first line of Table 17.2.

2 The next step is to find the lowest investment alternative that has an overall rate of return of at least 10%. Table 17.2 indicates a rate of return of 9.63% for Location C, resulting in its elimination from further consideration. The next alternative, Location A, has an i of 10.49% and replaces "do nothing" as the defender.

3 The incremental investment between alternatives must now be considered. Since all locations have a 30-year life, the relation used to find the incremental i is

$$0 = \text{incremental cost} + \text{incremental CFAT } (P/A,i\%,30) \qquad (17.1)$$

where i is found by trial and error. Note that $(P/A,10\%,30) = 9.4269$; thus any P/A value resulting from Eq. (17.1) greater than 9.4269 indicates the return is less than 10% and, therefore, is unacceptable. Comparing B incrementally to Location A, using Eq. (17.1), results in the equation $0 = -75,000 + 13,000(P/A,i\%,30)$. A rate of return of 17.28% on the extra investment justifies Location B, thereby eliminating Location A.

4 With B as the defender and D the challenger, the incremental investment yields 8.55%, which is less than 10% and eliminates Location D. Only Locations A and B are justified and B is selected, since it requires the larger investment.

Table 17.2   COMPUTATION OF INCREMENTAL RATE OF RETURN FOR MUTUALLY EXCLUSIVE EQUAL–LIVED PROJECTS

	C	A	B	D
Building cost	$—190,000	$—200,000	$—275,000	$—350,000
Annual CFAT	19,500	22,000	35,000	42,000
Projects compared	C to none	A to none	B to A	D to B
Incremental cost	$—190,000	$—200,000	$ —75,000	$ —75,000
Incremental CFAT	19,500	22,000	13,000	7,000
$(P/A,i\%,30)$	9.7436	9.0909	5.7692	10.7143
Incremental i	9.63%	10.49%	17.28%	8.55%
Increment justified?	No	Yes	Yes	No
Project selected	None	A	B	B

COMMENT We should mention here again, just as a word of warning, that an alternative should *always* be compared with an acceptable alternative, noting that the "do nothing" alternative may be the acceptable one. Since C was not justified, Location A was *not* compared to C. Thus, *if* the B-to-A comparison had not indicated that B was incrementally justified, then the comparison D-to-A would have been made, instead of D-to-B.

It is important to understand the use of incremental rate-of-return selection because if it is not properly applied in mutually-exclusive-alternative evaluation, the wrong alternatives may be selected. If the overall rate of return of each alternative is computed, the results are

Location	C	A	B	D
Overall i	9.63%	10.49%	12.40%	11.59%

If we now apply *only* the first criterion stated earlier, that is, make the largest investment that has a MARR of 10% or more, we would choose Location D. But, as shown above, this is the wrong selection because the extra investment of $75,000 between Locations B and D will not earn the MARR. In fact, it will earn only 8.55% (Table 17.2). Remember, therefore, that incremental analysis is necessary for selection of one alternative from several when the rate-of-return evaluation method is used.

/////

When the alternatives under consideration consist of disbursements only, the "income" is the difference between costs for two alternatives. In this case, there is no need to compare any of the alternatives again the "do nothing" alternative. The lowest-investment-cost alternative is the defender against the next-lowest-investment-cost alternative (challenger). This procedure is illustrated in Example 17.2.

Example 17.2 Four machines can be used for a certain stamping operation. The costs for each machine are shown in Table 17.3. Determine which machine should be selected if the company's MARR is 12%.

Table 17.3 FOUR MUTUALLY EXCLUSIVE ALTERNATIVES

	Machine			
	1	2	3	4
First cost	$-5,000	$-6,500	$-10,000	$-15,000
Annual operating cost	-3,500	-3,200	-3,000	-1,400
Salvage value	+500	+900	+700	+1,000
Life, years	8	8	8	8

Table 17.4   COMPUTATION OF INCREMENTAL RATE OF RETURN

	Machine			
	1	2	3	4
Initial investment	$-5,000	$-6,500	$-10,000	$-15,000
Annual operating cost	-3,500	-3,200	-3,000	-1,400
Salvage value	+500	+900	+700	+1,000
Plans compared	–	2 to 1	3 to 2	4 to 2
Incremental investment	–	$-1,500	$ -3,500	$ -8,500
Incremental annual savings	–	+300	+200	+1,800
Incremental salvage	–	+400	-200	+100
Incremental i	–	14.6%	<0%	13.7%
Increment justified?	–	Yes	No	Yes
Alternative selected	–	2	2	4

SOLUTION   The machines are already ranked according to increasing investment cost and since no incomes are involved, the incremental comparisons can be made because all machines are assumed to be acceptable. Comparing Machine 2 (challenger) to Machine 1 (defender) on an incremental basis,

$$0 = -1,500 + 300(P/A,i\%,8) + 400(P/F,i\%,8)$$

Solution of the equation yields $i = 14.6\%$. Therefore, eliminate Machine 1 from further consideration. (If you had trouble obtaining the rate-of-return equation above, prepare a tabulation of cash flow for Machines 1 and 2 as in Sec. 10.1.) The remaining calculations are summarized in Table 17.4. When Machine 3 is compared to Machine 2, the rate of return on the increment is less than 0%; therefore, Machine 3 is eliminated. The comparison of Machine 4 to Machine 2 shows that the rate of return on the increment is greater than the MARR, favoring Machine 4. Since no additional alternatives are available, Machine 4 represents the optimum selection.

COMMENT   You should recognize that when incomes are not considered in the analysis, it is implied that one of the machines *must* be selected. This situation could arise when the alternatives under consideration are part of a larger project which has been shown to be economical regardless of which alternative is selected. You should now calculate the present worth and equivalent uniform annual cost of each machine to satisfy yourself that Machine 4 would be selected by all the evaluation methods.

                                                                                    ////

When the alternatives under consideration have different lives, it is necessary to make the comparison over the least common multiple of years between *all* alternatives when using the present-worth method; however, as shown in the Solved Examples (Example 17.6), if the incremental investment rate-of-return method is used, the incremental rate of return is found using the least common multiple between the *two* alternatives compared. Since this is not necessary in an EUAC analysis, this method is generally preferred when the alternatives have different lives. Example 17.6 in the Solved Examples section illustrates the calculations for alternatives having different lives. Of course, you can always use incremental present-worth or incremental EUAC analysis at the MARR to solve the problem but, here too, it is necessary to make the comparison over the least common multiple of lives.

Often the lives of the alternatives are so long that they can be considered to be infinite, in which case the capitalized-cost method is used. See Solved Examples for this special case.

Examples 17.6, 17.7
Problems P17.3–P17.11

17.3 Selection Using Incremental Benefit/Cost Ratio

The B/C ratio (Chap. 11) can also be used as an evaluation technique for mutually exclusive alternatives. An incremental B/C ratio can be computed in a fashion similar to that used for the incremental rate of return. The project that has the incremental B/C $\geqslant$ 1.0 *and* requires the largest *justified* investment is selected. The procedure to be followed is similar to that used for rate-of-return analysis. However, in a B/C analysis, it is generally convenient to compute an overall B/C ratio for each alternative, since the total present-worth or EUAC values must be computed in preparation for the incremental analysis. Those alternatives that have an overall B/C $<$ 1.0 can be eliminated immediately and need not be considered in the incremental analysis. Example 17.3 presents a complete application of the incremental B/C ratio to mutually exclusive alternatives.

Example 17.3 Using the four alternatives of Example 17.1 (Table 17.1), apply incremental B/C ratio analysis to select the most advantageous alternative (MARR = 10%).

SOLUTION The alternatives are first ranked from smallest to largest initial investment cost. The next step is to calculate the overall B/C ratio and eliminate those alternatives that have B/C $<$ 1.0. As shown in Table 17.5, Location C can be eliminated on the basis of its overall B/C ratio (0.97). All other alternatives are

Table 17.5 INCREMENTAL B/C RATIO ANALYSIS FOR MUTUALLY EXCLUSIVE ALTERNATIVES, EXAMPLE 17.3

	C	A	B	D
First cost	$-190,000	$-200,000	$-275,000	$-350,000
Annual CFAT	19,500	22,000	35,000	42,000
Life, years	30	30	30	30
Present worth of CFAT	$ 183,826	$ 207,394	$ 329,945	$ 395,934
Overall B/C	0.97	1.03	1.20	1.13
Comparison	–	–	B to A	D to B
Incremental benefit	–	–	$ 122,551	$ 65,989
Incremental cost	–	–	75,000	75,000
Incremental B/C	–	–	1.64	0.88
Alternative selected	–	–	B	B

acceptable and, therefore, must be compared on an incremental basis. The incremental benefits and costs can be determined as follows:

Incremental benefits: increase in present worth of
CFAT between alternatives
Incremental cost: increase in first cost between alternatives

A summary of the incremental B/C analysis is presented in Table 17.5. Using the acceptable alternative that has the lowest investment cost as the defender (A) and the next lowest acceptable alternative (B) as the challenger, the incremental B/C ratio is 1.64, indicating that Location B should be selected over Location A (therefore eliminating A from further consideration). Using B as the defender and D as the challenger, the incremental analysis yields incremental B/C = 0.88, favoring Location B. Since Location B has an incremental ratio greater than 1.0 *and* is the largest justified investment, it is selected; this, of course, is the same conclusion reached with the incremental-investment rate-of-return method.

COMMENT   You should recognize that alternative selection should not be made on the basis of the overall B/C ratio, even though Location B would still be selected, *coincidentally in this case*. The incremental investment must also be justified in order to select the best alternative.

Although present-worth values were used in this example, EUAC values can also be used to compare the investments; in fact, this method is generally simpler if lives are unequal.   ////

Example 17.8
Problems P17.12–P17.16

17.4   The Capital-Budgeting Problem

Most corporations have capital-investment projects including alternatives that are not mutually exclusive. If a company is attempting to solve a problem with the following characteristics, it must solve a *capital-budgeting problem* (sometimes referred to as *portfolio selection*):

1 Several projects are available that are *independent* of each other, that is, selection of a particular project does *not* preclude selection of any other of the projects.
2 A budgetary constraint restricts the total investment possible in the projects.
3 The objective of financial investment for the corporation is to maximize the value of the investments.

The field of capital budgeting is being thoroughly researched at this time. We will investigate relatively simple problems that include a budget constraint and equal-risk projects with only one initial investment required. The projects will have future cash flows that are assumed to be deterministically predictable. We will look briefly at two solution techniques—rate-of-return and present-worth (discounted cash flow). This material is fundamental, and current research on it can be found in articles in such journals as *The Engineering Economist, The Journal of Finance, The Journal of Business,* and *Managerial Planning.*

In the typical capital-budgeting problem, there are several independent alternatives, each with a first cost, life, series of yearly independent cash flows, and possibly a salvage value. If any of the alternatives are mutually exclusive, this must be specifically accounted for. Likewise, if any are dependent, that is, must be done in conjunction with some other alternative, the dependent components should be combined into *one independent* alternative.

Problem P17.17

17.5 Rate-of-Return Solution to Capital-Budgeting Problems

There is a general procedure that may be applied when solving a capital-budgeting problem by the rate-of-return method. The steps in this procedure are the following:

1 For each alternative available for investment, compute the overall rate of return (before- or after-tax, as appropriate).

2 Rank the alternatives in decreasing order starting with the alternative having the greatest return and stopping with the alternative with a return that just exceeds the MARR. Any project with a return less than the minimum is discarded.

3 Select alternatives for investment from the ranked alternatives until the budgetary constraint is met as closely as possible, without exceeding it.

If a portion of the budget is unspent because of the pattern of ranked returns and first costs, it is necessary to determine which combination of acceptable alternatives would result in maximization of the rate of return. This requires that an estimate be made of the rate of return obtainable on the unspent portion of the original budget. If the MARR is considered a *reasonable* estimate, the return obtainable on a safe investment (such as interest from a conventional savings account) would be regarded as *pessimistic*, while an estimate somewhat greater than the MARR would be considered *optimistic*. The actual estimate used would be dependent on the past experience and financial philosophy of the company. Example 17.4 illustrates the rate-of-return solution to a capital-budgeting problem.

Example 17.4 Consider the six independent projects outlined in the first four columns of Table 17.6. If the MARR before taxes is 14% and an investment budget of $35,000 has been set, select the alternatives using the rate-of-return criteria.

Table 17.6 INDEPENDENT ALTERNATIVES CONSIDERED FOR INVESTMENT

Alternative	Investment	Life, years	Annual CFBT	Rate of return	Rank
A	$10,000	6	$2,870	18%	3
B	15,000	9	2,930	13	No
C	8,000	5	2,680	20	2
D	21,000	3	9,500	17	4
E	13,000	10	2,600	15	5
F	6,000	4	2,540	25	1

SOLUTION The returns computed by the EUAC method are shown and ranked in Table 17.6. Alternative B is unacceptable since 13% < 14%, the required return. The ranked alternatives and cumulative investment figures are tabulated below:

Ranked alternative	Rate of return	Investment	Cumulative investment
F	25%	$ 6,000	$ 6,000
C	20	8,000	14,000
A	18	10,000	24,000
D	17	21,000	45,000
E	15	13,000	58,000

Using the budgetary restriction of $35,000 allowed for total investment and selecting alternatives on the basis of overall rate of return, F, C, and A are selected. The overall rate of return (RR) for these alternatives, assuming a MARR of 14% on the unspent portion ($11,000), is

$$RR = \frac{1}{35,000}[6,000(0.25) + 8,000(0.20) + 10,000(0.18)$$
$$+ 11,000(0.14)]$$
$$= 0.184 \quad (18.4\%)$$

It should be recognized that other combinations of acceptable alternatives can be formulated that will not violate the budgetary constraint. Some of these combinations and the overall rate of return are the following:

Alternative	Total investment	Overall return
F, C, D	$35,000	19.1%
C, A, E	31,000	16.9
F, A, E	29,000	17.4

Thus, the Alternatives F, C, and D should be selected to maximize the overall rate of return if the unallocated portion of the budget can earn the MARR of 14%.

COMMENT Generally the budget amount is not a rigidly set value; rather, a range is placed on the total amount that can be invested. Because the budget is usually not an inflexible value, adjustments in this amount are often made in order to select the alternatives in ranked order to obtain a higher rate of return.   ////

Example 17.9
Problems P17.18–P17.21

17.6 Present-Worth Solution to Capital-Budgeting Problems

Another commonly accepted method for solving capital-budgeting problems is the present-worth, or discounted cash-flow, method. The purpose of the present-worth solution to capital-budgeting problems is to maximize the cash flow, and if the present worth of each alternative is found using the MARR, all alternatives with a positive present worth will return at least the MARR. When the alternatives under consideration have different lives and if the present worth is computed for each alternative for its life (rather than for the least common multiple between them or over a preselected planning horizon), the alternatives selected may be different than those selected on the basis of the rate-of-return method of the preceding section.[1-5] This difference in answers is brought about in part by the fact that the failure to equalize the lives of the alternatives assumes that the *rate of return* for the shorter-lived alternative is *zero* between the end of its life and that of all longer-lived alternatives. The assumption of a zero rate of return after the first life cycle is contradictory to even conservative investment policy for two reasons:

1 It is *always* possible for a company to make an investment that will yield some minimum rate of return (safe investment).

2 A company's MARR is established on the basis of past and present investment opportunities and, therefore, reflects the probable return on future investments.

The assumption of a zero interest rate can be partially negated by calculating the EUAC at the MARR of each alternative over its respective life and then finding the present worth at the MARR of each alternative, using the life (planning horizon) of the longest-lived alternative. This assumes that investments could be made that will make a rate of return equal to the MARR through the length of the planning horizon. It should be recognized that the ranking of alternatives by the rate-of-return and present-worth methods may still be different. This difference is possible, due to two reasons:

1 Implicit in the *rate-of-return method* is the assumption that the rate of return obtainable for any alternative continues indefinitely at the rate projected for the first life cycle.

2 The *present-worth method* explicitly uses the same interest rate (MARR) for all alternatives regardless of their actual rates of return through their first life cycle. The MARR may be considerably less than the actual rate of return obtainable on each alternative.

Using a planning horizon equal to the longest-lived alternative, the present-worth solution to capital-budgeting problems is as follows:

1 Compute the EUAC at the MARR of each alternative over its respective life with the equation

$$\text{EUAC} = \left[\sum_{j=1}^{n} (CF)_j (P/F,i\%,j) - \text{investment} \right] (A/P,i\%,n)$$

where $(CF)_j$ is the cash flow for year j ($j = 1, 2, \ldots, n$).

2 Compute the present worth at the MARR for each alternative over the planning horizon (n_1) using the relation

$$\text{PW} = \text{EUAC}(P/A,i\%,n_1) \qquad (17.2)$$

3 Rank the alternatives by present-worth value from largest to smallest.

4 Select alternatives for investment from the ranked alternatives by maximizing the present worth and not violating the budgetary constraint.

Example 17.5 illustrates the above procedure for unequal-lived alternatives.

Example 17.5 Refer to the projects presented in Table 17.6 (Example 17.4). If the budget is $35,000 and the MARR is 14%, select the best portfolio of alternatives by maximizing total present worth.

SOLUTION The first step is to compute the EUAC at 14% of each alternative over its respective life. For example, the EUAC of Alternative A is

$$\text{EUAC}_A = 2,870 - 10,000(A/P,14\%,6)$$
$$= \$298$$

The EUAC values for each alternative are given in column (4) of Table 17.7.

The next step is to calculate the present worth using Eq. (17.2) at the MARR of 14% for each alternative for a planning horizon of $n_1 = 10$ years. For Alternative A,

Table 17.7 **CAPITAL BUDGETING FOR UNEQUAL-LIVED ALTERNATIVES FOR EXAMPLE 17.5**

Alternative	Initial investment (1)	Life, years (2)	Annual CFBT (3)	EUAC (4)	PW* (5)	PW rank (6)
A	$10,000	6	$2,870	$+298	$+1,556	4
B	15,000	9	2,930	−103	−535	No
C	8,000	5	2,680	+350	+1,826	3
D	21,000	3	9,500	+455	2,372	2
E	13,000	10	2,600	+108	+562	5
F	6,000	4	2,540	+481	+2,508	1

*PW = EUAC$(P/A,14\%,10)$.

Table 17.8 ALTERNATIVES RANKED BY
PRESENT-WORTH VALUE FOR
EXAMPLE 17.5

Ranked alternative	Present worth	Investment	Cumulative investment
F	$+2,508	$ 6,000	$ 6,000
D	+2,372	21,000	27,000
C	+1,826	8,000	35,000
A	+1,556	10,000	45,000
E	+562	13,000	58,000

$$PW_A = 298(P/A,14\%,10) = \$1,556$$

Column (5) of Table 17.7 presents the PW values, which are ranked in column (6) according to decreasing PW. Alternative B is removed from further consideration because it has a negative PW value.

Table 17.8 shows the alternatives in ranked order with the cumulative investment. With a budget of $35,000, the present worth is maximized when Alternatives F, D, and C are selected.

COMMENT The alternatives selected above are the same as those selected by maximizing the overall rate of return. You should recognize that the initial ranking of alternatives is different by the two methods, as is shown below:

	Rank	
Alternative	RR	PW
A	3	4
B	No	No
C	2	3
D	4	2
E	5	5
F	1	1

Furthermore, if the budget had been different than $35,000, the alternatives selected would likely be different by each method regardless of whether the rates of return or the present-worth values are maximized. In this event, it is necessary to make a judgment about whether the return obtainable after the first life cycle of each alternative is more likely to approximate the MARR (which will favor PW ranking) or the actual alternative rates of return (which will favor the rate-of-return ranking).   ////

Example 17.10
Problems P17.22–P17.29

Table 17.9 THREE MUTUALLY EXCLUSIVE
DIFFERENT–LIVED
ALTERNATIVES

	A	B	C
Initial cost	$-6,000	$-7,000	$-9,000
Salvage value	0	+200	+300
CFBT per year	+2,000	+3,000	+3,000
Life, years	3	4	6

SOLVED EXAMPLES

Example 17.6 Three mutually exclusive alternatives are available, as shown in Table 17.9. If a 15% before-tax return is the minimum accepted, use (a) EUAC analysis and (b) incremental rate-of-return analysis to select the most attractive alternative.

SOLUTION

(a) The EUAC for the respective life of each alternative must first be computed. For Alternative A,

$$\text{EUAC}_A = -6,000(A/P,15\%,3) + 2,000 = \$-628$$

Similarly, $\text{EUAC}_B = \$588$ and $\text{EUAC}_C = \$656$. Therefore, select Alternative C.
(b) The first step in the incremental rate-of-return analysis is to compare A to the "do nothing" alternative. The equation for this comparison is

$$0 = -6,000 + 2,000(P/A,i\%,3)$$

which is satisfied at $i = 0\%$. Therefore, delete Alternative A. Comparing B to "do nothing" yields $i = 26.4\%$, which justifies Alternative B, because $26.4\% > \text{MARR} = 15\%$. Comparing C to B incrementally requires a 12-year evaluation period. Table 17.10 is a cash-flow tabulation for the incremental analysis of C to B. The incremental rate of return for the net cash-flow column is $i = 19.4\%$, which is greater than the MARR. Thus, Alternative C is selected, as in the EUAC analysis of part (a).

COMMENT When comparing different-lived alternatives by incremental analysis it is necessary to use the least common multiple of years between only the *two* alternatives being compared, and *not* the common multiple of all alternative lives.

////

Sec. 17.2

Example 17.7 The U.S. Army Corps of Engineers wants to construct a dam on the Sacochsi River. Six different sites have been suggested. The construction and average annual dollar benefits (income) to the area are tabulated below. If a MARR of

Table 17.10   NET CASH–FLOW TABULA-
TION FOR EXAMPLE 17.6

Year	Cash flow B	Cash flow C	Net cash flow (C − B)
0	$ −7,000	$ −9,000	$−2,000
1	+3,000	+3,000	0
2	+3,000	+3,000	0
3	+3,000	+3,000	0
4	−3,800	+3,000	+6,800
5	+3,000	+3,000	0
6	+3,000	−5,700	−8,700
7	+3,000	+3,000	0
8	−3,800	+3,000	+6,800
9	+3,000	+3,000	0
10	+3,000	+3,000	0
11	+3,000	+3,000	0
12	+3,200	+3,300	+100
	$+15,600	$+18,600	$+3,000

6% is required and dam life is long enough to be considered infinite for analysis purposes, select the best location from the economic viewpoint.

Site	Construction cost (millions)	Annual income
A	$ 6	$350,000
B	8	420,000
C	3	125,000
D	10	400,000
E	5	350,000
F	11	700,000

SOLUTION After ranking the projects by first cost, we can use the capitalized-cost equation, Eq. (8.3), $P = A/i$, in the form

$$0 = \frac{A}{i} - P \qquad (17.3)$$

to determine if the incremental investment is justified. If the righthand side of Eq. (17.3) is greater than zero, the incremental investment is justified. Table 17.11 indicates that only Site E is justified, so this is the most economical dam site.   ////

Sec. 17.2

Example 17.8 Compare the dam sites described in Example 17.7 using incremental B/C ratio analysis.

Table 17.11 CAPITALIZED COST COMPARISON OF MUTUALLY EXCLUSIVE DAM SITES

	C	E	A	B	D	F
P(million	$ 3	$ 5	$ 6	$ 8	$ 10	$ 11
A($1,000)	125	350	350	420	400	700
Comparison	C to none	E to none	A to E	B to E	D to E	F to E
ΔP(million)*	$ 3	$ 5	$ 1	$ 3	$ 5	$ 6
ΔA($1,000)	125	350	0	70	50	350
ΔA/i − ΔP(million)	−0.92	0.83	−1.0	−1.83	−4.17	−0.17
Site selected	None	E	E	E	E	E

*Δ = incremental.

SOLUTION We make use of the capitalized-cost equation, Eq. (8.2),

$$A = Pi$$

to obtain EUAC values for capital recovery (cost), as shown in the first row of Table 17.12. Since Site E is justified and has the largest investment, it is selected.

COMMENT Suppose Site G is added with a construction cost of $10 million and an annual benefit of $700,000. What site should G be compared to? What is the ΔB/C ratio? If you determine that a comparison of G to E is to be made and ΔB/C = 1.17 in favor of G, you are correct! Now Site F must be incrementably evaluated with G, but since the annual benefits are the same ($700,000), the ΔB/C ratio is zero and the added investment is not justified. Therefore, Site G is chosen.   ////

Sec. 17.3

Example 17.9 Mrs. I. M. Able has $2,000 to invest. She has singled out four possibilities, as detailed in Table 17.13. Which investments should she make if the MARR is 10% and the maximization of rate of return is used as a criteria?

Table 17.12 USE OF INCREMENTAL B/C RATIO ANALYSIS FOR EXAMPLE 17.8

	C	E	A	B	D	F
Capital recovery ($1,000)	$180	$300	$360	$480	$600	$660
Annual benefits ($1,000)	125	350	350	420	400	700
Comparison	C to none	E to none	A to E	B to E	D to E	F to E
ΔCapital recovery*	$180	$300	$ 60	$180	$300	$360
ΔAnnual benefits	125	350	0	70	50	350
ΔB/C ratio	0.70	1.17	0	0.39	0.17	0.97
Site selected	None	E	E	E	E	E

*Δ = incremental.

Table 17.13 POSSIBLE INVESTMENTS FOR
EXAMPLE 17.9

Plan	Investment	CFAT for year		
		1	2	3
A	$1,000	$ 0	$ 0	$1,500
B	1,000	1,100	100	–
C	1,000	400	400	400
D	1,000	1,200	–	–

SOLUTION Using the EUAC method for computation of actual rate of return and then ranking by i gives us the results below.

Plan	Rate of return	Rank
A	14.5%	3
B	18.5	2
C	9.3	No
D	20.0	1

Since Plans A, B, and D have a return of at least MARR, with a budget of $2,000 select Plans B and D for investment.

COMMENT Since all alternatives have the same initial investment, maximization of the rate of return is assured without considering other combinations of alternatives.

////

Sec. 17.5

Example 17.10 (*a*) Solve the capital-budgeting problem in Example 17.9 using the maximization of present worth as a criterion (MARR = 10%). (*b*) Compare the answer in (*a*) with the portfolio selected using the rate-of-return criterion and state why the answers are different.

SOLUTION

(*a*) The PW values using Eq. (17.2) for $n_1 = 3$ with $i = 10\%$ are shown below.

Plan	PW	Rank
A	$127	2
B	118	3
C	−5	No
D	249	1

With a $2,000 budget, select Plans D and A.

(*b*) The answer using the return criteria is selection of Plans B and D. The two criteria give different answers because lives are different and the assumptions about the rate of return obtainable after the first life cycle are different between the two methods, as described in Sec. 17.6. ////

Sec. 17.6

REFERENCES

1. **G. Fleisher**, Two Major Issues Associated with the Rate of Return Method for Capital Allocation: The "Ranking Error" and "Preliminary Selection," *J. Ind. Eng.*, April, 1966, pp. 202–208.
2. **J. H. Lorie** and **L. J. Savage**, Three Problems in Rationing Capital, *J. Bus.*, October, 1955, pp. 229–239.
3. **G. D. Quinin**, "The Capital Expenditure Decision," Richard D. Irwin, Inc., Homewood, Ill., 1967, pp. 175–197.
4. **C. S. Rowley**, Methods of Capital Project Selection, *Manag. Plan.*, March–April, 1973.
5. **E. Solomon**, The Arithmetic of Capital-budgeting Decisions, *J. Bus.*, April, 1956, pp. 124–129.

BIBLIOGRAPHY

Barish, pp. 231, 233–235.
DeGarmo and **Canada**, pp. 206–208, 289–291, 303–304.
Emerson and **Taylor**, pp. 8-3 – 8-8.
Fabrycky and **Thuesen**, pp. 85–91, 105–110, 195–198, 347–353.
Grant and **Ireson**, pp. 193–195, 227–235, 517–531.
Riggs, pp. 240–243, 245–248.
Smith, pp. 99–100, 133–139, 499–526.
Taylor, p. 117.
Thuesen, **Fabrycky**, and **Thuesen**, pp. 128–165.

PROBLEMS

P17.1 What is the basic difference between an engineering-economy problem that requires selection from several mutually exclusive alternatives and one that requires selection from independent projects?

P17.2 What criteria are used to select a mutually exclusive alternative by the B/C ratio method?

P17.3 Five different methods can be used for recovering by-product heavy metals from a waste stream. The investment costs and incomes associated with each method are shown below. Assuming all methods have a ten year life with zero some salvage value and the company's MARR is 15%, determine which one should be selected by (a) the EUAC method and (b) the incremental rate-of-return method.

	Method				
	1	2	3	4	5
First cost	$15,000	$18,000	$25,000	$35,000	$52,000
Salvage value	+1,000	+2,000	−500	−700	+4,000
Annual income	5,000	6,000	7,000	9,000	12,000

P17.4 If Method 2 in problem P17.3(b) has a life of five years and Method 3 has a life of 15 years, which alternative should be selected?

P17.5 Select the best alternative using the incremental rate-of-return method from the proposals shown below if the MARR is 10% and the projects will have a useful life of 15 years. Assume the cost of the land will be recovered when the project is terminated.

	Proposal						
	1	2	3	4	5	6	7
Land cost	$ 50,000	$ 40,000	$ 70,000	$ 80,000	$ 90,000	$ 65,000	$ 75,000
Construction cost	200,000	150,000	170,000.	185,000	165,000	175,000	190,000
Annual maintenance	15,000	16,000	14,000	17,000	18,000	13,000	12,000
Annual income	52,000	49,000	68,000	50,000	81,000	77,000	45,000

P17.6 Any one of five machines can be used in a certain phase of a canning operation. The costs of the machines are shown below and all are expected to have a ten-year life. If the company's minimum attractive rate of return is 18%, determine which machine should be selected using (a) the present-worth method and (b) the incremental rate-of-return method.

	Machine				
	1	2	3	4	5
First cost	$28,000	$33,000	$22,000	$51,000	$46,000
Annual operating cost	20,000	18,000	25,000	12,000	14,000

P17.7 An oil and gas company is considering five sizes of pipe for a new pipeline. The costs for each size are shown below. Assuming all pipes will last 15 years and the company's MARR is 8%, which size of pipe should be used according to (a) the present-worth method and (b) the incremental rate-of-return method?

	Pipe size, inches				
	14	16	20	24	30
Initial investment	$9,180	$10,510	$13,180	$15,850	$30,530
Installation cost	600	800	1,400	1,500	2,000
Annual operating cost	6,000	5,800	5,200	4,900	4,800

P17.8   An independent dirt contractor is trying to determine which size dump truck to buy. The contractor knows that as the bed size increases, the net income increases but it is uncertain whether the incremental expenditure required in the larger trucks could be justified. The cash flows associated with each size truck are shown below. If the contractor's MARR is 8% and all trucks are expected to have a useful life of eight years, determine which size truck should be purchased using (a) the EUAC method and (b) the incremental rate-of-return method.

	Truck size, yards				
	8	10	15	20	40
Initial investment	$10,000	$12,000	$18,000	$24,000	$33,000
Annual operating cost	5,000	5,500	7,000	11,000	16,000
Salvage value	2,000	2,500	3,000	3,500	4,500
Annual income	9,000	4,000	10,500	12,500	14,500

P17.9   Five processes can be used for producing a certain part. If the company's MARR is 15%, determine which process should be selected by (a) the present-worth method and (b) the incremental rate-of-return method.

	Process				
	1	2	3	4	5
First cost	$15,000	$22,000	$27,000	$31,000	$42,000
Annual operating cost	6,000	5,000	4,500	3,000	2,000
Salvage value	500	1,000	1,100	600	3,000
Life, years	3	4	6	6	6

P17.10   One phase of a meat-packing operation requires the use of separate machines for the following functions: pressing, slicing, weighing, and wrapping. All machines under consideration are expected to have a life of six years with no salvage value. There are two alternatives for each of the functions as follows:

	Alternative 1		Alternative 2	
	First cost	Annual cost	First cost	Annual cost
Pressing	$ 5,000	$13,000	$10,000	$11,000
Slicing	4,000	10,000	17,000	4,000
Weighing	12,000	15,000	15,000	13,000
Wrapping	3,000	9,000	11,000	7,000

(a) If the company's MARR is 20%, use the incremental rate-of-return method to determine which machine should be selected for each function (identify them as Pressing 1, Pressing 2, Slicing 1, etc.). (b) For the machines selected in part (a), determine the total investment and operating cost for the entire operation.

P17.11 A third alternative can be added in problem P17.10: one machine to do the pressing and slicing and another machine to do the weighing and wrapping. The machine that will do the pressing and slicing (identified as Pressing-Slicing 3) will cost $29,000 and will have an annual operating cost of $9,000. The machine that will do the weighing and wrapping (identified as Weighing-Wrapping 3) will cost $26,000 and will have an annual operating cost of $18,000. (a) Which machines should be selected for the entire operation? (b) Determine the total investment and operating cost for the entire operation. (c) If a fourth alternative can be considered—a single machine to perform all four functions (identified as Machine 4) having an initial cost of $45,000 and an annual operating cost of $32,000—which machine(s) should be selected?

P17.12 Work problem P17.5 using the benefit/cost method of evaluation.

P17.13 Use the benefit/cost method in evaluating the alternatives of problem P17.6.

P17.14 (a) Rework problem P17.7 by the benefit/cost method. (b) If a MARR of 15% is required instead of 8%, use the B/C method to select the most economical alternative.

P17.15 If the life of the 30-inch pipe in problem P17.7 can be extended to 30 years by corrosion protection costing $200 per year, use the benefit/cost method to determine which size should be selected.

P17.16 Work problem P17.9 by the benefit/cost method.

P17.17 State three characteristics of a capital-budgeting problem.

P17.18 Select, from the following table, projects for investment using the rate-of-return criterion for (a) a MARR of 12% and a budget of $60,000 and (b) a MARR of 15% and a budget of $85,000.

Alternative	Investment	Rate of return	Alternative	Investment	Rate of return
1	$12,000	16%	5	$25,000	23%
2	18,000	21	6	11,000	19
3	9,000	13	7	18,000	14
4	3,000	8			

P17.19 The following five investment opportunities are being considered by the executive committee of the Smoky Ashtray Company.

Alternative	Investment	Life, years	Annual CFBT
A	$125,000	10	$16,000
B	78,000	12	14,500
C	50,000	8	10,100
D	90,000	10	16,000
E	10,000	4	4,500

(*a*) Select alternatives by rate of return using a MARR of 10% and a budget of $190,000. (*b*) Compute the overall rate of return if it is assumed that uninvested capital will return the MARR.

P17.20 Is the selection of alternatives different in problem P17.19 if the budget is $170,000? What percent change in overall rate of return takes place with this budget?

P17.21 The Hoof-Power Horse Cart Company can invest in three of five alternatives this year. All plans will require an initial investment of $10,000 this year and will return money for a varying number of years as shown below. Determine which alternatives should be selected for investment if rate of return is to be maximized and the MARR is 10%.

	CFBT for year				
Alternative	1	2	3	4	5
1	$1,000	$1,700	$ 2,400	$ 3,100	$ 3,800
2	500	500	500	500	10,500
3	5,000	5,000	2,000	–	–
4	0	0	0	15,000	–
5	0	3,000	12,000	–	–

P17.22 Select alternatives from those described in problem P17.19 by maximizing the present worth for the respective alternative life at a MARR of 8% and a budget value of (*a*) $190,000 and (*b*) $170,000. (*c*) Are the answers to (*a*) and (*b*) the same as those to problems P17.19 and P17.20 where the rate of return was maximized? Why?

P17.23 Select alternatives from those described in problem P17.19 by maximizing the present worth over the planning horizon of 12 years and a budget value of (*a*) $190,000 and (*b*) $170,000. (*c*) Compare these answers with those of problem P17.22.

P17.24 Rework problem P17.21 by maximizing present worth over the life of the longest-lived alternative at a MARR value of 10%.

P17.25 Rework problem P17.24 at a MARR of 15%. Compare the ranking of alternatives for this problem and P17.24; comment on any differences.

P17.26 Solve the following capital-budgeting problem for a MARR of 10% and a budget of $40,000 by (*a*) maximizing the rate of return and (*b*) maximizing the present worth. Compare your answers.

Alternative	Investment	Life, years	Annual CFBT
A	$10,000	3	$3,950
B	12,000	8	2,400
C	18,000	5	5,750
D	22,000	12	3,550

P17.27 Rework problem P17.26 if all alternatives have a life of eight years. Will the answers for the rate-of-return and present-worth methods always be the same? Why?

P17.28 Explain clearly the basic reason why the rate-of-return and present-worth methods can result in different ranking of alternatives when solving a capital-budgeting problem.

P17.29 Compute and rank the present-worth values for the planning horizon of three years of each alternative for Example 17.9 using the following MARR values: 5%, 10%, 15%, 18%, 20%, 25%. What MARR value(s) seem to be critical to the changing ranks of present-worth values?

18

ESTABLISHING THE MINIMUM ATTRACTIVE RATE OF RETURN

This chapter introduces you to some of the problems and methods, both quantitative and subjective, in setting the minimum attractive rate of return (MARR) used in engineering-economy analysis.

The problems of an uncertain future, inflation, risk, subjectivity, and altered management policy make the establishing of a realistic MARR quite difficult. While the quantitative models illustrated here give approximate answers, the degree of error is not calculable. Nevertheless, the setting of MARR is very important to a good economy study and sound financial decisions.

CRITERIA

To complete this chapter, you must be able to do the following:

1 Define the two separate sources of capital financing available to a corporation.
2 Define the term *cost of capital* and state why the minimum attractive rate of return is greater than the cost of capital.
3 State five reasons why the minimum attractive rate of return can vary.

4 Compute the before- and after-tax cost of capital for debt financing, given the type of financing, period of financing, interest rate, repayment scheme, and tax rate.

5 Compute the before- and after-tax cost of capital for equity financing using the methods below, given market price, book value, earnings and dividend for a share of common stock, and tax rate:

   (*a*)   dividend method
   (*b*)   earnings/price ratio method
   (*c*)   Gordon-Shapiro method
   (*d*)   opportunity-cost method

6 Compute the average cost of capital before and after taxes given the before- or after-tax cost of capital for both debt and equity financing, the proportion of each type of financing, and the tax rate.

7 Compute the debt-to-equity ratio given the proportion of debt and equity financing, and state what happens to the cost of capital as this ratio changes.

EXPLANATION OF MATERIAL

18.1 Types of Capital Financing

A corporation will accumulate capital by three different methods. These three are categorized into two sources, debt and equity financing, to correspond with the balance sheet sections of liabilities and owner's equity, respectively. The types of financing are defined as follows:

> *Debt financing* Capital borrowed from others that will be paid back at a stated interest rate by a certain specified date. The original owner (lender) takes no direct risk on the return of the funds and interest nor does he share in the profits the borrowing firm makes on the funds. Debt financing includes borrowing via bonds, mortgages, and loans and may be classed as long-term or short-term liabilities.
>
> *Equity financing* Capital owned by the corporation used to make a profit for the corporation. There are two types of equity financing: *owner's funds*, which are funds obtained from stock sales and may include funds from the company owners, if the firm is small or not a stock-issuing concern; *retained earnings*, which are sometimes referred to as plowback funds. These funds have been previously retained by the firm for investment and expansion purposes; they are owned by the stockholders, not the corporation per se.

In actual computations dealing with the setting of MARR, the cost to the corporation of each type of financing, debt or equity, is computed independently of other types. This assumption of independence will be used throughout this chapter. The proportion of debt and equity financing that should be used by a corporation is a very difficult problem to solve, a problem we will discuss only briefly in this chapter.

Problem P18.1

18.2   The Cost of Capital

The actual interest rate paid by the corporation in developing investment capital is called the *cost of capital* (CC). Since most firms use a combination of debt and equity financing and since these two types have different interest rates, the CC is an intermediate rate. As mentioned earlier, independence between types of financing is assumed in computing the CC. If we assume a project will be financed by a $100,000 bond issue (debt financing) and the actual rate of interest paid is 8% per year, the CC is 8%. In other words, the CC is a minimum cutoff of the return required on an investment. Thus, if the $100,000 investment will return 6%, money will be lost.

Commonly, CC < MARR, the latter being the return criterion used in economy studies. If the CC is 8% and an added 7% return is expected, then the MARR is 15%. As discussed in Sec. 1.5, the overall return on a project must be at least equal to the MARR to consider the project for funding (Fig. 1.1). If a company performs its present-worth and EUAC analyses at the CC rate, obviously the company is satisfied to *just* recover the investment; since this is generally not the case, the economic analysis is usually performed at a MARR > CC. Determination of *actual* CC is virtually impossible. While the quantitative methods available can give an approximate value, when the MARR is established, subjective judgment and experience are vital. Moreover, accuracy in determining CC is justified in terms of neither time nor economics because of a fluctuating economy, inflation, risks, and changing management policies.

It is important to remember that as the demand for capital exceeds its supply, the MARR will far exceed the CC because of the great selectivity that is possible and necessary.

Problem P18.2

18.3   Variations in MARR

The MARR is not a set, nonvarying value. Rather, it is altered by corporations for different types of projects. For example, a firm may use a MARR of 15% for depreciable assets and a MARR of 20% for diversification investment, that is, purchasing smaller companies, land, etc.

The MARR varies from one project to another and through time because of the following:

1 Project risk The more risk that is *judged* to be associated with a proposed project, the higher the MARR and, for that matter, the higher the CC for the project.

2 Sensitivity of project area If management is determined to diversify (or invest) in a certain area, the MARR may be lowered to encourage investment with the hope of recovering lost profit in other investment areas. This subjective reaction to investment opportunity can create much havoc with an economy study.

3 Tax structure If taxes are increasing due to increased profits, capital gains from retired assets, and increasing local taxes, the MARR will be increased. An after-tax study would eliminate this reason for a fluctuating MARR.

4 Capital-financing methods As capital becomes limited, the MARR is increased and management begins to look closely at the service life of the project (Chap. 11). As the demand for the limited capital exceeds supply, the MARR is further increased.

5 Rates used by other firms If the rates of other firms that are used as a standard increase, a company may alter its MARR upward in response. A typical standard may be that of the firm called the government, even though the MARR for government projects varies drastically and is set in a nonquantitative fashion, since profit is not a requisite of government investment.

Problem P18.3

18.4 Cost of Capital for Debt Financing

Debt financing includes borrowing by bond and mortgages. The interest (or dividend) paid on borrowed money can be used to reduce taxes; therefore, in computing the CC for debt financing, an after-tax approach should be taken. The two examples below illustrate determination of CC for debt financing.

Example 18.1 A bond face value of $500,000 will be raised by issuing 500 $1,000, 8% a year, ten-year bonds. If the effective tax rate is 50% and the bonds are discounted 2% for quick sale, compute the cost of capital (*a*) before taxes and (*b*) after taxes.

SOLUTION

(*a*) The annual dividend payment is $1,000(0.08) = $80 and the discount sale price of $980. Using Eq. (10.4), we find the *i* at which

$$0 = 980 - 80(P/A,i\%,10) - 1,000(P/F,i\%,10)$$

Trial and error results in a value $i = 8.31\%$, which is the before-tax CC for the $500,000.

(*b*) With the allowance to reduce taxes by deducting the interest on borrowed money, a tax savings of $80(0.5) = $40 per year is realized and actual annual dividend outlay is $80 - $40 = $40. Using Eq. (10.4), the after-tax CC value is now 4.26%.   ////

Example 18.2 A company plans to purchase a certain asset for $20,000 with a zero salvage value and anticipated life of ten years. Management has decided to put $10,000 down now and borrow $10,000 at an interest rate of 6% on the unpaid balance. The repayment scheme will be $600 interest each year and the $10,000 principal in year 10. (*a*) What is the loan's ATCC if the tax rate is 50%? (*b*) If the asset

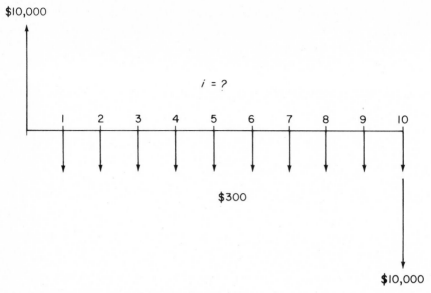

$10,000

i = ?

$300

$10,000

FIG. 18.1   Cash flow for a loan of $10,000 at 6%.

will have a cash flow before taxes (CFBT) of $5,000 per year and will be straight-line depreciated, is the investment justified?

SOLUTION

(a) Figure 18.1 presents the cash flow for repayment of the $10,000 loan only with a tax credit for $0.50(\$600) = \$300$ on the annual interest. We can set up the following relation for the loan only

$$0 = 10{,}000 - 300(P/A, i\%, 10) - 10{,}000(P/F, i\%, 10)$$

to find the after-tax CC of 3%.

(b) The CFAT (cash flow after taxes) value for each year of asset ownership is computed using Eqs. (16.7) and (16.8). (See Example 16.8 for a quick review.)

$$\text{CFBT} = \$5{,}000$$
$$\text{Taxable income} = 5{,}000 - (20{,}000/10) - 600 = \$2{,}400$$
$$\text{Taxes} = 0.5(2{,}400) = \$1{,}200$$
$$\text{CFAT} = 5{,}000 - 1{,}200 - 600 = \$3{,}200 \quad \text{(years 1–9)}$$
$$= 5{,}000 - 1{,}200 - 600 - 10{,}000$$
$$= \${-}6{,}800 \quad \text{(year 10)}$$

The after-tax rate of return is $i = 26.5\%$, which satisfies the relation

$$0 = -10{,}000 + 3{,}200(P/A, i\%, 9) - 6{,}800(P/F, i\%, 10)$$

The investment is justified for all values of 26.5% > MARR > 3.0%.

COMMENT In part (*a*), realize that the CC is *not* the loan rate of 6%, since this is not the rate paid on the entire $20,000 first cost.　　　　　　////

Problems P18.4–P18.8

18.5 Cost of Capital for Equity Financing

The use of equity funds—stocks or retained earnings—involves *no tax advantage*. If the cost of capital (CC) is computed before taxes, the desired income must include earnings and a dollar amount sufficient to cover the required tax payment. There are many ways to compute CC, some quantitative, some subjective. We briefly look at four methods, the first three quantitative, the last a subjective method that can be easily applied by the corporation analyst or the individual. All methods use the same example to permit a comparison of answers. Realize that, since the CC is computed independently for debt or equity financing, the methods presented here have no effect on the CC computation using debt financing.

18.5.1 Dividend method If a company is primarily interested in the dividend paid to stockholders and not in retained earnings as a source of equity financing, the after-tax cost of capital (ATCC) is the stated dividend rate. The before-tax cost of capital (BTCC), with taxes accounted for, is computed as

$$BTCC = \frac{\text{after-tax dividend rate}}{1 - \text{tax rate}} \qquad (18.1)$$

Example 18.3 A total of $50,000 new capital is to be raised by selling 2,500 common stocks at the market value of $20 per share. If a dividend rate of 5% is anticipated and the tax rate is 60%, compute (*a*) ATCC and (*b*) BTCC.

SOLUTION

(*a*) ATCC is stated as 5% by management.
(*b*) BTCC may be computed using Eq. (18.1) as

$$BTCC = \frac{0.05}{1 - 0.60} = 0.125$$

COMMENT In part (*b*), the actual earnings needed to cover the dividend will be $50,000(0.125) = $6,250 per year.　　　　　　////

18.5.2 Earnings/price ratio method A corporation is usually more interested in total earnings per share, which includes dividends *and* retained earnings, that is,

$$\text{Earnings} = \text{dividend} + \text{retained earnings}$$

Here the BTCC is computed as

$$BTCC = \frac{\text{earnings}}{\text{market price}(1 - \text{tax rate})} \qquad (18.2)$$

Example 18.4 Using the details of Example 18.3 and assuming an earning rate of 10% per share of market price, compute (*a*) ATCC and (*b*) BTCC.

SOLUTION

(*a*) ATCC is 10%, as stated.

(*b*) Using Eq. (18.2) with earnings of $20(0.10) = $2 per share

$$\text{BTCC} = \frac{2}{20(1 - 0.6)} = 0.25$$

COMMENT Of course BTCC can be computed using Eq. (18.1) as BTCC = 0.10/ (1 − 0.6) = 0.25; however, dividends are often stated in dollars per share not as a percentage of market price, which justifies the inclusion of Eq. (18.2). Note that doubling the earnings per share has doubled the CC values of Example 18.3.   ////

18.5.3 Gordon-Shapiro method

This method is presented without proof here.[1] Both stocks and retained-earnings equity financing are included in the CC formula. After conversion from a discrete to a continuous time scale, the ATCC is computed as

$$\text{ATCC} = \frac{D}{P} + \frac{Y - D}{\text{BV}} \qquad (18.3)$$

where D = current dividend per share
$\quad P$ = current market value per share
$\quad Y$ = current earnings per share
$\quad \text{BV}$ = current book value per share
Then,

$$\text{BTCC} = \frac{\text{ATCC}}{1 - \text{tax rate}} \qquad (18.4)$$

A simplifying assumption in this formula is that the earnings/book value ratio (Y/BV) remains constant at all times. Therefore, a value of $Y/\text{BV} = 0.20$ now is the same five or seven years from now. Of course, this value actually fluctuates, but it is hoped not chaotically. Also worthy of mention is that the first term in Eq. (18.3), (D/P), is the same as the ATCC that could be computed using the *dividend method* above. For a $20-per-share stock with a 5% dividend ($1 per share), ATCC = D/P = 1/20 = 0.05, the same as the ATCC for the dividend method. Thus, with addition of the retained earnings/book value ratio, ($Y{-}D$)/BV, both types of equity financing are accounted for.

Example 18.5 Using the details of Examples 18.3 and 18.4 and assuming BV is $15 per share, use the Gordon-Shapiro method to compute (*a*) ATCC and (*b*) BTCC.

SOLUTION

(*a*) Using Eq. (18.3)

$$\text{ATCC} = \frac{1}{20} + \frac{2-1}{15} = 0.117$$

(*b*) By Eq. (18.4)

$$\text{BTCC} = \frac{0.117}{1 - 0.6} = 0.292$$

COMMENT A review of these three quantitative, progressively more complex methods shows BTCC to progress from 12.5% (dividend only) to 29.2% (dividend and retained earnings). While all are only approximate, they do give an idea of the cost of capital. ////

In summary, the selected methods presented here would be utilized under the following conditions:

1 If the consistent payment of a specified dividend rate is of primary concern to company management, then the dividend method should be used.

2 If both the dividend rate and the percent of earnings retained in the company are of primary interest to management, use the earning/price ratio method.

3 When, in addition to the above, the book value of the stock is to be considered, use the Gordon-Shapiro method.

4 If quantitative analysis is not desired and subjective evaluation will suffice, use the opportunity-cost method discussed below.

Other methods usable in obtaining the cost of capital are summarized in Reisman.[2]

18.5.4 Opportunity-cost method If a firm (or individual) considers all present and future investments of about the same risk level as the one presently contemplated, a subjectively determined (and usually experienced) cost of capital can be stated. This rate is the opportunity cost of capital. For example, if a company can make a consistent 15% rate of return before taxes by investing in mining operations, this 15% may well be considered the CC and the MARR without resorting to quantitative methods of determining a cost of capital. Although this method is subjective, it is quite useful when applied in a rational, timely manner.

Problems P18.9–P18.14

18.6 Computation of an Average Cost of Capital

Most project capital is not obtained via only debt or equity financing; rather, it is taken from a common pool of capital. Since most firms do not operate on total equity capital and since it is impossible to continue to operate on 100% financing, an average CC value should be computed using the weightings of debt and equity financing of the firm. If a specific project breakdown is definitely known, these weightings should be used. Example 18.6 demonstrates the computation of an average cost of capital.

Table 18.1 COMPUTATION OF THE AVERAGE COST OF CAPITAL
BEFORE AND AFTER TAXES FOR EXAMPLE 18.6

Source of funds	BTCC	ATCC	Proportion of funds	Average BTCC	Average ATCC
Equity	30%	12%	60%	18%	7.2%
Debt	10	4	40	4	1.6
				22%	8.8%

Example 18.6 A firm must raise a total of $500,000 of new capital for a particular project. If the company raises capital in the proportion 60% equity and 40% debt and if, for this project, equity capital would cost 12% after taxes and debt capital would cost 10% before taxes, find the total earnings necessary (a) before taxes and (b) after taxes. The tax rate is 60%.

SOLUTION First we must compute the average ATCC and average BTCC values for both equity and debt portions of financing. Refer to Table 18.1 for results of the following computations. For *equity* capital no tax advantage is realizable. Using Eq. (18.1),

$$\text{BTCC} = \frac{0.12}{1 - 0.6} = 0.30$$

The average BTCC is found by weighting BTCC with the proportion of funds from each type of financing to obtain

$$\text{Average BTCC} = 0.3(0.6) + 0.1(0.4) = 0.22$$

For *debt* financing a tax advantage can be claimed:

$$\text{ATCC} = \text{BTCC}(1 - \text{tax rate}) = 0.10(1 - 0.6) = 0.04$$

Then,

$$\text{Average ATCC} = 0.12(0.6) + 0.04(0.4) = 0.088$$

(a) Earnings before taxes = $500,000(0.22) = $110,000 per year.
(b) Earnings after taxes = $500,000(0.088) = $44,000 per year.   ////

Often the MARR is set using a CC value representative of only one type of financing, even though the capital comes from a common fund including both types of capital. If the MARR is determined as though funds were obtained from only one method of financing when in reality both types of capital are used, an unrealistic MARR may result. When the actual type of financing is known, the MARR should be determined for this source; however, if the funds are obtained from a capital pool that includes both types of financing, a compromise MARR may be computed using the overall company split of debt and equity funds.

Example 18.7
Problems P18.15–P18.21

18.7   Effect of the Debt/Equity Ratio on the Cost of Capital

The value of the ratio

$$D/E = \frac{\text{proportion of debt capital}}{\text{proportion of equity capital}}$$

is called the debt/equity ratio. Of course, this ratio varies with the company and type of industry. The range is $0 < D/E < \infty$. For example, if debt capital represents 60% of the capital and equity 40%, $D/E = 60/40 = 1.5$. As the ratio increases, the average CC value decreases, as shown in the Solved Examples. Since CC decreases as D/E goes up, a likely question is, why not use 100% debt financing? You may remember that this question was first posed for *you* in Example 16.9. As debt financing for a firm increases, the lenders take a larger risk and require a greater interest rate on loans. It is impossible to maintain a business and solvency without a healthy share of equity financing. In other words, if the company does not own some part of itself, it would not be able to obtain operating or investment capital.

The advantage that debt financing allows is referred to as *leverage*. More debt capital releases other equity funds for use, but in the long run discourages potential stockholders from investing in the company because of the high D/E ratio.

Example 18.8
Problems P18.22–P18.26

SOLVED EXAMPLES

Example 18.7   The Hearty Food chain wants to purchase a fleet of 15 new delivery trucks for $150,000. Each truck has a salvage value of $1,000 after ten years and will be straight-line depreciated. There are two methods of capital financing available—equity and debt. Equity financing would involve the selling of stocks at the market value of $15. These stocks would pay a dividend of $0.50 per share and have an anticipated earning rate of 5% of market value. The present book value per share is $12.

Maximum debt financing approved by the bank is 50% of the sum needed. The loan will be for ten years and repayment will be at 8% interest, to be paid in ten equal annual installments. If the fleet is expected to produce an annual CFBT of $30,000 and the effective tax rate is 50%, which method of financing is more advantageous? Assume the company's MARR is two times the cost of capital.

SOLUTION

100% equity financing   We first compute ATCC using the Gordon-Shapiro method, Eq. (18.3):

$$\text{ATCC} = \frac{0.50}{15} + \frac{0.75 - 0.50}{12} = 0.0542$$

Therefore, MARR = 2(5.42) = 10.84%. For 100% equity financing, no interest tax credit is allowed. To find the actual rate of return, we need the CFAT values. For each of the ten years,

$$\text{CFBT} = \$30,000$$
$$\text{Depreciation} = \frac{150,000 - 15,000}{10} = \$13,500$$
$$\text{Taxes} = 0.5(\text{TI}) = 0.5(30,000 - 13,500) = \$8,250$$
$$\text{CFAT} = 30,000 - 8,250 = \$21,750$$

Now, we solve the rate-of-return equation:

$$0 = -150,000(A/P,i\%,10) + 15,000(A/F,i\%,10) + 21,750$$

A return of $i = 8.4\%$ satisfies this equation. Since $8.4\% < 10.84\%$, equity financing is not advisable.

50% debt financing Since two types of financing are involved here, we must find an average ATCC. The ATCC for 50% equity financing is still 5.42%, since all values remain the same. For the $75,000 loan, we must compute the equal annual payment (A) as follows:

$$A = 75,000(A/P,8\%,10) = \$11,177$$

For ease of calculation, we will assume that the principal is reduced uniformly by an amount of $75,000/10 = \$7,500$ per year. Therefore, the annual interest payment is approximated as $\$11,177 - \$7,500 = \$3,677$. The annual tax credit for the interest is $0.5(\$3,677) = \$1,839$. Therefore, the CC value for debt financing is approximated as the return which satisfies

$$0 = 75,000(A/P,i\%,10) - (11,177 - 1,839)$$

The value of $(A/P,i\%,10) = 0.12451$ is correct for $i = 4.2\%$, which is the ATCC value for debt financing. Note that the 50% debt financing effectively reduced the cost of capital from the loan rate of 8% to 4.2% because of the tax advantage due to interest.

Weighting the equity and debt ATCC values by 50% gives

$$\text{Average ATCC} = 0.5(5.42) + 0.5(4.2) = 4.81\%$$

The 50% debt financing has reduced the ATCC from 5.42% for 100% equity financing to 4.81%. Now, MARR = 2(4.81) = 9.62%, compared to the previous value of 10.84%. We are ready to compute the actual rate of return on the $75,000 put in by the company. The annual CFAT value is computed from Eq. (16.8) as follows:

$$\text{CFBT} = \$30,000$$
$$\text{Depreciation} = \$13,500$$
$$\text{Taxes} = 0.5(30,000 - 13,500 - 3,677) = \$6,412$$
$$\text{CFAT} = 30,000 - 6,412 - 3,677 - 7,500 = \$12,411$$

The rate-of-return equation may be written

$$0 = -75,000(A/P,i\%,10) + 15,000(A/F,i\%,10) + 12,411$$

which is satisfied at $i = 11.95\%$. Since $11.95\% > \text{MARR} = 9.62\%$, this method of financing is advised.

COMMENT For the 50% debt financing we have used the average annual interest and principal values of \$3,677 and \$7,500, respectively, to compute $i = 11.95\%$. Actually, the changing interest and principal values for each year should be used, since the portion of the \$11,177 annual payments applied toward the loan principal increases each year. For example, in the first year

$$\text{Interest} = 75,000(0.08) = \$6,000$$
$$\text{TI} = 30,000 - 13,500 - 6,000 = \$10,500$$
$$\text{Taxes} = 0.5(10,500) = \$5,250$$
$$\text{CFAT} = 30,000 - 5,250 - 6,000 - 5,177 = \$13,573$$

However, for year 2

$$\text{Interest} = (75,000 - 5,177)(0.08) = \$5,586$$
$$\text{TI} = 30,000 - 13,500 - 5,586 = \$10,914$$
$$\text{Taxes} = 0.5(10,914) = \$5,457$$
$$\text{CFAT} = 30,000 - 5,457 - 5,586 - 5,591 = \$13,366$$

Therefore, even though an annual \$11,177 payment is made, due to the 8% interest on the declining balance, the actual rate-of-return equation should be written

$$0 = -75,000(A/P,i\%,10) + 15,000(A/F,i\%,10)$$

$$+ \left[\sum_{j=1}^{n} (\text{CFAT})_j (P/F,i\%,j) \right] (A/P,i\%,10)$$

However, the additional calculations are not worth the trouble since the approximation of the actual rate of return is usually quite good. ////

Secs. 18.4, 18.5, and 18.6

Example 18.8 Mr. Billy N. Aire is making plans to invest in common stock. He has records of three electronics companies. The total asset value and ownership is given in Table 18.2. For each company, answer the following questions for the particular situation explained:

Table 18.2 FINANCING PROFILE FOR THREE COMPANIES*

| Firm | Asset value | Financing | |
		Debt	Equity
A	\$5.0	\$1.0	\$4.0
B	4.0	2.0	2.0
C	6.0	5.0	1.0

*All \$ values in millions.

(a) Compute the present debt/equity ratio (D/E).

(b) If a decrease of 20% in asset value takes place, compute the new debt/equity ratio.

(c) If revenue is $1.5 million for each company, compute the rate of return on the issued common stock. Assume interest on debt financing averages 6% for Company A, 8% for B, and 10% for C. Compare the returns and comment about their relative magnitudes.

SOLUTION

(a) The present D/E ratios (values divided by 1 million) are

$$(D/E)_A = \frac{1}{4} = 0.25$$

$$(D/E)_B = \frac{2}{2} = 1.00$$

$$(D/E)_C = \frac{5}{1} = 5.00$$

(b) Table 18.3 presents computations of D/E ratios after a 20% decrease in asset value. The company debt remains the same after the decrease; thus, the equity share must decrease and D/E ratios increase. The zero equity value for Company C is used, even though the actual loss of $1.2 million would result in a $-200,000 stock ownership. In this case, the stockholders of Company C are completely wiped out. You can see that when a company suffers an asset devaluation, because of obsolescence or some other reason, the stockholder is hurt and the D/E ratio increases.

(c) Table 18.4 gives a summary of rate of return on issued stock values using the given interest rates. These interest rates increase as the D/E ratio increases, due to the greater *risk* taken by the lender. The return values increase from 36% to 100%. The great increase in return as D/E increases is called *leverage*. However, don't be fooled. This situation is by no means a panacea, since such a high return as 100% is based on a present equity value of zero and a $5 million

Table 18.3 D/E RATIOS AFTER A 20% DECREASE IN ASSET VALUE*

| Firm | Asset value | | | Financing | | D/E ratio |
	Old	Reduction	New	Debt	Equity	
A	$5.0	$1.0	$4.0	$1.0	$3.0	0.33
B	4.0	0.8	3.2	2.0	1.2	1.67
C	6.0	1.2	4.8	5.0	0.0	∞

*All $ values in millions.

Table 18.4 RATE OF RETURN ON ISSUED STOCKS*

Firm	Revenue (1)	Debt (2)	Rate (3)	Amount (4) = (3) (2)	Net income (5) = (1) — (4)	Stock value (6)	Rate of return (7) = (5)/(6)
A	$1.5	$1.0	0.06	$0.06	$1.44	$4.00	0.36
B	1.5	2.0	0.08	0.16	1.34	2.00	0.67
C	1.5	5.0	0.10	0.50	1.00	1.00	1.00

*All $ values in millions.

debt. In short, confidence in a company with a very large D/E ratio is bad, no matter how large the return on stock value.   ////

Sec. 18.7

REFERENCES

1. **M. J. Gordon** and **E. Shapiro**, Capital Equipment Analysis: The Required Rate of Return, *Mngmt. Sci.*, October, 1959.
2. **A. Reisman**, "Managerial and Engineering Economics," Allyn & Bacon, Inc., Boston, 1971.

BIBLIOGRAPHY

Barish, pp. 209–216, 225–226.
DeGarmo and Canada, pp. 186–192, 198–206.
Emerson and Taylor, pp. 13-1 – 13-19.
Grant and Ireson, pp. 196–204, 517–518.
Park, pp. 111–118.
Reisman, pp. 174–175, 283–299, 325–328, 340–341.
Riggs, pp. 223–230.
Smith, pp. 455–467, 474–476.
Taylor, pp. 159–176.
Thuesen, Fabrycky, and Thuesen, pp. 135–136.

PROBLEMS

P18.1   State whether each of the following is in the category of debt or equity financing: (*a*) short-term note from the bank; (*b*) $5,000 taken from a co-owner's savings account to pay a company bill; (*c*) a $150,000 bond issue; (*d*) an issue of preferred stock worth $55,000; (*e*) Ms. Broke borrows $50,000 from her brother at 3% interest to run her business. The brother is not a co-owner in the company.

P18.2   If the owner of the Dollar Daze Variety Store has computed an overall cost of capital of $5\frac{1}{2}\%$ and is about to evaluate the purchase of a new cash register using a MARR of 6%, what return is expected on the investment? Does this seem reasonable? Why?

P18.3   Will each of the following cause the MARR to be raised or lowered? State why. (*a*) Investment in a chain of quick-food stores is contemplated, but the president is very leery of such an undertaking. (*b*) The Crooked Nail Construction Company built a 250-unit apartment house three years ago and still retains ownership. Due to the risk, when the project was undertaken a 12% return was required; however, because of the favorable outcome management feels this is safer than some other types of investments.

P18.4   A large auto-manufacturing firm requires $2 million in new funds. The financial consultants of the company recommend that the firm sell 12-year, 8% semiannual bonds to a brokerage firm at a 4% discount. (*a*) What is the total face value of the bond issue? (*b*) Compute the after-tax cost of capital obtained in this manner, if the tax rate is 50%.

P18.5   The Barely-Making-It Swimming Suit Company has to raise $50,000 in new capital. Two methods of debt financing are available. The first is to borrow $50,000 from a bank. The company will pay an effective 8% per year for eight years to the bank. The other method would require the issuing of 50 $1,000, ten-year bonds which will pay a 6% annual dividend. If you assume that the principal on the loan is reduced uniformly for the eight years, with the remainder going toward interest, which method of financing would you recommend after taxes are accounted for? Assume the tax rate is 52%.

P18.6   Is the answer the same for problem P18.5 if a before-tax analysis is made?

P18.7   Purchase of some new equipment for $75,000 is contemplated by the Goat and Cow Dairy. The equipment will last five years and can be salvaged for $15,000. The company has $25,000 in available money and hopes to borrow the remainder from a bank for a five-year period. The equipment is expected to increase cash flow before taxes by $18,000 per year. The new equipment will be straight-line depreciated and a 50% tax rate is applicable. If the economist for the firm estimates that the taxes for this endeavor will be $1,500 per year, what is the (*a*) stated interest rate paid on the loan and (*b*) effective interest rate on the loan after taxes provided the MARR of 10% is realized? Assume uniform reduction in principal over the life of the loan.

P18.8   Compute the after-tax rate of return for problem P18.7 if the stated loan rate were reduced by 10%.

P18.9   The common stock that is outstanding for Bottleneck Contractors earned an average of $0.75 per share last year. If the market price averaged $11.50 per share and the tax rate was 47.5%, what was (*a*) the before-tax cost of capital and (*b*) the after-tax cost of capital?

P18.10 The owners of a grocery store plan to construct a laundromat next door. They will use 100% equity financing. It will cost $22,000 to build the facility, which will be straight-line depreciated over a 15-year life using a salvage value of $7,000. They have the equity funds invested at the present time and make 10% per year. If the annual CFBT is expected to be $5,000 and the tax rate is 48%, is the venture expected to be profitable?

P18.11 Use the Gordon-Shapiro method to compute the BTCC for the following situation. The AZ Company has $1,500,000 in common stocks outstanding. Last year a dividend of $1.25 per share was paid. Earnings were reported as $1.37 per share on a market value of $13.75. Book value is 75% of market value, and the tax rate was 51%.

P18.12 The AZ Company of problem P18.11 would like to reduce its BTCC on equity financing to 15% but pay the same dividend. What earnings per share would be necessary to obtain a 15% BTCC? How much would be left per share for retained earnings?

P18.13 Mr. Snoozer, the president of Westfall Mattresses, sees an opportunity to invest $100,000 in a new mattress line. He anticipates an annual net income of $37,800 per year for the next seven years, which will be taxed at a rate of 50%. No depreciable assets are involved. The president hopes to raise the investment capital by selling stocks at a market value of $5.80. What percent of market value would have to be earned in order to keep his ATCC at one-half his rate of return on the investment?

P18.14 An investor is interested in a certain company. She finds that stock has a market value of $28.50 per share and has paid an 8% dividend for the last five years. An accountant friend who works for this company tells her that this company retains only 50% of its earnings to keep the ATCC very low. If the accountant also tells the investor that the current book value of stock is 60% of market value, what is the ATCC? Does this ATCC seem high or low?

P18.15 A large company would like to purchase a small firm which has been a supplier for many years. A cost of $780,000 has been placed on the small firm. The purchasing company does not know exactly how to finance the purchase to obtain an average ATCC as low as that for other ventures. The average BTCC is presently 10%. Two schemes of financing are available. The first requires that the company invest 50% equity funds at 8% and borrow the balance at 11% per year. The second scheme requires only 25% equity funds and the balance can be borrowed at 9%. Which scheme will require the smaller earnings?

P18.16 If ATCC is 10% in problem P18.15, what rate of interest could be paid on the debt capital for the two schemes?

P18.17 Rework Example 18.7 by computing the correct interest value each year for the 50% debt financing. Will the decision be altered due to this more exact treatment of debt financing?

P18.18 The Pure Trash Company has always used 100% equity financing in the past. A good opportunity is now offered that will require the raising of $250,000. The owner can supply the money from personal investments, which earn at an ATCC of $8\frac{1}{2}\%$ per year. The annual CFAT is expected to be $30,000 for the next 15 years at a tax rate of 50%. Alternatively, 60% of the required amount can be borrowed at 5% per year for 15 years. If it is assumed that the principal is uniformly reduced and an average annual interest is paid, use a MARR of 1.2 times ATCC to determine which plan is better.

P18.19 The Sno-Plow Company uses a MARR value that is 1.5 times the cost of capital. Three plans for raising $50,000 are available. These are detailed below.

Type of	Plan		
financing	1	2	3
Equity	90%	60%	20%
Debt	10	40	80

At present the before-tax cost for equity capital is 10% and for debt capital, 12%. If the project is expected to yield $10,000 for five years, do a before-tax analysis to determine the return for each plan and which plan is best.

P18.20 A conscientious couple devised a plan to buy certain types of groceries now for $600 in order to save a total of 25% in the next six months. However, they are not sure how to finance this plan. The couple wants to make a return of 50% more than the financing methods will cost over the six-month period. The financing plans are as follows: (a) Take $600 from a savings account now and put the monthly savings of $125 in as they are received. This account pays 6% per year compounded quarterly. (b) Borrow $600 now from the credit union at an effective 1% per month and repay the loan at $103.54 per month for six months and put the difference between the payment and the amount saved each month in the 6% compounded quarterly savings account. (c) Use the extra $300 from this month's budget, borrow $300 at 1% per month, and repay at the rate of $51.77 per month for six months. Again, the difference between the payments and the savings would be saved at 6% compounded quarterly.

Perform a before-tax, average cost-of-capital analysis to determine which financing plan is the most profitable. Assume there is no interperiod interest paid on the savings account.

P18.21 The Hack-Away Cough Syrup Company has a total of 153,000 shares of stock outstanding at a market value of $28 per share. Earnings are 12.5% of market and a 48% tax rate is used by the accountant for Hack-Away. Stocks are sufficient to finance only 50% of the company's undertakings. The remaining is financed by bonds and funds borrowed from a bank. Thirty percent of the debt financing is by $1,285,000 worth of $10,000, 6% per year, 15-year bonds, which were sold at a 2% discount for rapid sale. The remaining 70% of debt financing is by loans which are repaid at an effective after-tax rate of 9% per year. Using just the information above, determine (a) the average BTCC and (b) the average ATCC.

P18.22 Compute the debt-to-equity ratio for all plans in the following problems: (a) P18.19, (b) P18.20, and (c) P18.21.

P18.23 Why is an extremely large D/E value not healthy for a company?

P18.24 Compute the new rate-of-return values for the three firms described in Example 18.8(c) if the D/E ratio for each firm is 1.00.

P18.25 Company A has a total asset value of $2,500,000 and a D/E ratio value of 0.40 while Company B has total assets valued at $1,600,000 and D/E = 2.5. (a) Compute the dollar amount that is from equity and debt sources for each company. (b) If asset value is reduced by 15%, compute the new D/E values for both firms.

P18.26 Assume that the revenue for each company in problem P18.25(a) is $500,000 and that Company A pays 7% for debt financing and B pays 9%, due to the greater risk taken by the lender because of the larger D/E value. Compute the rate of return on common stock for each company, assuming that common stock is the sole source of equity financing.

19
SENSITIVITY AND RISK ANALYSIS

In this chapter you will learn the use of sensitivity analysis in economy studies and be introduced to the idea of probabilistic engineering economy.

Probabilistic analysis is, in realistic terms, largely an uncharted course for the engineering economist. The computations and decision-making steps are relatively simple, *once the input data are known*. However, to obtain good, real-world probabilistic data is quite difficult and expensive. In this chapter we take only a brief look at risk analysis.

CRITERIA

To complete this chapter you must be able to do the following:

1 State the purpose of sensitivity analysis and how it differs from breakeven analysis.
2 Determine the sensitivity of factor(s) for one or more projects, given the factor(s) to be studied, possible variation of each factor, and project details for all other factors.

3 Select the most economic of two or more projects using EUAC analysis, given three estimates (pessimistic, reasonable, and optimistic) for important factors of each project and values for the nonvarying factors.

4 Compute the expected value of a variable, given the variable values and associated probabilities.

5 Determine the desirability of a project using expected value computations, given the time of the project, the cash flows, and/or the interest rate.

EXPLANATION OF MATERIAL

19.1 The Approach of Sensitivity Analysis

Since the workplace of engineering economists is the *future*, the estimates they use can possibly be in error. Sensitivity analysis is a study to see how the economic decision will be altered if certain factors are varied. For example, variation in the minimum attractive rate of return (MARR) may not alter a decision when all compared proposals would return far more than the MARR; thus, the decision is relatively insensitive to MARR. However, if a change in the economic life is critical, the decision is sensitive to life estimates.

Usually the variations in values of life, annual costs, or incomes, etc., result from variations in selling price, operation at different levels of capacity, inflation, etc. For example, if 75% capacity is compared to 50% capacity for a contemplated proposal, operating costs and revenue will increase, but anticipated life will probably decrease. Usually several important factors are studied to learn how the uncertainty of estimates will affect the economic study.

Plotting the sensitivity of present worth, EUAC, or rate of return versus the factor studied is quite illustrative. Two projects can be compared with respect to a given factor and the *breakeven point* computed, as in Sec. 12.5. This is the variable value at which the two proposals are economically equivalent. However, the breakeven chart represents only one factor per chart. Thus, several charts, one for each factor, must be constructed and independence of each factor assumed. In previous uses of breakeven analysis, we computed two values and connected the points with a straight line. However, if a factor generates sensitive results, several intermediary points should be used to be conscious of the sensitivity. This fact is illustrated in this chapter.

If several factors are to be studied, as is the usual case, they may be studied one at a time using manual computation. However, to get an idea of how several factors affect the sensitivity, a computer program should be written using general formulas with the varying factors expressed as unknowns. The computer easily allows more than one basis of comparison to be employed, for example, present-worth and rate-of-return analysis. In addition, the computer can plot the results, giving a rapid visual display of sensitivity.

19.2 Determination of the Sensitivity of Estimates

There is a general procedure that should be followed when conducting a sensitivity analysis. The steps in this procedure are as follows:

1 Determine which factor(s) are most likely to vary from the estimated value.
2 Select the probable range and increment of variation for each factor.
3 Select an evaluation method, such as present-worth, EUAC, or rate-of-return, that will be used to evaluate each factor's sensitivity.
4 Compute and, if desired, plot the results from the evaluation method selected in step *3*.

The results of the sensitivity analysis will show those factors that should be carefully estimated by collecting more information when possible. Example 19.1 illustrates sensitivity analysis for one project.

Example 19.1   The ACQ Company is contemplating the purchase of a new piece of automatic machinery for $80,000 with zero salvage value and an anticipated before-tax cash flow of $27,000 - 2,000k$ $(k = 1, 2, \ldots, n)$ per year. Figure 19.1 is a cash-flow diagram of the asset purchase. The MARR for the company has varied from 10% to 25% for different types of investments. The economic life of similar machinery normally varies from eight to 12 years. Use EUAC analysis to investigate the sensitivity of varying (*a*) the MARR using $n = 10$ and (*b*) the economic life, assuming a MARR of 15%.

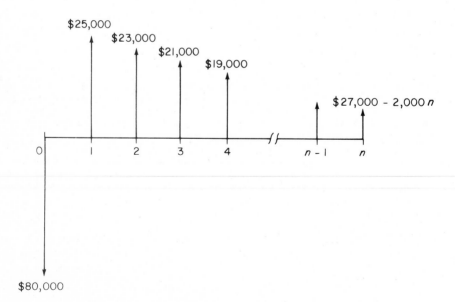

FIG. 19.1   Cash-flow diagram used for sensitivity analysis of *i* and *n*, Example 19.1.

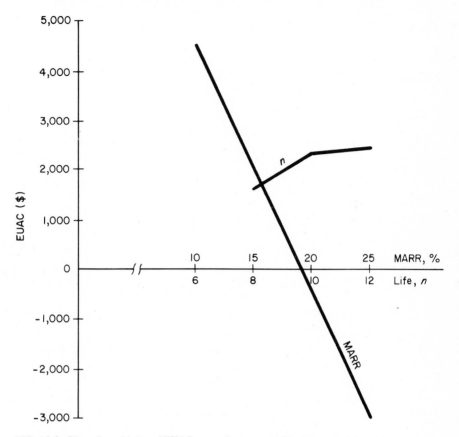

FIG. 19.2 Plot of sensitivity of EUAC versus interest and life, Example 19.1.

SOLUTION

(*a*) Allowing *i* to change by a 5% increment should be sufficient for sensivity purposes. For $i = 10\%$,

$$P = -80,000 + 25,000(P/A,10\%,10) - 2,000(P/G,10\%,10)$$
$$= \$27,830$$
$$\text{EUAC} = P(A/P,10\%,10) = \$4,529$$

Similarly, other results are

i	P	EUAC
15%	$ 11,512	$ 2,294
20	−962	−229
25	−10,711	−3,000

A plot of MARR versus EUAC is shown in Fig. 19.2. The steep negative slope of EUAC indicates that the decision is quite sensitive to variations in MARR. If it is

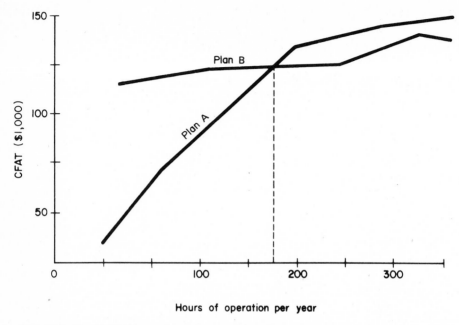

FIG. 19.3 Plot of CFAT versus hours for two alternatives.

likely that the MARR will be set toward the upper end of the company's MARR range, the investment would not be attractive.

(*b*) Using an increment of two years, the *P* and EUAC values for $8 \leqslant n \leqslant 12$ at $i = 15\%$ are as follows:

n	*P*	EUAC
8	$ 7,221	$1,609
10	11,511	2,294
12	13,145	2,425

Figure 19.2 presents the characteristic nonlinear relation of EUAC versus *n*. Since the EUAC is positive for all values of *n*, the decision to invest would not be affected by the economic life of the machine.

COMMENT Note that after $n = 10$, the EUAC curve seems to level out and become quite insensitive. This insensitivity to changes in cash flow in the distant future is the expected trait because, when discounted to time zero, the present-worth value gets smaller as *n* increases.   ////

If two projects are compared and the sensitivity for one factor is to be determined, the actual plot may show quite nonlinear results. Take a look at the general form of the plots in Fig. 19.3. We won't be concerned with the actual computations, but the graph shows that the CFAT of each plan is a nonlinear function of hours of operation. Plan A is extremely sensitive in the range of 50 to 200 hours but is comparatively insensitive above 200 hours. We mentioned in Sec. 19.1 that sensitive plans should have intermediary points plotted for a more accurate sensitivity analysis to be made. If a considerable variation in the hours of operation is expected and if Plans A and B are both economically justified, Plan B should be selected due to its relative insensitivity to hours of operation. This will provide some assurance of a relatively stable CFAT value.

Example 19.6
Problems P19.1–P19.14

19.3 Sensitivity of Alternatives Using Three Estimates of Factors

We can thoroughly study the economic decision between two or more projects by borrowing from the field of project scheduling the concept of making three estimates for pertinent factors: a pessimistic, a reasonable, and an optimistic estimate. This allows us to study decision sensitivity for each factor, thereby obtaining different decisions depending on the factor considered. After the sensitizing, the analyst should accept the decision representing the economic situation as he or she best understands it. This involves a subjective weighing of the sensitized factors.

Example 19.2 You are an engineering economist attempting to evaluate three alternatives for which pessimistic (P), reasonable (R), and optimistic (O) estimates are made for the life, salvage value, and annual operating costs (Table 19.1). Determine the most economic alternative using EUAC analysis and a before-tax MARR of 12%.

Table 19.1 COMPETING ALTERNATIVES WITH THREE ESTIMATES FOR n, SV, AND AOC

	First cost, P	Salvage, SV	AOC	Life, n
Alternative A				
P	$20,000	0	$11,000	3
R	20,000	0	9,000	5
O	20,000	0	5,000	8
Alternative B				
P	$15,000	$ 500	$ 4,000	2
R	15,000	1,000	3,500	4
O	15,000	2,000	2,000	7
Alternative C				
P	$30,000	$3,000	$ 8,000	3
R	30,000	3,000	7,000	7
O	30,000	3,000	3,500	9

Table 19.2   EUAC VALUES FOR THREE ALTERNATIVES
WITH VARYING FACTOR ESTIMATES,
EXAMPLE 19.2

	Alternatives		
Strategy	A	B	C
P	$19,327	$12,640	$19,601
R	14,548	8,229	13,276
O	9,026	5,089	8,927

SOLUTION For each alternative description in Table 19.1 we must compute EUAC. For example, using the pessimistic (P) estimates for Alternative A, compute

$$EUAC = 20,000(A/P, 12\%,3) + 11,000 = \$19,327$$

Table 19.2 presents EUAC values at 12% for all situations. Figure 19.4 is a plot of EUAC versus the three life estimates for each alternative. Since the EUAC calculated from the "reasonable" estimates for Alternative B is less than even the optimistic EUAC values for A and C, Alternative B is clearly favored.

COMMENT While the alternative that should be selected in this example is obvious, this may not always be the case. For example, in Table 19.2, if the pessimistic

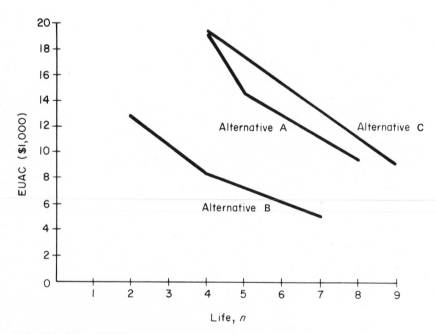

FIG. 19.4   Plot of EUAC versus three life estimates, Example 19.2.

EUAC for Alternative B were $21,000 and the optimistic EUAC values for Alternatives A and C were less than $5,089, selection of B would not be as apparent. In this case, it would be necessary to decide which strategy (P, R, or O) would control the decision. In determining which strategy should be used, other factors, such as the project MARR, the availability of capital, and the economic stability of the company, would have to be taken into consideration. ////

Problems P19.15–P19.20

19.4 Economic Uncertainty and the Expected Value

The use of probability and its computations by the engineering economist are not very common; the reason for this is not because the computations are difficult to perform or understand, but because realistic probability values are difficult to obtain. The economist must deal with *uncertain future* monetary and life values. Often reliance on past data, if any exist, for future cash-flow values is incorrect. However, experience and wise judgment can often be used in conjunction with the expected value to evaluate the desirability of a project. The *expected value* can be interpreted as a long-run average outcome if the project were repeated many times. However, even for a single purchase, the expected value is meaningful.

The expected value $E(X)$ may be computed using the relation

$$E(X) = \sum_{i=1}^{m} XP(X) \quad (m = 1, 2, \ldots, n) \qquad (19.1)$$

where X = specific variable value

$P(X)$ = probability a specific value will occur

A much more detailed explanation of probability, expected values, and statistics can be found in any text on probability and statistics.

Example 19.3 You expect to be mentioned in your favorite uncle's will. You anticipate being willed $5,000 with a probability of 0.50, $50,000 with a 0.45 chance, and a value of zero dollars with a 0.05 chance. What is your expected inheritance?

SOLUTION Let

$$X = \text{inheritance value}$$
$$P(X) = \text{associated, subjectively evaluated probability}$$

Your expected inheritance is

$$E(X) = 5,000(0.50) + 50,000(0.45) + 0(0.05) = \$25,000$$

COMMENT As in all probability statements, the sum of all $P(X)$ values must be 1.0. If the $P(X) = 0.05$ for $X = \$0$ were not stated, it would have to be assumed. Of course, it makes no computational difference but is necessary for completeness.

If actual cash flows are the X values, some can be negative, as for the first cost. If the expected value is positive, then the overall outcome is expected to be a cash inflow. Therefore, $E(\text{cash flow}) = \$-1,500$ in present worth indicates a losing proposition. The expected-value concept can be used in engineering-economy studies, as discussed below.   ////

Problems P19.21–P19.24

19.5 Expected Value of Economy Alternatives

The expected value $E(X)$ can be used in many ways and the probabilities used in its computation can be stated in different fashions. Example 19.4 below is a simple $E(X)$ computation, while Example 19.5 involves the use of PW for an asset with the possibility of different cash-flow sequences.

Example 19.4 A utility company is experiencing a difficult time in obtaining natural gas for electric generation and distribution. Monthly expenses are now averaging $7,750,000. The economist for this city-owned utility has collected average revenue figures for the past two years using the categories "fuel plentiful," "less than 30% other fuel purchased," "greater than 30% other fuel" (Table 19.3). "Other" is a fuel other than natural gas, purchased at a premium, the cost of which is transferable directly to the consumer. Can the utility expect to meet future expenses based on the two years of data?

SOLUTION Using the 24-month data-collection period, probabilities may be computed as

Fuel	$P(X)$
Plentiful	$12/24 = 0.50$
$< 30\%$	$6/24 = 0.25$
$\geqslant 30\%$	$6/24 = 0.25$

Probabilities add up to 1.0, so Eq. (19.1) results in

$$E(X) = E(\text{revenue}) = 5,270,000(0.50)$$
$$+ 7,850,000(0.25) + 12,130,000(0.25)$$
$$= \$7,630,000$$

With average monthly expenses of $7,750,000, the utility will have to raise rates in the near future.   ////

Table 19.3 FUEL DATA FOR A MUNICIPAL UTILITY

Fuel situation	Months in past 2 years	Average revenue per month, X
Plentiful	12	$ 5,270,000
$< 30\%$ other	6	7,850,000
$\geqslant 30\%$ other	6	12,130,000

Table 19.4 EQUIPMENT CASH FLOW PROBABILITIES
FOR EXAMPLE 19.5

	Economy		
Year	Receding (Prob. = 0.2)	Stable (Prob. = 0.6)	Expanding (Prob. = 0.2)
0	$-5,000	$-5,000	$-5,000
1	+2,500	+2,000	+2,000
2	+2,000	+2,000	+3,000
3	+1,000	+2,000	+3,500

Example 19.5 The Tule Company has had experience with automatic reaming equipment. A certain piece of equipment will cost $5,000 and have a life of three years. Suspected actual cash flows and probability of each are listed in Table 19.4, depending on a receding, stable, or expanding economy. Using expected-present-worth analysis, determine if the equipment should be purchased at a 15% rate of return.

SOLUTION The first step is to find the present worth under each type of anticipated economy and then find the $E(\text{PW})$ using Eq. (19.1). Letting the subscript R represent a receding economy, S a stable, and E an expanding economy,

$$PW_R = 4,344 - 5,000 = \$ \ -656$$
$$PW_S = 4,566 - 5,000 = \ -434$$
$$PW_E = 6,309 - 5,000 = \ +1,309$$

$$E(\text{PW}) = \sum_{j=R,S}^{E} PW_j[P(j)]$$
$$= -656(0.2) - 434(0.6) + 1,309(0.2)$$
$$= \$-310$$

Since $E(\text{PW}) < 0$, the project is not expected to be a paying proposition at a 15% required rate of return. ////

The probability analysis presented in this chapter is extremely basic and intended only as an introduction to the field of probabilistic engineering economy. Current research into the area can be found in recent issues of *The Engineering Economist, Management Science, The Journal of Finance,* and *The Journal of Business.*

Example 19.7
Problems P19.25–P19.29

SOLVED EXAMPLES

Example 19.6 The city of Booney has a three-mile stretch of heavily traveled highway to resurface. The Ajax Construction Company offers two methods of resurfacing. The first is a concrete surface for a cost of $150,000 and an annual maintenance charge of $1,000.

The second method is an asphalt covering with a first cost of $100,000 and a yearly service charge of $2,000. However, Ajax would also request that every third year the highway be "touched up" at a cost of $2,500 per mile.

The city uses the interest rate on revenue bonds, 6% in this case, as the MARR. (a) Determine the breakeven number of years of the two methods. If the city expects an interstate to replace this stretch of highway in 20 years, which method should be selected? (b) If touch-up cost increases by $500 per mile every three years, is the decision sensitive to this cost?

SOLUTION

(a) We can set up the equation

$$\text{EUAC of concrete} = \text{EUAC of asphalt}$$

and compute the breakeven n value. The breakeven equation is

$$150,000(A/P,6\%,n) + 1,000 = 100,000(A/P,6\%,n) + 2,000$$

$$+ 7,500 \left[\sum_j (P/F,6\%,j) \right] (A/P, 6\%,n) \qquad (19.2)$$

where $j = 3, 6, 9, \ldots, n$, which may be rewritten as

$$50,000(A/P,6\%,n) - 1,000 - 7,500 \left[\sum_j (P/F,6\%,j) \right] (A/P,6\%,n) = 0 \qquad (19.3)$$

A value of $n = 39.6$ satisfies the equation, so a life of approximately 40 years is required to break even at 6%. Since the road is required for only 20 years, the asphalt surface should be constructed.

(b) The total touch-up cost will increase by $1,500 every three years. Equation (19.3) may now be written as

$$50,000(A/P,6\%,n) - 1,000$$

$$- \sum_j \left[7,500 + 1,500 \left(\frac{j-3}{3} \right) (P/F,6\%,j) \right] (A/P, 6\%,n) = 0$$

where $j = 3, 6, 9, \ldots, n$. The breakeven point is now found to be approximately 21 years, which still favors the asphalt surface. The decision is therefore insensitive to possible increases in the touch-up cost.   ////

Sec. 19.2

Table 19.5   RAINFALL AND SUPPORT WALL COST

Rainfall (in.)	Probability of greater rainfall occurring	Cost of support wall to carry rainfall
2.0	0.300	$10,000
2.5	0.100	15,000
3.0	0.050	22,000
3.5	0.010	30,000
4.0	0.005	42,000

Example 19.7   The Holdup Construction Company plans to build an apartment complex close to the edge of a partially leveled hill. Support of the cliff below the complex will assure that no damage will occur to the buildings or occupants. The amount of rainfall at any one time can cause varying amounts of damage. Table 19.5 itemizes the probability of certain rainfalls within a period of a few hours and the first cost to construct a support wall to insure protection against the corresponding amount of water.

The project of wall construction will be financed by a 30-year 9% loan. Data on record show that an average of $20,000 damage has often occurred when a heavy rain fell. Without taking the intangible (and extremely important) fact of human life into account, what size support wall is most economic?

SOLUTION   We will use EUAC analysis to find the most economic plan. For each rainfall level

$$EUAC = \text{annual loan cost} + \text{expected annual damage}$$
$$= \text{cost}(A/P,9\%,30) + 20,000(\text{probability of greater rainfall})$$

The $20,000 damage figure is used because this has been the previous damage on the *average*. We assume the probabilities are a yearly value. Computations in Table 19.6 show that a $30,000 wall to protect against a 3.5-inch rainfall is most economic. A wall to protect against a 3.0-inch rain is a close second to the most economic plan.

Table 19.6   EUAC FOR DIFFERENT SUPPORT WALLS

Rainfall (in.)	Support wall cost	Annual loan cost	Expected annual damage	EUAC
2.0	$10,000	$ 973	$6,000	$6,973
2.5	15,000	1,460	2,000	3,460
3.0	22,000	2,141	1,000	3,141
3.5	30,000	2,920	200	3,120
4.0	42,000	4,088	100	4,188

COMMENT Usually a rather hefty safety factor is automatically added when people are endangered. Pure economic analysis is not used alone as the decision maker. By building to protect to a greater degree than usually needed, the probabilities of damage are lowered, by how much we don't know, but it makes us *feel safer!*   ////

Sec. 19.5

BIBLIOGRAPHY

Barish, pp. 19–20, 307–314 (risk).
DeGarmo and **Canada,** pp. 253–266 (sensitivity and risk).
Fabrycky and **Thuesen,** pp. 213–220 (sensitivity); pp. 227–231 (risk).
Grant and **Ireson,** pp. 251–260, 266–270 (sensitivity); pp. 279–288 (risk).
Park, pp. 66–69 (sensitivity); pp. 69–74, 222–226 (risk).
Smith, pp. 315–317 (sensitivity); pp. 309–315, 317–320 (risk).
Taylor, pp. 443–447 (sensitivity); pp. 297–303 (risk).
Thuesen, Fabrycky, and **Thuesen,** pp. 312–314 (sensitivity); pp. 320–326 (risk).

PROBLEMS

P19.1   The A-1 Salvage Company is contemplating the purchase of a new crane equipped with a magnetic pick-up device to be used in moving scrap metal around the yard. The complete crane will cost $62,000, have an eight-year life, and a salvage value of $1,500. Annual maintenance, fuel, and overhead costs are estimated at $0.50 per ton. Labor cost will be $8 per hour for regular wages and $12 for overtime. A total of 25 tons can be moved in an eight-hour period. The salvage yard has, in the past, handled anywhere from ten to 30 tons of scrap per day. If the company uses a MARR of 10%, plot the sensitivity of the present worth of costs versus annual volume moved assuming the operator is paid for 200 days of work per year. Use a five-ton increment for the graph.

P19.2   A new piece of equipment is being economically evaluated by three engineers for the Zoot-Suit Electric Utility. The first cost will be $77,000 and the life is estimated at six years with a salvage value of $10,000 at disposal time. The engineers disagree, however, on the annual net income to be credited to this new equipment. Engineer A has given an estimate of $14,000 per year. Engineer B states that this is too low and estimates $15,000, while C estimates $18,000 per year. If the MARR is 8%, use present-worth analysis to determine if these three estimates will change the decision to buy the equipment.

P19.3   Perform the same analysis as in problem P19.2, except make it an after-tax consideration using straight-line depreciation and a 52% effective tax rate. Assume an annual equipment-operating cost of $1,000 and an after-tax rate of return of 5%.

P19.4   In problem P8.8 (p. 144) you made a present-worth comparison between building and leasing storage space. Determine the sensitivity of your decision to the following situations: (*a*) construction costs go up 10% and lease costs go down to $0.125 per square foot per month; (*b*) lease costs remain at $0.15 per square foot per month, but construction costs vary from $5 to $9 per square foot.

P19.5   For the situation presented in problem P10.8 (p. 179), plot the sensitivity of the rate of return to the amount of the income gradient. Perform the computations for values of the *negative* gradient from $300 to $700 in increments of $100. If the company would like a return of at least 40%, would variation in this income gradient have affected the decision to buy the dump trucks?

P19.6   Consider the two air-conditioning systems detailed below.

	System 1	System 2
First cost	$10,000	$17,000
Annual operating cost	200	150
Salvage value	−100	−300
New compressor and motor cost at midlife	1,750	3,000
Life, years	8	12

Use an EUAC analysis to determine the sensitivity of the economic decision to MARR values of 8%, 10%, 12%, and 15%.

P19.7   Reread problem P19.6. If MARR is 10%, plot the EUAC for each system for life values from four to eight for System 1 and six to 12 for System 2. Assume the salvage values and annual operating costs are the same for each life value. Further, assume that the compressor is replaced at midlife. Plot EUAC for even-numbered years only. Which EUAC is more sensitive to a varying-life estimate?

P19.8   A city couple, Joan and John Pollution, would like to buy a small section of land in the woods to be used as a weekend vacation home. Alternatively, they have thought of buying a travel trailer and four-wheel-drive vehicle to pull the trailer for vacations.

   The Pollutions have found a five-acre tract with a cabin, well, etc., 25 miles from their home. It will cost them $30,000, but they expect to sell the acreage for $45,000 in ten years when their children are grown. The insurance, upkeep, etc., costs are estimated at $500 per year, but this weekend site is expected to save the family $50 every day they don't go on a traveling vacation. The Pollutions estimate that, even though the cabin is only 25 miles from home, they will travel 50 miles a day when at the cabin while working on it, visiting neighbors, etc. The Pollution car averages 15 miles per gallon of gasoline.

   The trailer and van combination would cost $11,000 and could be sold for $2,000 in ten years. Insurance, maintenance, etc., costs will be $750 per year, but the trailer is expected to save $25 per vacation day. On a normal vacation, the Pollutions travel 300 miles each day. Mileage per gallon for the van is estimated at 60% that of the family car.

Gas costs $0.60 per gallon and the Pollution family want a return of fun and 10% from either investment. Plot the sensitivity of EUAC for each plan if the Pollutions' vacation time in the past has been from six to 14 days per year. Also compute the breakeven number of days per year for the two plans.

P19.9 Suppose the Pollution family of problem P19.8 plan to purchase the acreage with the cabin and still go on a four-day traveling vacation in the car at a cost of $65 per day. (a) Was the decision to buy the land still the better decision? (b) Does the breakeven point seem to be sensitive to the type of vacation plans that combine the use of the acreage and traveling?

P19.10 Plot the sensitivity of rate of return versus the life of the bond for problem P13.15 (p. 234). Use life values of 10, 12, 15, 18, and 20.

P19.11 The Charley Horse Company has been offered an investment opportunity that will require a cash outlay of $30,000 now and a cash inflow of $3,500 for each year of investment. However, the company must state now the number of years it will retain the investment. If the investment is kept for six years, $25,000 can be gotten for the company's share, but after ten years the resale value will be only $12,000. If money is worth 8%, is the decision sensitive to the retention period?

P19.12 The person who did the analysis in problem P12.31 (p. 223) is concerned with the sensitivity of the breakeven point to the expected life of the project. Determine the sensitivity for (a) 15 years and (b) 25 years.

P19.13 Determine the sensitivity of the most economic life value of problem P12.23 (p. 221) to the cost gradient. Investigate the gradient values of $60 to $140 in increments of $20 and plot the results.

P19.14 Rework problem P19.13 at an interest rate of $i = 5\%$ and plot the results on the graph used in P19.13.

P19.15 Reread problem P9.7 (p. 158). The time of overhaul can vary from two to four years for the used machine and from four to six years for the new machine. Plot the EUAC values for these three estimates and determine if they will alter the decision of P9.7.

P19.16 If the spray method of problem P9.12 (p. 159) is used, the amount of water used can vary from an optimistic value of 15 gallons to a pessimistic value of 30 gallons with 20 gallons being a reasonable figure. The immersion technique always takes four gallons per ham. How will this varying use of water for the spray method affect the economic decision?

P19.17 An engineer is trying to decide between two ways to pump concrete up to the top floors of a seven-story office building now under construction. Plan 1 requires the purchase of equipment costing $6,000 and costing between $0.40 and $0.75 per ton to operate with an expected cost of $0.50 per ton. The asset is able to pump 100 tons per day. If purchased, the asset will last for five years, have no salvage value, and be used from 50 to 100 days per year. Plan 2 is an equipment-leasing option and is expected to cost the company $2,500 per year for equipment with an optimistic cost of $1,800 and a pessimistic value of $3,200 per year. In addition, a $5-per-hour labor cost will be incurred for operation of the leased equipment.

Plot the EUAC of each plan versus total annual operating or lease cost at $i = 12\%$. Determine which plan the engineer should select, using the reasonable estimates for a use of (a) 50 and (b) 100 days per year.

P19.18 When the country's economy is expanding, the AB Investment Company is optimistic and uses a MARR of 8% on all new investments. However, in a receding economy a return of 15% on investments is required. Normally a 10% return is required. Similarly, an expanding economy causes the estimates of asset life to go down about 20% and a receding economy makes the n values increase about 10%. Plot the sensitivity of present worth versus (a) MARR and (b) life values for the two plans detailed below using the reasonable estimate for other varying factors.

	Plan M	Plan Q
Initial investment	$-100,000	$-110,000
Annual CFBT	+15,000	+19,000
Life, years	20	20

P19.19 Rework problem P19.18 except use the plans detailed in problem P9.18 (p. 160).

P19.20 When is it necessary to select a particular strategy of pessimistic, reasonable, or optimistic and make an economic decision on the basis of the selected strategy?

P19.21 The variable X can take on the values $X = 5, 10, 15, 20$ with a probability of 0.40, 0.30, 0.233, 0.067, respectively. Compute the expected value of X.

P19.22 The AOC value for a plan can take on one of two values. Your office partner told you that the high value is $2,800 per year. If her computations show a probability of 0.75 for the high value and an expected AOC of $2,575, what is the lower AOC value used in the computation of the average AOC value?

P19.23 Find the expected present worth of the following series of payments if each series is expected to be realized with the probability shown at the head of each column (assume $i = 20\%$).

	Annual cash flow		
Year	$P = 0.5$	$P = 0.2$	$P = 0.3$
0	$-5,000	$-6,000	$-4,000
1	+1,000	+500	+3,000
2	+1,000	+1,500	+1,200
3	+1,000	+2,000	-800

P19.24 Compute the expected EUAC value for the cash flows of problem P19.23.

P19.25 The officers of a resort country club are thinking of constructing an additional 18-hole golf course. Due to the northerly location of the resort there is a 60% chance of a 120-day golf season, a 20% chance of 150 days of golfing weather, and a 20% chance of a 165-day season. The course will be used by an estimated 350 golfers each day of the four-month season, but by only 100 per day for each extra day in the golfing season. The course will cost $375,000 to construct and will require a $25,000 rework cost after four years. Annual maintenance cost will be $36,000 and the green fees will be $4.25 per person. If a life of ten years is anticipated before a major rework is required and a 12% return is required, determine if the course should be constructed.

P19.26 The owners of the Dial-A-Hole Roofing Company want to invest $10,000 in new equipment. A life of six years and a salvage value of 12% is anticipated. The annual income will depend upon the state of the housing and construction industry. The income is expected to be $2,000 per year; however, a current slump in the industry is given a 50% chance of lasting three years and a 20% chance of continuing for three additional years. However, if the outlook of the depressed market does improve, either during the first or second three-year period, the annual income of the investment is expected to be $3,500. Can the company expect to make 8% on its investment?

P19.27 The Way-Too-High Construction Company is building an apartment complex in the arid Southwest on the top of a partially leveled hill. The road that winds around the hill to the apartment complex entrance needs retaining walls for support above and below the road surface. The probability of a rain shower greater than a given amount and the associated damage and retaining wall construction costs are shown below.

Rainfall (in.)	Probability of greater rainfall occurring	Retaining wall cost to carry rainfall	Expected annual damage for specified rainfall
1.0	0.6	$15,000	$ 1,000
2.0	0.3	16,000	1,500
2.5	0.1	18,000	2,000
3.0	0.02	21,000	5,000
3.5	0.005	28,000	9,000
4.0	0.001	35,000	14,000

Determine which plan will result in the lowest annual cost over a 25-year period at an interest rate of 10%.

P19.28 Rework problem P19.27 using an after-tax analysis, assuming the tax rate is 50% and the retaining-wall construction cost will be secured by an 8% 25-year loan. Assume the principal amount is reduced an equal amount each year with the remainder of the payment applied to interest.

P19.29 A private citizen has $5,000 to invest. If he puts the money in a savings and loan account, he is assured of receiving an effective 6.35% per year on the principal. If he puts the money in stocks he has a fifty-fifty chance of one of the following cash-flow sequences for the next five years.

Year	Stock 1	Stock 2
0	$-5,000	$-5,000
1-4	+250	+600
5	+6,800	+5,400

Finally, he can invest his $5,000 in improved property for five years with the following outcomes and probabilities P.

	Cash flow		
Year	$P = 0.3$	$P = 0.5$	$P = 0.2$
0	$-5,000	$-5,000	$-5,000
1	-425	0	+500
2	-425	0	+600
3	-425	0	+700
4	-425	0	+800
5	+9,500	+7,200	+5,200

Which of the three investments—savings, stocks, or property— is the best?

APPENDIXES

Appendix	Subject
A	Compound interest factors
B	Continuous compounding
C	Answers to problems

Tabulated here are values of interest factors in which interest is compounded once each period (Tables A-1 – A-24). If interest is compounded more or less frequently than once each period, refer to Chap. 4 and Appendix B. The computational form of the formulas is given below.

Factor	Notation	Formula
Single-payment compound amount	$(F/P,i\%,n)$	$(1+i)^n$
Single-payment present worth	$(P/F,i\%,n)$	$\dfrac{1}{(1+i)^n}$
Sinking fund	$(A/F,i\%,n)$	$\dfrac{i}{(1+i)^n-1}$
Uniform-series compound amount	$(F/A,i\%,n)$	$\dfrac{(1+i)^n-1}{i}$
Capital recovery	$(A/P,i\%,n)$	$\dfrac{i(1+i)^n}{(1+i)^n-1}$
Uniform-series present worth	$(P/A,i\%,n)$	$\dfrac{(1+i)^n-1}{i(1+i)^n}$

A few useful computational relations between factors are given below. (The $i\%$ and n are omitted from the notation when possible simply for the sake of brevity.)

$$(P/F) = \frac{1}{(F/P)} \quad (F/A) = \frac{1}{(A/F)} \quad (P/A) = \frac{1}{(A/P)}$$

$$(P/A) = (F/A)(P/F) \quad (A/P) = (A/F)(F/P)$$

$$(P/A) = \sum_{j=1}^{n} (P/F, i\%, j) \quad F/A = \sum_{j=1}^{n} (F/P, i\%, j)$$

Table A-25 and A-26 present factors that convert a uniform gradient (G) of \$1 per period to a present-worth or equivalent uniform annual series, respectively. Computational formulas are as follows:

Factor	Notation	Formula
Uniform-gradient present worth	$(P/G, i\%, n)$	$\frac{1}{i}\left[\frac{(1+i)^n - 1}{i(1+i)^n} - \frac{n}{(1+i)^n}\right]$
Uniform-gradient capital recovery	$(A/G, i\%, n)$	$\frac{1}{i} - \frac{n}{(1+i)^n - 1}$

Useful gradient relations are

$$(P/G) = (A/G)(P/A) \quad (A/G) = (P/G)(A/P)$$

INTEREST TABLES

TABLE A-1

0.25% COMPOUND INTEREST FACTORS

	SINGLE PAYMENTS		UNIFORM SERIES PAYMENTS				
N	COMPOUND AMOUNT F/P	PRESENT WORTH P/F	SINKING FUND A/F	COMPOUND AMOUNT F/A	CAPITAL RECOVERY A/P	PRESENT WORTH P/A	N
1	1.0025	0.9975	1.00017	0.9998	1.00267	0.9973	1
2	1.0050	0.9950	0.49951	2.0020	0.50201	1.9920	2
3	1.0075	0.9925	0.33259	3.0067	0.33509	2.9843	3
4	1.0100	0.9901	0.24912	4.0142	0.25162	3.9743	4
5	1.0126	0.9876	0.19905	5.0240	0.20155	4.9616	5
6	1.0151	0.9851	0.16566	6.0364	0.16816	5.9466	6
7	1.0176	0.9827	0.14181	7.0515	0.14431	6.9293	7
8	1.0202	0.9802	0.12393	8.0688	0.12643	7.9093	8
9	1.0227	0.9778	0.11002	9.0889	0.11252	8.8870	9
10	1.0253	0.9753	0.09890	10.1112	0.10140	9.8619	10
11	1.0278	0.9729	0.08979	11.1366	0.09229	10.8350	11
12	1.0304	0.9705	0.08221	12.1643	0.08471	11.8053	12
13	1.0330	0.9681	0.07579	13.1943	0.07829	12.7730	13
14	1.0356	0.9657	0.07029	14.2273	0.07279	13.7386	14
15	1.0382	0.9632	0.06552	15.2626	0.06802	14.7017	15
16	1.0408	0.9608	0.06135	16.3006	0.06385	15.6623	16
17	1.0434	0.9584	0.05767	17.3412	0.06017	16.6207	17
18	1.0460	0.9561	0.05439	18.3842	0.05689	17.5763	18
19	1.0486	0.9537	0.05147	19.4301	0.05397	18.5300	19
20	1.0512	0.9513	0.04883	20.4784	0.05133	19.4811	20
22	1.0565	0.9466	0.04428	22.5830	0.04678	21.3762	22
24	1.0617	0.9418	0.04049	24.6982	0.04299	23.2619	24
25	1.0644	0.9395	0.03882	25.7599	0.04132	24.2013	25
26	1.0671	0.9372	0.03728	26.8242	0.03978	25.1384	26
28	1.0724	0.9325	0.03453	28.9604	0.03703	27.0052	28
30	1.0778	0.9278	0.03215	31.1077	0.03465	28.8630	30
32	1.0832	0.9232	0.03006	33.2657	0.03256	30.7116	32
34	1.0886	0.9186	0.02822	35.4343	0.03072	32.5508	34
35	1.0913	0.9163	0.02738	36.5227	0.02988	33.4669	35
36	1.0940	0.9140	0.02659	37.6137	0.02909	34.3807	36
38	1.0995	0.9095	0.02512	39.8041	0.02762	36.2016	38
40	1.1050	0.9050	0.02381	42.0055	0.02631	38.0136	40
45	1.1189	0.8937	0.02103	47.5574	0.02353	42.5040	45
50	1.1329	0.8827	0.01880	53.1792	0.02130	46.9388	50
55	1.1472	0.8717	0.01699	58.8711	0.01949	51.3182	55
60	1.1616	0.8609	0.01547	64.6347	0.01797	55.6434	60
65	1.1762	0.8502	0.01419	70.4708	0.01669	59.9151	65
70	1.1910	0.8397	0.01309	76.3801	0.01559	64.1338	70
75	1.2059	0.8292	0.01214	82.3639	0.01464	68.3002	75
80	1.2211	0.8190	0.01131	88.4228	0.01381	72.4149	80
85	1.2364	0.8088	0.01058	94.5576	0.01308	76.4785	85
90	1.2519	0.7988	0.00992	100.7694	0.01242	80.4917	90
95	1.2676	0.7889	0.00934	107.0595	0.01184	84.4552	95
100	1.2836	0.7791	0.00882	113.4281	0.01132	88.3692	100

TABLE A-2

0.50% COMPOUND INTEREST FACTORS

	SINGLE PAYMENTS			UNIFORM SERIES PAYMENTS			
N	COMPOUND AMOUNT F/P	PRESENT WORTH P/F	SINKING FUND A/F	COMPOUND AMOUNT F/A	CAPITAL RECOVERY A/P	PRESENT WORTH P/A	N
1	1.0050	0.9950	1.00017	0.9998	1.00517	0.9949	1
2	1.0100	0.9901	0.49885	2.0046	0.50385	1.9847	2
3	1.0151	0.9852	0.33174	3.0144	0.33674	2.9696	3
4	1.0201	0.9803	0.24818	4.0293	0.25318	3.9497	4
5	1.0252	0.9754	0.19805	5.0493	0.20305	4.9250	5
6	1.0304	0.9705	0.16463	6.0743	0.16963	5.8953	6
7	1.0355	0.9657	0.14076	7.1045	0.14576	6.8608	7
8	1.0407	0.9609	0.12285	8.1400	0.12785	7.8217	8
9	1.0459	0.9561	0.10893	9.1805	0.11393	8.7775	9
10	1.0511	0.9514	0.09779	10.2263	0.10279	9.7288	10
11	1.0564	0.9466	0.08867	11.2772	0.09367	10.6753	11
12	1.0617	0.9419	0.08108	12.3333	0.08608	11.6169	12
13	1.0670	0.9372	0.07466	13.3947	0.07966	12.5540	13
14	1.0723	0.9326	0.06915	14.4617	0.07415	13.4865	14
15	1.0777	0.9279	0.06438	15.5338	0.06938	14.4143	15
16	1.0831	0.9233	0.06020	16.6113	0.06520	15.3374	16
17	1.0885	0.9187	0.05652	17.6941	0.06152	16.2559	17
18	1.0939	0.9142	0.05324	18.7824	0.05824	17.1700	18
19	1.0994	0.9096	0.05031	19.8761	0.05531	18.0794	19
20	1.1049	0.9051	0.04768	20.9753	0.05268	18.9843	20
22	1.1160	0.8961	0.04312	23.1903	0.04812	20.7807	22
24	1.1271	0.8872	0.03933	25.4274	0.04433	22.5593	24
25	1.1328	0.8828	0.03766	26.5543	0.04266	23.4419	25
26	1.1384	0.8784	0.03612	27.6869	0.04112	24.3201	26
28	1.1498	0.8697	0.03337	29.9690	0.03837	26.0635	28
30	1.1614	0.8611	0.03098	32.2741	0.03598	27.7896	30
32	1.1730	0.8525	0.02890	34.6022	0.03390	29.4986	32
34	1.1848	0.8440	0.02706	36.9537	0.03206	31.1907	34
35	1.1907	0.8398	0.02622	38.1384	0.03122	32.0305	35
36	1.1966	0.8357	0.02543	39.3288	0.03043	32.8659	36
38	1.2086	0.8274	0.02396	41.7276	0.02896	34.5245	38
40	1.2208	0.8192	0.02265	44.1505	0.02765	36.1667	40
45	1.2516	0.7990	0.01987	50.3147	0.02487	40.2012	45
50	1.2832	0.7793	0.01766	56.6344	0.02266	44.1362	50
55	1.3156	0.7601	0.01584	63.1136	0.02084	47.9744	55
60	1.3488	0.7414	0.01434	69.7565	0.01934	51.7182	60
65	1.3828	0.7232	0.01306	76.5669	0.01806	55.3696	65
70	1.4177	0.7053	0.01197	83.5495	0.01697	58.9312	70
75	1.4535	0.6880	0.01102	90.7082	0.01602	62.4050	75
80	1.4902	0.6710	0.01020	98.0477	0.01520	65.7933	80
85	1.5279	0.6545	0.00947	105.5726	0.01447	69.0982	85
90	1.5664	0.6384	0.00883	113.2874	0.01383	72.3217	90
95	1.6060	0.6227	0.00825	121.1970	0.01325	75.4659	95
100	1.6465	0.6073	0.00773	129.3061	0.01273	78.5325	100

TABLE A-3

0.75% COMPOUND INTEREST FACTORS

	SINGLE PAYMENTS			UNIFORM SERIES PAYMENTS			
N	COMPOUND AMOUNT F/P	PRESENT WORTH P/F	SINKING FUND A/F	COMPOUND AMOUNT F/A	CAPITAL RECOVERY A/P	PRESENT WORTH P/A	N
1	1.0075	0.9926	1.00004	1.0000	1.00754	0.9925	1
2	1.0151	0.9852	0.49818	2.0073	0.50568	1.9775	2
3	1.0227	0.9778	0.33086	3.0224	0.33836	2.9554	3
4	1.0303	0.9706	0.24722	4.0450	0.25472	3.9259	4
5	1.0381	0.9633	0.19703	5.0753	0.20453	4.8892	5
6	1.0458	0.9562	0.16358	6.1133	0.17108	5.8453	6
7	1.0537	0.9490	0.13968	7.1592	0.14718	6.7944	7
8	1.0616	0.9420	0.12176	8.2128	0.12926	7.7363	8
9	1.0696	0.9350	0.10782	9.2743	0.11532	8.6712	9
10	1.0776	0.9280	0.09668	10.3438	0.10418	9.5991	10
11	1.0857	0.9211	0.08756	11.4213	0.09506	10.5202	11
12	1.0938	0.9142	0.07996	12.5070	0.08746	11.4344	12
13	1.1020	0.9074	0.07353	13.6008	0.08103	12.3419	13
14	1.1103	0.9007	0.06801	14.7027	0.07551	13.2425	14
15	1.1186	0.8940	0.06324	15.8130	0.07074	14.1364	15
16	1.1270	0.8873	0.05906	16.9315	0.06656	15.0237	16
17	1.1354	0.8807	0.05538	18.0584	0.06288	15.9044	17
18	1.1440	0.8742	0.05210	19.1938	0.05960	16.7785	18
19	1.1525	0.8677	0.04917	20.3377	0.05667	17.6461	19
20	1.1612	0.8612	0.04653	21.4902	0.05403	18.5073	20
22	1.1787	0.8484	0.04198	23.8211	0.04948	20.2104	22
24	1.1964	0.8358	0.03819	26.1873	0.04569	21.8883	24
25	1.2054	0.8296	0.03652	27.3836	0.04402	22.7178	25
26	1.2144	0.8234	0.03498	28.5890	0.04248	23.5413	26
28	1.2327	0.8112	0.03223	31.0267	0.03973	25.1697	28
30	1.2513	0.7992	0.02985	33.5013	0.03735	26.7741	30
32	1.2701	0.7873	0.02777	36.0132	0.03527	28.3546	32
34	1.2892	0.7757	0.02593	38.5628	0.03343	29.9117	34
35	1.2989	0.7699	0.02509	39.8519	0.03259	30.6815	35
36	1.3086	0.7642	0.02430	41.1508	0.03180	31.4457	36
38	1.3283	0.7528	0.02284	43.7777	0.03034	32.9569	38
40	1.3483	0.7417	0.02153	46.4442	0.02903	34.4457	40
45	1.3997	0.7145	0.01877	53.2875	0.02627	38.0719	45
50	1.4529	0.6883	0.01656	60.3913	0.02406	41.5650	50
55	1.5082	0.6630	0.01476	67.7655	0.02226	44.9301	55
60	1.5657	0.6387	0.01326	75.4203	0.02076	48.1718	60
65	1.6252	0.6153	0.01200	83.3666	0.01950	51.2946	65
70	1.6871	0.5927	0.01092	91.6153	0.01842	54.3030	70
75	1.7513	0.5710	0.00998	100.1779	0.01748	57.2009	75
80	1.8180	0.5501	0.00917	109.0667	0.01667	59.9927	80
85	1.8872	0.5299	0.00845	118.2937	0.01595	62.6821	85
90	1.9590	0.5105	0.00782	127.8719	0.01532	65.2728	90
95	2.0336	0.4917	0.00726	137.8147	0.01476	67.7685	95
100	2.1110	0.4737	0.00675	148.1360	0.01425	70.1727	100

TABLE A-4

1.00% COMPOUND INTEREST FACTORS

	SINGLE PAYMENTS			UNIFORM SERIES PAYMENTS			
N	COMPOUND AMOUNT F/P	PRESENT WORTH P/F	SINKING FUND A/F	COMPOUND AMOUNT F/A	CAPITAL RECOVERY A/P	PRESENT WORTH P/A	N
1	1.0100	0.9901	1.00007	0.9999	1.01007	0.9900	1
2	1.0201	0.9803	0.49757	2.0098	0.50757	1.9702	2
3	1.0303	0.9706	0.33005	3.0298	0.34005	2.9407	3
4	1.0406	0.9610	0.24630	4.0601	0.25630	3.9017	4
5	1.0510	0.9515	0.19606	5.1005	0.20606	4.8530	5
6	1.0615	0.9420	0.16256	6.1515	0.17256	5.7950	6
7	1.0721	0.9327	0.13864	7.2129	0.14864	6.7277	7
8	1.0829	0.9235	0.12070	8.2851	0.13070	7.6512	8
9	1.0937	0.9143	0.10675	9.3678	0.11675	8.5654	9
10	1.1046	0.9053	0.09559	10.4613	0.10559	9.4706	10
11	1.1157	0.8963	0.08646	11.5659	0.09646	10.3669	11
12	1.1268	0.8875	0.07886	12.6815	0.08886	11.2543	12
13	1.1381	0.8787	0.07242	13.8083	0.08242	12.1329	13
14	1.1495	0.8700	0.06691	14.9462	0.07691	13.0028	14
15	1.1610	0.8614	0.06213	16.0956	0.07213	13.8641	15
16	1.1726	0.8528	0.05795	17.2565	0.06795	14.7169	16
17	1.1843	0.8444	0.05426	18.4290	0.06426	15.5612	17
18	1.1961	0.8360	0.05099	19.6132	0.06099	16.3972	18
19	1.2081	0.8278	0.04806	20.8092	0.05806	17.2248	19
20	1.2202	0.8196	0.04542	22.0172	0.05542	18.0443	20
22	1.2447	0.8034	0.04087	24.4696	0.05087	19.6591	22
24	1.2697	0.7876	0.03708	26.9713	0.04708	21.2420	24
25	1.2824	0.7798	0.03541	28.2409	0.04541	22.0217	25
26	1.2952	0.7721	0.03387	29.5232	0.04387	22.7937	26
28	1.3213	0.7569	0.03113	32.1264	0.04113	24.3149	28
30	1.3478	0.7419	0.02875	34.7820	0.03875	25.8061	30
32	1.3749	0.7273	0.02667	37.4909	0.03667	27.2679	32
34	1.4025	0.7130	0.02484	40.2542	0.03484	28.7009	34
35	1.4166	0.7059	0.02401	41.6567	0.03401	29.4068	35
36	1.4307	0.6989	0.02322	43.0732	0.03322	30.1057	36
38	1.4595	0.6852	0.02176	45.9487	0.03176	31.4828	38
40	1.4888	0.6717	0.02046	48.8820	0.03046	32.8327	40
45	1.5648	0.6391	0.01771	56.4761	0.02771	36.0925	45
50	1.6446	0.6081	0.01551	64.4573	0.02551	39.1939	50
55	1.7285	0.5786	0.01373	72.8456	0.02373	42.1449	55
60	1.8166	0.5505	0.01225	81.6619	0.02225	44.9527	60
65	1.9093	0.5238	0.01100	90.9277	0.02100	47.6242	65
70	2.0067	0.4983	0.00993	100.6663	0.01993	50.1660	70
75	2.1090	0.4742	0.00902	110.9015	0.01902	52.5845	75
80	2.2166	0.4511	0.00822	121.6588	0.01822	54.8856	80
85	2.3296	0.4292	0.00752	132.9648	0.01752	57.0751	85
90	2.4485	0.4084	0.00690	144.8475	0.01690	59.1583	90
95	2.5734	0.3886	0.00636	157.3362	0.01636	61.1404	95
100	2.7046	0.3697	0.00587	170.4620	0.01587	63.0263	100

TABLE A-5

1.50% COMPOUND INTEREST FACTORS

	SINGLE PAYMENTS			UNIFORM SERIES PAYMENTS			
N	COMPOUND AMOUNT F/P	PRESENT WORTH P/F	SINKING FUND A/F	COMPOUND AMOUNT F/A	CAPITAL RECOVERY A/P	PRESENT WORTH P/A	N
1	1.0150	0.9852	1.00004	1.0000	1.01504	0.9852	1
2	1.0302	0.9707	0.49631	2.0149	0.51131	1.9558	2
3	1.0457	0.9563	0.32840	3.0451	0.34340	2.9121	3
4	1.0614	0.9422	0.24446	4.0907	0.25946	3.8542	4
5	1.0773	0.9283	0.19410	5.1520	0.20910	4.7824	5
6	1.0934	0.9145	0.16053	6.2293	0.17553	5.6970	6
7	1.1098	0.9010	0.13656	7.3226	0.15156	6.5979	7
8	1.1265	0.8877	0.11859	8.4325	0.13359	7.4856	8
9	1.1434	0.8746	0.10461	9.5589	0.11961	8.3602	9
10	1.1605	0.8617	0.09344	10.7022	0.10844	9.2218	10
11	1.1779	0.8489	0.08430	11.8627	0.09930	10.0707	11
12	1.1956	0.8364	0.07668	13.0406	0.09168	10.9071	12
13	1.2135	0.8240	0.07024	14.2362	0.08524	11.7311	13
14	1.2317	0.8119	0.06473	15.4497	0.07973	12.5429	14
15	1.2502	0.7999	0.05995	16.6814	0.07495	13.3428	15
16	1.2690	0.7880	0.05577	17.9315	0.07077	14.1307	16
17	1.2880	0.7764	0.05208	19.2005	0.06708	14.9071	17
18	1.3073	0.7649	0.04881	20.4884	0.06381	15.6720	18
19	1.3269	0.7536	0.04588	21.7957	0.06088	16.4256	19
20	1.3468	0.7425	0.04325	23.1225	0.05825	17.1680	20
22	1.3875	0.7207	0.03871	25.8363	0.05371	18.6202	22
24	1.4295	0.6996	0.03493	28.6321	0.04993	20.0297	24
25	1.4509	0.6892	0.03327	30.0615	0.04827	20.7189	25
26	1.4727	0.6790	0.03173	31.5124	0.04673	21.3979	26
28	1.5172	0.6591	0.02900	34.4797	0.04400	22.7260	28
30	1.5631	0.6398	0.02664	37.5368	0.04164	24.0151	30
32	1.6103	0.6210	0.02458	40.6862	0.03958	25.2663	32
34	1.6590	0.6028	0.02276	43.9308	0.03776	26.4809	34
35	1.6838	0.5939	0.02193	45.5897	0.03693	27.0748	35
36	1.7091	0.5851	0.02115	47.2735	0.03615	27.6598	36
38	1.7608	0.5679	0.01972	50.7172	0.03472	28.8042	38
40	1.8140	0.5513	0.01843	54.2650	0.03343	29.9150	40
45	1.9542	0.5117	0.01572	63.6107	0.03072	32.5514	45
50	2.1052	0.4750	0.01357	73.6786	0.02857	34.9987	50
55	2.2679	0.4409	0.01183	84.5246	0.02683	37.2705	55
60	2.4431	0.4093	0.01039	96.2088	0.02539	39.3793	60
65	2.6319	0.3799	0.00919	108.7960	0.02419	41.3368	65
70	2.8353	0.3527	0.00817	122.3559	0.02317	43.1539	70
75	3.0545	0.3274	0.00730	136.9637	0.02230	44.8406	75
80	3.2905	0.3039	0.00655	152.7004	0.02155	46.4064	80
85	3.5448	0.2821	0.00589	169.6533	0.02089	47.8598	85
90	3.8187	0.2619	0.00532	187.9163	0.02032	49.2089	90
95	4.1139	0.2431	0.00482	207.5906	0.01982	50.4613	95
100	4.4318	0.2256	0.00437	228.7855	0.01937	51.6239	100

TABLE A-6

2.00% COMPOUND INTEREST FACTORS

	SINGLE PAYMENTS		UNIFORM SERIES PAYMENTS				
N	COMPOUND AMOUNT F/P	PRESENT WORTH P/F	SINKING FUND A/F	COMPOUND AMOUNT F/A	CAPITAL RECOVERY A/P	PRESENT WORTH P/A	N
1	1.0200	0.9804	1.00002	1.0000	1.02002	0.9804	1
2	1.0404	0.9612	0.49507	2.0199	0.51507	1.9415	2
3	1.0612	0.9423	0.32677	3.0603	0.34677	2.8838	3
4	1.0824	0.9238	0.24263	4.1215	0.26263	3.8076	4
5	1.1041	0.9057	0.19216	5.2039	0.21216	4.7133	5
6	1.1262	0.8880	0.15853	6.3079	0.17853	5.6013	6
7	1.1487	0.8706	0.13452	7.4341	0.15452	6.4718	7
8	1.1717	0.8535	0.11651	8.5827	0.13651	7.3253	8
9	1.1951	0.8368	0.10252	9.7543	0.12252	8.1620	9
10	1.2190	0.8204	0.09133	10.9494	0.11133	8.9824	10
11	1.2434	0.8043	0.08218	12.1684	0.10218	9.7866	11
12	1.2682	0.7885	0.07456	13.4117	0.09456	10.5751	12
13	1.2936	0.7730	0.06812	14.6799	0.08812	11.3481	13
14	1.3195	0.7579	0.06260	15.9735	0.08260	12.1060	14
15	1.3459	0.7430	0.05783	17.2929	0.07783	12.8490	15
16	1.3728	0.7285	0.05365	18.6387	0.07365	13.5774	16
17	1.4002	0.7142	0.04997	20.0115	0.06997	14.2916	17
18	1.4282	0.7002	0.04670	21.4117	0.06670	14.9917	18
19	1.4568	0.6864	0.04378	22.8399	0.06378	15.6782	19
20	1.4859	0.6730	0.04116	24.2966	0.06116	16.3511	20
22	1.5460	0.6468	0.03663	27.2981	0.05663	17.6577	22
24	1.6084	0.6217	0.03287	30.4209	0.05287	18.9136	24
25	1.6406	0.6095	0.03122	32.0293	0.05122	19.5231	25
26	1.6734	0.5976	0.02970	33.6698	0.04970	20.1207	26
28	1.7410	0.5744	0.02699	37.0500	0.04699	21.2809	28
30	1.8113	0.5521	0.02465	40.5668	0.04465	22.3961	30
32	1.8845	0.5306	0.02261	44.2256	0.04261	23.4679	32
34	1.9606	0.5100	0.02082	48.0322	0.04082	24.4982	34
35	1.9999	0.5000	0.02000	49.9928	0.04000	24.9982	35
36	2.0399	0.4902	0.01923	51.9926	0.03923	25.4884	36
38	2.1223	0.4712	0.01782	56.1130	0.03782	26.4402	38
40	2.2080	0.4529	0.01656	60.3999	0.03656	27.3551	40
45	2.4378	0.4102	0.01391	71.8901	0.03391	29.4897	45
50	2.6915	0.3715	0.01182	84.5762	0.03182	31.4232	50
55	2.9717	0.3365	0.01014	98.5827	0.03014	33.1744	55
60	3.2809	0.3048	0.00877	114.0468	0.02877	34.7605	60
65	3.6224	0.2761	0.00763	131.1205	0.02763	36.1971	65
70	3.9994	0.2500	0.00667	149.9712	0.02667	37.4982	70
75	4.4157	0.2265	0.00586	170.7839	0.02586	38.6767	75
80	4.8752	0.2051	0.00516	193.7626	0.02516	39.7442	80
85	5.3827	0.1858	0.00456	219.1331	0.02456	40.7109	85
90	5.9429	0.1683	0.00405	247.1440	0.02405	41.5866	90
95	6.5614	0.1524	0.00360	278.0698	0.02360	42.3797	95
100	7.2443	0.1380	0.00320	312.2148	0.02320	43.0981	100

TABLE A-7

3.00% COMPOUND INTEREST FACTORS

	SINGLE PAYMENTS		UNIFORM SERIES PAYMENTS				
N	COMPOUND AMOUNT F/P	PRESENT WORTH P/F	SINKING FUND A/F	COMPOUND AMOUNT F/A	CAPITAL RECOVERY A/P	PRESENT WORTH P/A	N
1	1.0300	0.9709	1.00001	1.0000	1.03001	0.9709	1
2	1.0609	0.9426	0.49262	2.0300	0.52262	1.9134	2
3	1.0927	0.9151	0.32353	3.0909	0.35353	2.8286	3
4	1.1255	0.8885	0.23903	4.1836	0.26903	3.7171	4
5	1.1593	0.8626	0.18836	5.3091	0.21836	4.5797	5
6	1.1940	0.8375	0.15460	6.4683	0.18460	5.4171	6
7	1.2299	0.8131	0.13051	7.6624	0.16051	6.2302	7
8	1.2668	0.7894	0.11246	8.8922	0.14246	7.0196	8
9	1.3048	0.7664	0.09843	10.1590	0.12843	7.7860	9
10	1.3439	0.7441	0.08723	11.4637	0.11723	8.5301	10
11	1.3842	0.7224	0.07808	12.8077	0.10808	9.2526	11
12	1.4258	0.7014	0.07046	14.1919	0.10046	9.9539	12
13	1.4685	0.6810	0.06403	15.6176	0.09403	10.6349	13
14	1.5126	0.6611	0.05853	17.0861	0.08853	11.2960	14
15	1.5580	0.6419	0.05377	18.5987	0.08377	11.9378	15
16	1.6047	0.6232	0.04961	20.1566	0.07961	12.5610	16
17	1.6528	0.6050	0.04595	21.7613	0.07595	13.1660	17
18	1.7024	0.5874	0.04271	23.4142	0.07271	13.7534	18
19	1.7535	0.5703	0.03981	25.1166	0.06981	14.3237	19
20	1.8061	0.5537	0.03722	26.8701	0.06722	14.8774	20
22	1.9161	0.5219	0.03275	30.5364	0.06275	15.9368	22
24	2.0328	0.4919	0.02905	34.4260	0.05905	16.9354	24
25	2.0938	0.4776	0.02743	36.4588	0.05743	17.4131	25
26	2.1566	0.4637	0.02594	38.5526	0.05594	17.8768	26
28	2.2879	0.4371	0.02329	42.9304	0.05329	18.7640	28
30	2.4272	0.4120	0.02102	47.5748	0.05102	19.6004	30
32	2.5751	0.3883	0.01905	52.5020	0.04905	20.3887	32
34	2.7319	0.3660	0.01732	57.7294	0.04732	21.1317	34
35	2.8138	0.3554	0.01654	60.4612	0.04654	21.4871	35
36	2.8983	0.3450	0.01580	63.2751	0.04580	21.8322	36
38	3.0748	0.3252	0.01446	69.1584	0.04446	22.4924	38
40	3.2620	0.3066	0.01326	75.4002	0.04326	23.1147	40
45	3.7816	0.2644	0.01079	92.7184	0.04079	24.5186	45
50	4.3838	0.2281	0.00887	112.7951	0.03887	25.7297	50
55	5.0821	0.1968	0.00735	136.0693	0.03735	26.7743	55
60	5.8915	0.1697	0.00613	163.0505	0.03613	27.6755	60
65	6.8299	0.1464	0.00515	194.3290	0.03515	28.4528	65
70	7.9177	0.1263	0.00434	230.5895	0.03434	29.1234	70
75	9.1787	0.1089	0.00367	272.6250	0.03367	29.7018	75
80	10.6407	0.0940	0.00311	321.3557	0.03311	30.2007	80
85	12.3354	0.0811	0.00265	377.8479	0.03265	30.6311	85
90	14.3001	0.0699	0.00226	443.3379	0.03226	31.0024	90
95	16.5777	0.0603	0.00193	519.2583	0.03193	31.3226	95
100	19.2181	0.0520	0.00165	607.2710	0.03165	31.5989	100

TABLE A-8

4.00% COMPOUND INTEREST FACTORS

	SINGLE PAYMENTS			UNIFORM SERIES PAYMENTS			
N	COMPOUND AMOUNT F/P	PRESENT WORTH P/F	SINKING FUND A/F	COMPOUND AMOUNT F/A	CAPITAL RECOVERY A/P	PRESENT WORTH P/A	N
1	1.0400	0.9615	1.00000	1.000	1.04000	0.9615	1
2	1.0816	0.9246	0.49020	2.040	0.53020	1.8861	2
3	1.1249	0.8890	0.32035	3.122	0.36035	2.7751	3
4	1.1699	0.8548	0.23549	4.246	0.27549	3.6299	4
5	1.2167	0.8219	0.18463	5.416	0.22463	4.4518	5
6	1.2653	0.7903	0.15076	6.633	0.19076	5.2421	6
7	1.3159	0.7599	0.12661	7.898	0.16661	6.0021	7
8	1.3686	0.7307	0.10853	9.214	0.14853	6.7327	8
9	1.4233	0.7026	0.09449	10.583	0.13449	7.4353	9
10	1.4802	0.6756	0.08329	12.006	0.12329	8.1109	10
11	1.5395	0.6496	0.07415	13.486	0.11415	8.7605	11
12	1.6010	0.6246	0.06655	15.026	0.10655	9.3851	12
13	1.6651	0.6006	0.06014	16.627	0.10014	9.9857	13
14	1.7317	0.5775	0.05467	18.292	0.09467	10.5631	14
15	1.8009	0.5553	0.04994	20.024	0.08994	11.1184	15
16	1.8730	0.5339	0.04582	21.825	0.08582	11.6523	16
17	1.9479	0.5134	0.04220	23.697	0.08220	12.1657	17
18	2.0258	0.4936	0.03899	25.645	0.07899	12.6593	18
19	2.1068	0.4746	0.03614	27.671	0.07614	13.1339	19
20	2.1911	0.4564	0.03358	29.778	0.07358	13.5903	20
22	2.3699	0.4220	0.02920	34.248	0.06920	14.4511	22
24	2.5633	0.3901	0.02559	39.083	0.06559	15.2470	24
25	2.6658	0.3751	0.02401	41.646	0.06401	15.6221	25
26	2.7725	0.3607	0.02257	44.312	0.06257	15.9828	26
28	2.9987	0.3335	0.02001	49.968	0.06001	16.6631	28
30	3.2434	0.3083	0.01783	56.085	0.05783	17.2920	30
32	3.5081	0.2851	0.01595	62.701	0.05595	17.8735	32
34	3.7943	0.2636	0.01431	69.858	0.05431	18.4112	34
35	3.9461	0.2534	0.01358	73.652	0.05358	18.6646	35
36	4.1039	0.2437	0.01289	77.598	0.05289	18.9083	36
38	4.4388	0.2253	0.01163	85.970	0.05163	19.3679	38
40	4.8010	0.2083	0.01052	95.025	0.05052	19.7928	40
45	5.8412	0.1712	0.00826	121.029	0.04826	20.7200	45
50	7.1067	0.1407	0.00655	152.667	0.04655	21.4822	50
55	8.6463	0.1157	0.00523	191.159	0.04523	22.1086	55
60	10.5196	0.0951	0.00420	237.990	0.04420	22.6235	60
65	12.7987	0.0781	0.00339	294.968	0.04339	23.0467	65
70	15.5716	0.0642	0.00275	364.290	0.04275	23.3945	70
75	18.9452	0.0528	0.00223	448.630	0.04223	23.6804	75
80	23.0497	0.0434	0.00181	551.243	0.04181	23.9154	80
85	28.0435	0.0357	0.00148	676.088	0.04148	24.1085	85
90	34.1192	0.0293	0.00121	827.981	0.04121	24.2673	90
95	41.5112	0.0241	0.00099	1012.781	0.04099	24.3978	95
100	50.5048	0.0198	0.00081	1237.620	0.04081	24.5050	100

TABLE A-9

5.00% COMPOUND INTEREST FACTORS

	SINGLE PAYMENTS			UNIFORM SERIES PAYMENTS			
N	COMPOUND AMOUNT F/P	PRESENT WORTH P/F	SINKING FUND A/F	COMPOUND AMOUNT F/A	CAPITAL RECOVERY A/P	PRESENT WORTH P/A	N
1	1.0500	0.9524	1.00001	1.000	1.05001	0.9524	1
2	1.1025	0.9070	0.48781	2.050	0.53781	1.8594	2
3	1.1576	0.8638	0.31721	3.152	0.36721	2.7232	3
4	1.2155	0.8227	0.23202	4.310	0.28202	3.5459	4
5	1.2763	0.7835	0.18098	5.526	0.23098	4.3294	5
6	1.3401	0.7462	0.14702	6.802	0.19702	5.0756	6
7	1.4071	0.7107	0.12282	8.142	0.17282	5.7863	7
8	1.4774	0.6768	0.10472	9.549	0.15472	6.4631	8
9	1.5513	0.6446	0.09069	11.026	0.14069	7.1077	9
10	1.6289	0.6139	0.07951	12.578	0.12951	7.7216	10
11	1.7103	0.5847	0.07039	14.207	0.12039	8.3063	11
12	1.7958	0.5568	0.06283	15.917	0.11283	8.8632	12
13	1.8856	0.5303	0.05646	17.713	0.10646	9.3935	13
14	1.9799	0.5051	0.05103	19.598	0.10102	9.8985	14
15	2.0789	0.4810	0.04634	21.578	0.09634	10.3796	15
16	2.1828	0.4581	0.04227	23.657	0.09227	10.8377	16
17	2.2920	0.4363	0.03870	25.840	0.08870	11.2740	17
18	2.4066	0.4155	0.03555	28.132	0.08555	11.6895	18
19	2.5269	0.3957	0.03275	30.538	0.08275	12.0852	19
20	2.6533	0.3769	0.03024	33.065	0.08024	12.4621	20
22	2.9252	0.3419	0.02597	38.504	0.07597	13.1629	22
24	3.2250	0.3101	0.02247	44.501	0.07247	13.7985	24
25	3.3863	0.2953	0.02095	47.726	0.07095	14.0938	25
26	3.5556	0.2812	0.01956	51.112	0.06956	14.3751	26
28	3.9200	0.2551	0.01712	58.401	0.06712	14.8980	28
30	4.3218	0.2314	0.01505	66.437	0.06505	15.3724	30
32	4.7648	0.2099	0.01328	75.297	0.06328	15.8026	32
34	5.2532	0.1904	0.01176	85.064	0.06176	16.1928	34
35	5.5159	0.1813	0.01107	90.318	0.06107	16.3741	35
36	5.7917	0.1727	0.01043	95.833	0.06043	16.5468	36
38	6.3853	0.1566	0.00928	107.706	0.05928	16.8678	38
40	7.0398	0.1420	0.00828	120.796	0.05828	17.1590	40
45	8.9847	0.1113	0.00626	159.694	0.05626	17.7740	45
50	11.4670	0.0872	0.00478	209.340	0.05478	18.2559	50
55	14.6350	0.0683	0.00367	272.701	0.05367	18.6334	55
60	18.6784	0.0535	0.00283	353.567	0.05283	18.9292	60
65	23.8388	0.0419	0.00219	456.775	0.05219	19.1610	65
70	30.4249	0.0329	0.00170	588.497	0.05170	19.3427	70
75	38.8306	0.0258	0.00132	756.611	0.05132	19.4849	75
80	49.5585	0.0202	0.00103	971.171	0.05103	19.5964	80
85	63.2504	0.0158	0.00080	1245.009	0.05080	19.6838	85
90	80.7251	0.0124	0.00063	1594.502	0.05063	19.7522	90
95	103.028	0.0097	0.00049	2040.552	0.05049	19.8059	95
100	131.492	0.0076	0.00038	2609.835	0.05038	19.8479	100

TABLE A-10

6.00% COMPOUND INTEREST FACTORS

	SINGLE PAYMENTS		UNIFORM SERIES PAYMENTS				
N	COMPOUND AMOUNT F/P	PRESENT WORTH P/F	SINKING FUND A/F	COMPOUND AMOUNT F/A	CAPITAL RECOVERY A/P	PRESENT WORTH P/A	N
1	1.0600	0.9434	1.00001	1.000	1.06001	0.9434	1
2	1.1236	0.8900	0.48544	2.060	0.54544	1.8334	2
3	1.1910	0.8396	0.31411	3.184	0.37411	2.6730	3
4	1.2625	0.7921	0.22859	4.375	0.28859	3.4651	4
5	1.3382	0.7473	0.17740	5.637	0.23740	4.2123	5
6	1.4185	0.7050	0.14336	6.975	0.20336	4.9173	6
7	1.5036	0.6651	0.11914	8.394	0.17914	5.5823	7
8	1.5938	0.6274	0.10104	9.897	0.16104	6.2098	8
9	1.6895	0.5919	0.08702	11.491	0.14702	6.8017	9
10	1.7908	0.5584	0.07587	13.181	0.13587	7.3600	10
11	1.8983	0.5268	0.06679	14.971	0.12679	7.8868	11
12	2.0122	0.4970	0.05928	16.870	0.11928	8.3838	12
13	2.1329	0.4688	0.05296	18.882	0.11296	8.8526	13
14	2.2609	0.4423	0.04759	21.015	0.10759	9.2949	14
15	2.3965	0.4173	0.04296	23.276	0.10296	9.7122	15
16	2.5403	0.3936	0.03895	25.672	0.09895	10.1058	16
17	2.6927	0.3714	0.03545	28.212	0.09545	10.4772	17
18	2.8543	0.3503	0.03236	30.905	0.09236	10.8276	18
19	3.0256	0.3305	0.02962	33.759	0.08962	11.1581	19
20	3.2071	0.3118	0.02718	36.785	0.08718	11.4699	20
22	3.6035	0.2775	0.02305	43.392	0.08305	12.0415	22
24	4.0489	0.2470	0.01968	50.815	0.07968	12.5503	24
25	4.2918	0.2330	0.01823	54.864	0.07823	12.7833	25
26	4.5493	0.2198	0.01690	59.155	0.07690	13.0031	26
28	5.1116	0.1956	0.01459	68.527	0.07459	13.4061	28
30	5.7434	0.1741	0.01265	79.057	0.07265	13.7648	30
32	6.4533	0.1550	0.01100	90.888	0.07100	14.0840	32
34	7.2509	0.1379	0.00960	104.182	0.06960	14.3681	34
35	7.6860	0.1301	0.00897	111.433	0.06897	14.4982	35
36	8.1471	0.1227	0.00840	119.118	0.06839	14.6210	36
38	9.1541	0.1092	0.00736	135.901	0.06736	14.8460	38
40	10.2855	0.0972	0.00646	154.759	0.06646	15.0463	40
45	13.7643	0.0727	0.00470	212.738	0.06470	15.4558	45
50	18.4197	0.0543	0.00344	290.328	0.06344	15.7619	50
55	24.6496	0.0406	0.00254	394.160	0.06254	15.9905	55
60	32.9867	0.0303	0.00188	533.111	0.06188	16.1614	60
65	44.1435	0.0227	0.00139	719.059	0.06139	16.2891	65
70	59.0738	0.0169	0.00103	967.897	0.06103	16.3845	70
75	79.0539	0.0126	0.00077	1300.899	0.06077	16.4558	75
80	105.792	0.0095	0.00057	1746.529	0.06057	16.5091	80
85	141.573	0.0071	0.00043	2342.881	0.06043	16.5489	85
90	189.456	0.0053	0.00032	3140.934	0.06032	16.5787	90
95	253.534	0.0039	0.00024	4208.902	0.06024	16.6009	95
100	339.285	0.0029	0.00018	5638.082	0.06018	16.6175	100

TABLE A-11

7.00% COMPOUND INTEREST FACTCRS

	SINGLE PAYMENTS			UNIFORM SERIES PAYMENTS			
N	COMPOUND AMOUNT F/P	PRESENT WORTH P/F	SINKING FUND A/F	CCMPCUND AMOUNT F/A	CAPITAL RECOVERY A/P	PRESENT WORTH P/A	N
1	1.0700	0.9346	1.00000	1.000	1.C7000	C.9346	1
2	1.1449	0.8734	0.48310	2.070	0.55310	1.8J80	2
3	1.2250	0.8163	0.31105	3.215	0.38105	2.6243	3
4	1.3108	0.7629	0.22523	4.440	0.29523	3.3872	4
5	1.4C25	J.713J	0.17389	5.751	0.24389	4.1J02	5
6	1.5007	J.6663	0.13980	7.153	0.2C98C	4.7665	6
7	1.6058	0.6228	0.11555	8.654	0.18555	5.3893	7
8	1.7182	J.582J	J.09747	1J.260	0.16747	5.9713	8
9	1.8385	0.5439	0.08349	11.978	0.15349	6.5152	9
10	1.9671	0.5C84	0.07238	13.816	0.14238	7.0236	10
11	2.1J48	J.4751	0.06336	15.784	0.13336	7.4987	11
12	2.2522	0.4440	0.05590	17.888	0.1259C	7.9427	12
13	2.4C98	0.4150	J.04965	2J.141	0.11965	8.3576	13
14	2.5785	J.3878	0.04435	22.550	0.11435	8.7454	14
15	2.7590	0.3624	0.03979	25.129	0.10979	9.1079	15
16	2.9521	J.3387	J.03586	27.888	0.10586	9.4466	16
17	3.1588	0.3166	0.03243	30.840	0.10243	9.7632	17
18	3.3799	0.2959	0.02941	33.999	0.09941	10.0591	18
19	3.6165	0.2765	0.02675	37.379	0.C9675	10.3356	19
20	3.8697	0.2584	0.02439	40.995	0.09439	10.5940	20
22	4.4304	J.2257	J.02041	49.JC5	0.09041	11.0612	22
24	5.0723	0.1971	0.01719	58.176	0.08719	11.4693	24
25	5.4274	0.1843	0.01581	63.248	0.08581	11.6536	25
26	5.8073	J.1722	J.01456	68.676	0.08456	11.8258	26
28	6.6488	0.1504	0.01239	80.697	0.08239	12.1371	28
30	7.6122	0.1314	0.01059	94.460	0.08059	12.4090	30
32	8.7152	0.1147	0.00907	110.217	0.07907	12.6465	32
34	9.9780	0.1002	0.00780	128.257	0.0778C	12.8540	34
35	1C.6765	J.0937	J.00723	138.235	0.07723	12.9477	35
36	11.4238	0.0875	0.00672	148.912	0.07672	13.0352	36
38	13.0791	0.0765	0.00580	172.559	0.07580	13.1935	38
40	14.9743	J.0668	0.00501	199.633	0.07501	13.3317	40
45	21.0022	0.0476	0.00350	285.745	0.0735C	13.6055	45
50	29.4566	0.0339	0.00246	406.523	0.07246	13.8008	5J
55	41.3143	0.0242	0.00174	575.919	0.07174	13.9399	55
60	57.9454	0.0173	0.00123	813.506	0.07123	14.0392	60
65	81.2713	0.0123	0.00087	1146.734	0.07087	14.1099	65
70	113.987	0.0088	0.00062	1614.102	0.07062	14.1604	7C
75	159.873	0.0063	0.00044	2269.609	0.07044	14.1964	75
80	224.229	J.0045	0.00031	3188.990	0.07031	14.2220	80
85	314.493	0.0032	0.00022	4478.465	0.07022	14.2403	85
90	441.C92	0.0023	0.00016	6287.020	0.07016	14.2533	90
95	618.653	0.0016	0.00011	8823.613	0.07011	14.2626	95
100	867.691	0.0012	0.00008	12381.300	0.07008	14.2693	100

TABLE A-12

8.00% COMPOUND INTEREST FACTORS

| | SINGLE PAYMENTS | | UNIFORM SERIES PAYMENTS | | | |
N	COMPOUND AMOUNT F/P	PRESENT WORTH P/F	SINKING FUND A/F	COMPOUND AMOUNT F/A	CAPITAL RECOVERY A/P	PRESENT WORTH P/A	N
1	1.0800	0.9259	1.00000	1.000	1.08000	0.9259	1
2	1.1664	0.8573	0.48077	2.080	0.56077	1.7833	2
3	1.2597	0.7938	0.30803	3.246	0.38803	2.5771	3
4	1.3605	0.7350	0.22192	4.506	0.30192	3.3121	4
5	1.4693	0.6806	0.17046	5.867	0.25046	3.9927	5
6	1.5869	0.6302	0.13632	7.336	0.21632	4.6229	6
7	1.7138	0.5835	0.11207	8.923	0.19207	5.2064	7
8	1.8509	0.5403	0.09401	10.637	0.17401	5.7466	8
9	1.9990	0.5002	0.08008	12.488	0.16008	6.2469	9
10	2.1589	0.4632	0.06903	14.487	0.14903	6.7101	10
11	2.3316	0.4289	0.06008	16.645	0.14008	7.1390	11
12	2.5182	0.3971	0.05270	18.977	0.13270	7.5361	12
13	2.7196	0.3677	0.04652	21.495	0.12652	7.9038	13
14	2.9372	0.3405	0.04130	24.215	0.12130	8.2442	14
15	3.1722	0.3152	0.03683	27.152	0.11683	8.5595	15
16	3.4259	0.2919	0.03298	30.324	0.11298	8.8514	16
17	3.7000	0.2703	0.02963	33.750	0.10963	9.1216	17
18	3.9960	0.2502	0.02670	37.450	0.10670	9.3719	18
19	4.3157	0.2317	0.02413	41.446	0.10413	9.6036	19
20	4.6609	0.2145	0.02185	45.762	0.10185	9.8181	20
22	5.4365	0.1839	0.01803	55.457	0.09803	10.2007	22
24	6.3412	0.1577	0.01498	66.765	0.09498	10.5288	24
25	6.8485	0.1460	0.01368	73.106	0.09368	10.6748	25
26	7.3963	0.1352	0.01251	79.954	0.09251	10.8100	26
28	8.6271	0.1159	0.01049	95.339	0.09049	11.0511	28
30	10.0626	0.0994	0.00883	113.283	0.08883	11.2578	30
32	11.7371	0.0852	0.00745	134.213	0.08745	11.4350	32
34	13.6901	0.0730	0.00630	158.626	0.08630	11.5869	34
35	14.7853	0.0676	0.00580	172.316	0.08580	11.6546	35
36	15.9681	0.0626	0.00534	187.102	0.08534	11.7172	36
38	18.6252	0.0537	0.00454	220.315	0.08454	11.8289	38
40	21.7245	0.0460	0.00386	259.056	0.08386	11.9246	40
45	31.9203	0.0313	0.00259	386.504	0.08259	12.1084	45
50	46.9014	0.0213	0.00174	573.768	0.08174	12.2335	50
55	68.9136	0.0145	0.00118	848.920	0.08118	12.3186	55
60	101.257	0.0099	0.00080	1253.208	0.08080	12.3766	60
65	148.779	0.0067	0.00054	1847.240	0.08054	12.4160	65
70	218.605	0.0046	0.00037	2720.067	0.08037	12.4428	70
75	321.203	0.0031	0.00025	4002.534	0.08025	12.4611	75
80	471.952	0.0021	0.00017	5886.902	0.08017	12.4735	80
85	693.452	0.0014	0.00012	8655.652	0.08012	12.4820	85
90	1018.908	0.0010	0.00008	12723.850	0.08008	12.4877	90
95	1497.110	0.0007	0.00005	18701.380	0.08005	12.4917	95
100	2199.746	0.0005	0.00004	27484.320	0.08004	12.4943	100

TABLE A-13

9.00% COMPOUND INTEREST FACTORS

	SINGLE PAYMENTS			UNIFORM SERIES PAYMENTS			
N	COMPOUND AMOUNT F/P	PRESENT WORTH P/F	SINKING FUND A/F	COMPOUND AMOUNT F/A	CAPITAL RECOVERY A/P	PRESENT WORTH P/A	N
1	1.0900	0.9174	1.00001	1.000	1.09001	0.9174	1
2	1.1881	0.8417	0.47847	2.090	0.56847	1.7591	2
3	1.2950	0.7722	0.30506	3.278	0.39506	2.5313	3
4	1.4116	0.7084	0.21867	4.573	0.30867	3.2397	4
5	1.5386	0.6499	0.16709	5.985	0.25709	3.8896	5
6	1.6771	0.5963	0.13292	7.523	0.22292	4.4859	6
7	1.8280	0.5470	0.10869	9.200	0.19869	5.0329	7
8	1.9926	0.5019	0.09068	11.028	0.18068	5.5348	8
9	2.1719	0.4604	0.07680	13.021	0.16680	5.9952	9
10	2.3673	0.4224	0.06582	15.193	0.15582	6.4176	10
11	2.5804	0.3875	0.05695	17.560	0.14695	6.8052	11
12	2.8126	0.3555	0.04965	20.140	0.13965	7.1607	12
13	3.0658	0.3262	0.04357	22.953	0.13357	7.4869	13
14	3.3417	0.2992	0.03843	26.019	0.12843	7.7861	14
15	3.6424	0.2745	0.03406	29.360	0.12406	8.0607	15
16	3.9703	0.2519	0.03030	33.003	0.12030	8.3125	16
17	4.3276	0.2311	0.02705	36.973	0.11705	8.5436	17
18	4.7171	0.2120	0.02421	41.301	0.11421	8.7556	18
19	5.1416	0.1945	0.02173	46.018	0.11173	8.9501	19
20	5.6043	0.1784	0.01955	51.159	0.10955	9.1285	20
22	6.6585	0.1502	0.01591	62.872	0.10591	9.4424	22
24	7.9109	0.1264	0.01302	76.788	0.10302	9.7066	24
25	8.6229	0.1160	0.01181	84.699	0.10181	9.8226	25
26	9.3990	0.1064	0.01072	93.322	0.10072	9.9290	26
28	11.1669	0.0896	0.00885	112.966	0.09885	10.1161	28
30	13.2674	0.0754	0.00734	136.304	0.09734	10.2736	30
32	15.7630	0.0634	0.00610	164.033	0.09610	10.4062	32
34	18.7279	0.0534	0.00508	196.977	0.09508	10.5178	34
35	20.4134	0.0490	0.00464	215.705	0.09464	10.5668	35
36	22.2506	0.0449	0.00424	236.118	0.09424	10.6118	36
38	26.4359	0.0378	0.00354	282.621	0.09354	10.6908	38
40	31.4085	0.0318	0.00296	337.872	0.09296	10.7574	40
45	48.3257	0.0207	0.00190	525.841	0.09190	10.8812	45
50	74.3548	0.0134	0.00123	815.053	0.09123	10.9617	50
55	114.404	0.0087	0.00079	1260.041	0.09079	11.0140	55
60	176.024	0.0057	0.00051	1944.707	0.09051	11.0480	60
65	270.833	0.0037	0.00033	2998.146	0.09033	11.0701	65
70	416.708	0.0024	0.00022	4618.984	0.09022	11.0845	70
75	641.156	0.0016	0.00014	7112.840	0.09014	11.0938	75
80	986.494	0.0010	0.00009	10949.930	0.09009	11.0999	80
85	1517.837	0.0007	0.00006	16853.750	0.09006	11.1038	85
90	2335.372	0.0004	0.00004	25937.470	0.09004	11.1064	90
95	3593.246	0.0003	0.00003	39913.870	0.09002	11.1080	95
100	5528.633	0.0002	0.00002	61418.200	0.09002	11.1091	100

TABLE A-14

10.00% COMPOUND INTEREST FACTORS

	SINGLE PAYMENTS		UNIFORM SERIES PAYMENTS				
N	COMPOUND AMOUNT F/P	PRESENT WORTH P/F	SINKING FUND A/F	COMPOUND AMOUNT F/A	CAPITAL RECOVERY A/P	PRESENT WORTH P/A	N
1	1.1000	0.9091	1.00000	1.000	1.10001	0.9091	1
2	1.2100	0.8264	0.47619	2.100	0.57619	1.7355	2
3	1.3310	0.7513	0.30212	3.310	0.40212	2.4868	3
4	1.4641	0.6830	0.21547	4.641	0.31547	3.1698	4
5	1.6105	0.6209	0.16380	6.105	0.26380	3.7908	5
6	1.7716	0.5645	0.12961	7.716	0.22961	4.3552	6
7	1.9487	0.5132	0.10541	9.487	0.20541	4.8684	7
8	2.1436	0.4665	0.08744	11.436	0.18744	5.3349	8
9	2.3579	0.4241	0.07364	13.579	0.17364	5.7590	9
10	2.5937	0.3855	0.06275	15.937	0.16275	6.1445	10
11	2.8531	0.3505	0.05396	18.531	0.15396	6.4950	11
12	3.1384	0.3186	0.04676	21.384	0.14676	6.8137	12
13	3.4522	0.2897	0.04078	24.522	0.14078	7.1033	13
14	3.7975	0.2633	0.03575	27.975	0.13575	7.3667	14
15	4.1772	0.2394	0.03147	31.772	0.13147	7.6061	15
16	4.5949	0.2176	0.02782	35.949	0.12782	7.8237	16
17	5.0544	0.1978	0.02466	40.544	0.12466	8.0215	17
18	5.5599	0.1799	0.02193	45.599	0.12193	8.2014	18
19	6.1158	0.1635	0.01955	51.159	0.11955	8.3649	19
20	6.7274	0.1486	0.01746	57.274	0.11746	8.5136	20
22	8.1402	0.1228	0.01401	71.402	0.11401	8.7715	22
24	9.8496	0.1015	0.01130	88.496	0.11130	8.9847	24
25	10.8346	0.0923	0.01017	98.346	0.11017	9.0770	25
26	11.9180	0.0839	0.00916	109.180	0.10916	9.1609	26
28	14.4208	0.0693	0.00745	134.208	0.10745	9.3066	28
30	17.4491	0.0573	0.00608	164.491	0.10608	9.4269	30
32	21.1134	0.0474	0.00497	201.134	0.10497	9.5264	32
34	25.5472	0.0391	0.00407	245.472	0.10407	9.6086	34
35	28.1019	0.0356	0.00369	271.019	0.10369	9.6442	35
36	30.9121	0.0323	0.00334	299.121	0.10334	9.6765	36
38	37.4036	0.0267	0.00275	364.036	0.10275	9.7327	38
40	45.2583	0.0221	0.00226	442.583	0.10226	9.7791	40
45	72.8888	0.0137	0.00139	718.888	0.10139	9.8628	45
50	117.388	0.0085	0.00086	1163.878	0.10086	9.9148	50
55	189.054	0.0053	0.00053	1880.538	0.10053	9.9471	55
60	304.472	0.0033	0.00033	3034.720	0.10033	9.9672	60
65	490.354	0.0020	0.00020	4893.539	0.10020	9.9796	65
70	789.718	0.0013	0.00013	7887.180	0.10013	9.9873	70
75	1271.846	0.0008	0.00008	12708.460	0.10008	9.9921	75
80	2048.315	0.0005	0.00005	20473.160	0.10005	9.9951	80
85	3298.823	0.0003	0.00003	32978.240	0.10003	9.9970	85
90	5312.773	0.0002	0.00002	53117.770	0.10002	9.9981	90
95	8556.250	0.0001	0.00001	85552.500	0.10001	9.9988	95

TABLE A-15

12.00% COMPOUND INTEREST FACTORS

	SINGLE PAYMENTS		UNIFORM SERIES PAYMENTS				
N	COMPOUND AMOUNT F/P	PRESENT WORTH P/F	SINKING FUND A/F	COMPOUND AMOUNT F/A	CAPITAL RECOVERY A/P	PRESENT WORTH P/A	N
1	1.1200	0.8929	1.00000	1.000	1.12000	0.8929	1
2	1.2544	0.7972	0.47170	2.120	0.59170	1.6900	2
3	1.4049	0.7118	0.29635	3.374	0.41635	2.4018	3
4	1.5735	0.6355	0.20923	4.779	0.32923	3.0373	4
5	1.7623	0.5674	0.15741	6.353	0.27741	3.6048	5
6	1.9738	0.5066	0.12323	8.115	0.24323	4.1114	6
7	2.2107	0.4523	0.09912	10.089	0.21912	4.5638	7
8	2.4760	0.4039	0.08130	12.300	0.20130	4.9676	8
9	2.7731	0.3606	0.06768	14.776	0.18768	5.3283	9
10	3.1058	0.3220	0.05698	17.549	0.17698	5.6502	10
11	3.4785	0.2875	0.04842	20.655	0.16842	5.9377	11
12	3.8960	0.2567	0.04144	24.133	0.16144	6.1944	12
13	4.3635	0.2292	0.03568	28.029	0.15568	6.4236	13
14	4.8871	0.2046	0.03087	32.393	0.15087	6.6282	14
15	5.4736	0.1827	0.02682	37.280	0.14682	6.8109	15
16	6.1304	0.1631	0.02339	42.753	0.14339	6.9740	16
17	6.8660	0.1456	0.02046	48.884	0.14046	7.1196	17
18	7.6900	0.1300	0.01794	55.750	0.13794	7.2497	18
19	8.6127	0.1161	0.01576	63.440	0.13576	7.3658	19
20	9.6463	0.1037	0.01388	72.052	0.13388	7.4695	20
22	12.1003	0.0826	0.01081	92.502	0.13081	7.6446	22
24	15.1786	0.0659	0.00846	118.155	0.12846	7.7843	24
25	17.0000	0.0588	0.00750	133.334	0.12750	7.8431	25
26	19.0400	0.0525	0.00665	150.333	0.12665	7.8957	26
28	23.8838	0.0419	0.00524	190.698	0.12524	7.9844	28
30	29.9598	0.0334	0.00414	241.332	0.12414	8.0552	30
32	37.5816	0.0266	0.00328	304.847	0.12328	8.1116	32
34	47.1423	0.0212	0.00260	384.520	0.12260	8.1566	34
35	52.7994	0.0189	0.00232	431.662	0.12232	8.1755	35
36	59.1353	0.0169	0.00206	484.461	0.12206	8.1924	36
38	74.1794	0.0135	0.00164	609.828	0.12164	8.2210	38
40	93.0506	0.0107	0.00130	767.088	0.12130	8.2438	40
45	163.987	0.0061	0.00074	1358.225	0.12074	8.2825	45
50	289.000	0.0035	0.00042	2400.006	0.12042	8.3045	50

TABLE A-16

15.00% COMPOUND INTEREST FACTORS

	SINGLE PAYMENTS		UNIFORM SERIES PAYMENTS				
N	COMPOUND AMOUNT F/P	PRESENT WORTH P/F	SINKING FUND A/F	COMPOUND AMOUNT F/A	CAPITAL RECOVERY A/P	PRESENT WORTH P/A	N
1	1.1500	0.8696	1.00000	1.000	1.15000	0.8696	1
2	1.3225	0.7561	0.46512	2.150	0.61512	1.6257	2
3	1.5209	0.6575	0.28798	3.472	0.43798	2.2832	3
4	1.7490	0.5718	0.20027	4.993	0.35027	2.8550	4
5	2.0114	0.4972	0.14832	6.742	0.29832	3.3522	5
6	2.3131	0.4323	0.11424	8.754	0.26424	3.7845	6
7	2.6600	0.3759	0.09036	11.067	0.24036	4.1604	7
8	3.0590	0.3269	0.07285	13.727	0.22285	4.4873	8
9	3.5179	0.2843	0.05957	16.786	0.20957	4.7716	9
10	4.0455	0.2472	0.04925	20.304	0.19925	5.0188	10
11	4.6524	0.2149	0.04107	24.349	0.19107	5.2337	11
12	5.3502	0.1869	0.03448	29.002	0.18448	5.4206	12
13	6.1528	0.1625	0.02911	34.352	0.17911	5.5831	13
14	7.0757	0.1413	0.02469	40.505	0.17469	5.7245	14
15	8.1370	0.1229	0.02102	47.580	0.17102	5.8474	15
16	9.3576	0.1069	0.01795	55.717	0.16795	5.9542	16
17	10.7612	0.0929	0.01537	65.075	0.16537	6.0472	17
18	12.3754	0.0808	0.01319	75.836	0.16319	6.1280	18
19	14.2317	0.0703	0.01134	88.211	0.16134	6.1982	19
20	16.3664	0.0611	0.00976	102.443	0.15976	6.2593	20
22	21.6446	0.0462	0.00727	137.631	0.15727	6.3587	22
24	28.6249	0.0349	0.00543	184.166	0.15543	6.4338	24
25	32.9187	0.0304	0.00470	212.791	0.15470	6.4642	25
26	37.8565	0.0264	0.00407	245.710	0.15407	6.4906	26
28	50.0651	0.0200	0.00306	327.101	0.15306	6.5335	28
30	66.2111	0.0151	0.00230	434.741	0.15230	6.5660	30
32	87.5641	0.0114	0.00173	577.094	0.15173	6.5905	32
34	115.803	0.0086	0.00131	765.357	0.15131	6.6091	34
35	133.174	0.0075	0.00113	881.160	0.15113	6.6166	35
36	153.150	0.0065	0.00099	1014.334	0.15099	6.6231	36
38	202.541	0.0049	0.00074	1343.606	0.15074	6.6338	38
40	267.860	0.0037	0.00056	1779.067	0.15056	6.6418	40
45	538.761	0.0019	0.00028	3585.076	0.15028	6.6543	45
50	1083.639	0.0009	0.00014	7217.598	0.15014	6.6605	50

TABLE A-17

18.00% COMPOUND INTEREST FACTORS

	SINGLE PAYMENTS			UNIFORM SERIES PAYMENTS			
N	COMPOUND AMOUNT F/P	PRESENT WORTH P/F	SINKING FUND A/F	COMPOUND AMOUNT F/A	CAPITAL RECOVERY A/P	PRESENT WORTH P/A	N
1	1.1800	0.8475	1.00000	1.000	1.18000	0.8475	1
2	1.3924	0.7182	0.45872	2.180	0.63872	1.5656	2
3	1.6430	0.6086	0.27992	3.572	0.45992	2.1743	3
4	1.9388	0.5158	0.19174	5.215	0.37174	2.6901	4
5	2.2878	0.4371	0.13978	7.154	0.31978	3.1272	5
6	2.6995	0.3704	0.10591	9.442	0.28591	3.4976	6
7	3.1855	0.3139	0.08236	12.141	0.26236	3.8115	7
8	3.7588	0.2660	0.06524	15.327	0.24524	4.0776	8
9	4.4354	0.2255	0.05240	19.386	0.23239	4.3030	9
10	5.2338	0.1911	0.04251	23.521	0.22251	4.4941	10
11	6.1759	0.1619	0.03478	28.755	0.21478	4.6560	11
12	7.2875	0.1372	0.02863	34.931	0.20863	4.7932	12
13	8.5993	0.1163	0.02369	42.218	0.20369	4.9095	13
14	10.1472	0.0985	0.01968	50.818	0.19968	5.0081	14
15	11.9736	0.0835	0.01640	60.965	0.19640	5.0916	15
16	14.1289	0.0708	0.01371	72.938	0.19371	5.1624	16
17	16.6721	0.0600	0.01149	87.067	0.19149	5.2223	17
18	19.6730	0.0508	0.00964	103.739	0.18964	5.2732	18
19	23.2142	0.0431	0.00810	123.412	0.18810	5.3162	19
20	27.3927	0.0365	0.00682	146.626	0.18682	5.3527	20
22	38.1416	0.0262	0.00485	206.342	0.18485	5.4099	22
24	53.1083	0.0188	0.00345	289.490	0.18345	5.4510	24
25	62.6678	0.0160	0.00292	342.599	0.18292	5.4669	25
26	73.9479	0.0135	0.00247	405.266	0.18247	5.4804	26
28	102.9650	0.0097	0.00177	566.472	0.18177	5.5016	28
30	143.3683	0.0070	0.00126	790.935	0.18126	5.5168	30
32	199.6258	0.0050	0.00091	1103.477	0.18091	5.5277	32
34	277.9585	0.0036	0.00065	1538.660	0.18065	5.5356	34
35	327.9910	0.0030	0.00055	1816.617	0.18055	5.5386	35
36	387.0291	0.0026	0.00047	2144.608	0.18047	5.5412	36
38	538.899	0.0019	0.00033	2988.329	0.18033	5.5453	38
40	750.362	0.0013	0.00024	4163.121	0.18024	5.5482	40
45	1716.641	0.0006	0.00010	9531.344	0.18010	5.5523	45
50	3927.249	0.0003	0.00005	21812.500	0.18005	5.5541	50

TABLE A-18

20.00% COMPOUND INTEREST FACTORS

	SINGLE PAYMENTS			UNIFORM SERIES PAYMENTS			
N	COMPOUND AMOUNT F/P	PRESENT WORTH P/F	SINKING FUND A/F	COMPOUND AMOUNT F/A	CAPITAL RECOVERY A/P	PRESENT WORTH P/A	N
1	1.2000	0.8333	1.00000	1.000	1.20000	0.8333	1
2	1.4400	0.6944	0.45455	2.200	0.65455	1.5278	2
3	1.7280	0.5787	0.27473	3.640	0.47473	2.1065	3
4	2.0736	0.4823	0.18629	5.368	0.38629	2.5887	4
5	2.4883	0.4019	0.13438	7.442	0.33438	2.9906	5
6	2.9860	0.3349	0.10071	9.930	0.30071	3.3255	6
7	3.5832	0.2791	0.07742	12.916	0.27742	3.6046	7
8	4.2998	0.2326	0.06061	16.499	0.26061	3.8372	8
9	5.1598	0.1938	0.04808	20.799	0.24808	4.0310	9
10	6.1917	0.1615	0.03852	25.959	0.23852	4.1925	10
11	7.4301	0.1346	0.03110	32.150	0.23110	4.3271	11
12	8.9161	0.1122	0.02527	39.580	0.22526	4.4392	12
13	10.6993	0.0935	0.02062	48.497	0.22062	4.5327	13
14	12.8392	0.0779	0.01689	59.196	0.21689	4.6106	14
15	15.4070	0.0649	0.01388	72.035	0.21388	4.6755	15
16	18.4884	0.0541	0.01144	87.442	0.21144	4.7296	16
17	22.1861	0.0451	0.00944	105.930	0.20944	4.7746	17
18	26.6232	0.0376	0.00781	128.116	0.20781	4.8122	18
19	31.9479	0.0313	0.00646	154.740	0.20646	4.8435	19
20	38.3375	0.0261	0.00536	186.687	0.20536	4.8696	20
22	55.2059	0.0181	0.00369	271.030	0.20369	4.9094	22
24	79.4965	0.0126	0.00255	392.483	0.20255	4.9371	24
25	95.3958	0.0105	0.00212	471.979	0.20212	4.9476	25
26	114.4750	0.0087	0.00176	567.375	0.20176	4.9563	26
28	164.8439	0.0061	0.00122	819.220	0.20122	4.9697	28
30	237.3752	0.0042	0.00085	1181.877	0.20085	4.9789	30
32	341.8201	0.0029	0.00059	1704.102	0.20059	4.9854	32
34	492.2207	0.0020	0.00041	2456.105	0.20041	4.9898	34
35	590.6648	0.0017	0.00034	2948.327	0.20034	4.9915	35
36	708.7976	0.0014	0.00028	3538.992	0.20028	4.9929	36
38	1020.668	0.0010	0.00020	5098.344	0.20020	4.9951	38
40	1469.762	0.0007	0.00014	7343.816	0.20014	4.9966	40
45	3657.236	0.0003	0.00005	18281.190	0.20005	4.9986	45
50	9100.363	0.0001	0.00002	45496.870	0.20002	4.9995	50

TABLE A-19

25.00% COMPOUND INTEREST FACTORS

	SINGLE PAYMENTS		UNIFORM SERIES PAYMENTS				
	COMPOUND AMOUNT	PRESENT WORTH	SINKING FUND	COMPOUND AMOUNT	CAPITAL RECOVERY	PRESENT WORTH	
N	F/P	P/F	A/F	F/A	A/P	P/A	N
1	1.2500	0.8000	1.00000	1.000	1.25000	0.8000	1
2	1.5625	0.6400	0.44445	2.250	0.69445	1.4400	2
3	1.9531	0.5120	0.26230	3.812	0.5123C	1.9520	3
4	2.4414	0.4096	0.17344	5.766	0.42344	2.3616	4
5	3.0517	0.3277	0.12185	8.207	0.37185	2.6893	5
6	3.8147	0.2621	0.08882	11.259	C.33882	2.9514	6
7	4.7683	0.2097	0.06634	15.C73	0.31634	3.1611	7
8	5.9604	0.1678	0.05040	19.842	C.3004C	3.3289	8
9	7.4505	0.1342	0.03876	25.802	0.28876	3.4631	9
10	9.3132	0.1074	0.03007	33.253	0.28007	3.5705	10
11	11.6414	0.0859	0.02349	42.566	0.27349	3.6564	11
12	14.5518	0.0687	0.01845	54.207	0.26845	3.7251	12
13	18.1897	0.0550	0.01454	68.759	0.26454	3.7801	13
14	22.7371	0.0440	0.01150	86.949	0.26150	3.8241	14
15	28.4214	0.0352	0.00912	109.686	0.25912	3.8593	15
16	35.5267	0.0281	0.00724	138.107	0.25724	3.8874	16
17	44.4083	0.0225	0.00576	173.634	0.25576	3.9099	17
18	55.5104	0.0180	0.00459	218.042	0.25459	3.9279	18
19	69.3879	0.0144	0.00366	273.552	C.25366	3.9424	19
20	86.7348	0.0115	0.00292	342.939	0.25292	3.9535	20
22	135.5230	0.0074	0.00186	538.092	0.25186	3.9705	22
24	211.7543	0.0047	0.00119	843.018	C.25119	3.9811	24
25	264.6926	0.0038	0.00095	1054.771	C.25095	3.9849	25
26	330.8655	0.0030	0.00076	1319.463	C.25076	3.9879	26
28	516.9768	0.0019	0.00048	2063.909	0.25048	3.9923	28
30	807.7749	0.0012	0.00031	3227.103	0.25031	3.9951	30
32	1262.146	0.0008	0.00020	5044.590	C.25020	3.9968	32
34	1972.101	0.0005	0.00013	7884.406	C.25013	3.9980	34
35	2465.124	0.0004	0.00010	9856.504	0.25010	3.9984	35
36	3081.403	0.0003	0.00008	12321.620	C.25008	3.9987	36
38	4814.684	0.0002	0.00005	19254.750	0.25005	3.9992	38
40	7522.934	0.0001	0.00003	30087.750	C.25003	3.9995	40
45	22958.08	0.0000	0.00001	91828.370	C.25001	3.9998	45

TABLE A-20

30.00% COMPOUND INTEREST FACTORS

	SINGLE PAYMENTS		UNIFORM SERIES PAYMENTS				
N	COMPOUND AMOUNT F/P	PRESENT WORTH P/F	SINKING FUND A/F	COMPOUND AMOUNT F/A	CAPITAL RECOVERY A/P	PRESENT WORTH P/A	N
1	1.3000	0.7692	1.00000	1.000	1.30000	0.7692	1
2	1.6900	0.5917	0.43478	2.300	0.73478	1.3609	2
3	2.1970	0.4552	0.25063	3.990	0.55063	1.8161	3
4	2.8561	0.3501	0.16163	6.187	0.46163	2.1662	4
5	3.7129	0.2693	0.11058	9.043	0.41058	2.4356	5
6	4.8268	0.2072	0.07839	12.756	0.37839	2.6427	6
7	6.2748	0.1594	0.05687	17.583	0.35687	2.8021	7
8	8.1573	0.1226	0.04192	23.858	0.34192	2.9247	8
9	10.6044	0.0943	0.03124	32.015	0.33124	3.0190	9
10	13.7858	0.0725	0.02346	42.619	0.32346	3.0915	10
11	17.9215	0.0558	0.01773	56.405	0.31773	3.1473	11
12	23.2979	0.0429	0.01345	74.326	0.31345	3.1903	12
13	30.2873	0.0330	0.01024	97.624	0.31024	3.2233	13
14	39.3734	0.0254	0.00782	127.912	0.30782	3.2487	14
15	51.1854	0.0195	0.00598	167.285	0.30598	3.2682	15
16	66.5410	0.0150	0.00458	218.470	0.30458	3.2832	16
17	86.5033	0.0116	0.00351	285.011	0.30351	3.2948	17
18	112.4542	0.0089	0.00269	371.514	0.30269	3.3037	18
19	146.1904	0.0068	0.00207	483.968	0.30207	3.3105	19
20	190.0474	0.0053	0.00159	630.158	0.30159	3.3158	20
22	321.1797	0.0031	0.00094	1067.266	0.30094	3.3230	22
24	542.7930	0.0018	0.00055	1805.979	0.30055	3.3272	24
25	705.6306	0.0014	0.00043	2348.771	0.30043	3.3286	25
26	917.3191	0.0011	0.00033	3054.401	0.30033	3.3297	26
28	1550.268	0.0006	0.00019	5164.227	0.30019	3.3312	28
30	2619.949	0.0004	0.00011	8729.836	0.30011	3.3321	30
32	4427.707	0.0002	0.00007	14755.690	0.30007	3.3326	32
34	7482.816	0.0001	0.00004	24939.410	0.30004	3.3329	34
35	9727.660	0.0001	0.00003	32422.230	0.30003	3.3330	35

TABLE A-21

35.00% COMPOUND INTEREST FACTORS

	SINGLE PAYMENTS			UNIFORM SERIES PAYMENTS			
N	COMPOUND AMOUNT F/P	PRESENT WORTH P/F	SINKING FUND A/F	COMPOUND AMOUNT F/A	CAPITAL RECOVERY A/P	PRESENT WORTH P/A	N
1	1.3500	0.7407	1.00000	1.000	1.35000	0.7407	1
2	1.8225	0.5487	0.42553	2.350	0.77553	1.2894	2
3	2.4604	0.4064	0.23966	4.172	0.58966	1.6959	3
4	3.3215	0.3011	0.15076	6.633	0.50076	1.9969	4
5	4.4840	0.2230	0.10046	9.954	0.45046	2.2200	5
6	6.0534	0.1652	0.06926	14.438	0.41926	2.3852	6
7	8.1721	0.1224	0.04880	20.492	0.39880	2.5075	7
8	11.0324	0.0906	0.03489	28.664	0.38489	2.5982	8
9	14.8937	0.0671	0.02519	39.696	0.37519	2.6653	9
10	20.1065	0.0497	0.01832	54.590	0.36832	2.7150	10
11	27.1437	0.0368	0.01339	74.696	0.36339	2.7519	11
12	36.6440	0.0273	0.00982	101.840	0.35982	2.7792	12
13	49.4694	0.0202	0.00722	138.484	0.35722	2.7994	13
14	66.7836	0.0150	0.00532	187.953	0.35532	2.8144	14
15	90.1579	0.0111	0.00393	254.737	0.35393	2.8255	15
16	121.7131	0.0082	0.00290	344.895	0.35290	2.8337	16
17	164.3126	0.0061	0.00214	466.608	0.35214	2.8398	17
18	221.8219	0.0045	0.00158	630.920	0.35158	2.8443	18
19	299.4595	0.0033	0.00117	852.742	0.35117	2.8476	19
20	404.2700	0.0025	0.00087	1152.201	0.35087	2.8501	20
22	736.7817	0.0014	0.00048	2102.236	0.35048	2.8533	22
24	1342.783	0.0007	0.00026	3833.673	0.35026	2.8550	24
25	1812.757	0.0006	0.00019	5176.453	0.35019	2.8556	25
26	2447.221	0.0004	0.00014	6989.207	0.35014	2.8560	26
28	4460.055	0.0002	0.00008	12740.160	0.35008	2.8565	28
30	8128.445	0.0001	0.00004	23221.290	0.35004	2.8568	30
32	14814.08	0.0001	0.00002	42323.120	0.35002	2.8570	32
34	26998.64	0.0000	0.00001	77136.120	0.35001	2.8570	34
35	36448.14	0.0000	0.00001	104134.70	0.35001	2.8571	35

TABLE A-22

40.00% COMPOUND INTEREST FACTORS

	SINGLE PAYMENTS			UNIFORM SERIES PAYMENTS			
N	COMPOUND AMOUNT F/P	PRESENT WORTH P/F	SINKING FUND A/F	COMPOUND AMOUNT F/A	CAPITAL RECOVERY A/P	PRESENT WORTH P/A	N
1	1.4000	0.7143	1.00000	1.000	1.40000	0.7143	1
2	1.9600	0.5102	0.41667	2.400	0.81667	1.2245	2
3	2.7440	0.3644	0.22936	4.360	0.62936	1.5889	3
4	3.8416	0.2603	0.14077	7.104	0.54077	1.8492	4
5	5.3782	0.1859	0.09136	10.946	0.49136	2.0352	5
6	7.5295	0.1328	0.06126	16.324	0.46126	2.1680	6
7	10.5413	0.0949	0.04192	23.853	0.44192	2.2628	7
8	14.7579	0.0678	0.02907	34.395	0.42907	2.3306	8
9	20.6610	0.0484	0.02034	49.153	0.42034	2.3790	9
10	28.9254	0.0346	0.01432	69.814	0.41432	2.4136	10
11	40.4955	0.0247	0.01013	98.739	0.41013	2.4383	11
12	56.6937	0.0176	0.00718	139.234	0.40718	2.4559	12
13	79.3712	0.0126	0.00510	195.928	0.40510	2.4685	13
14	111.1196	0.0090	0.00363	275.299	0.40363	2.4775	14
15	155.5675	0.0064	0.00259	386.419	0.40259	2.4839	15
16	217.7944	0.0046	0.00185	541.986	0.40184	2.4885	16
17	304.9119	0.0033	0.00132	759.780	0.40132	2.4918	17
18	426.8767	0.0023	0.00094	1064.693	0.40094	2.4941	18
19	597.6272	0.0017	0.00067	1491.570	0.40067	2.4958	19
20	836.6780	0.0012	0.00048	2089.197	0.40048	2.4970	20
22	1639.888	0.0006	0.00024	4097.223	0.40024	2.4985	22
24	3214.178	0.0003	0.00012	8032.949	0.40012	2.4992	24
25	4499.848	0.0002	0.00009	11247.120	0.40009	2.4994	25
26	6299.785	0.0002	0.00006	15746.970	0.40006	2.4996	26
28	12347.57	0.0001	0.00003	30866.460	0.40003	2.4998	28
30	24201.23	0.0000	0.00002	60500.640	0.40002	2.4999	30
32	47434.39	0.0000	0.00001	118583.50	0.40001	2.4999	32
34	92971.31	0.0000	0.00000	232425.90	0.40000	2.5000	34
35	130159.8	0.0000	0.00000	325397.20	0.40000	2.5000	35

TABLE A-23

45.00% COMPOUND INTEREST FACTORS

	SINGLE PAYMENTS			UNIFORM SERIES PAYMENTS			
N	COMPOUND AMOUNT F/P	PRESENT WORTH P/F	SINKING FUND A/F	COMPOUND AMOUNT F/A	CAPITAL RECOVERY A/P	PRESENT WORTH P/A	N
1	1.4500	0.6897	1.00000	1.000	1.45000	0.6897	1
2	2.1025	0.4756	0.40816	2.450	C.85816	1.1653	2
3	3.0486	0.3280	0.21966	4.552	0.66966	1.4933	3
4	4.4205	0.2262	0.13156	7.601	C.58156	1.7195	4
5	6.4097	0.1560	0.08318	12.022	0.53318	1.8755	5
6	9.2941	0.1076	0.05426	18.431	0.50426	1.9831	6
7	13.4764	0.0742	0.03607	27.725	0.48607	2.0573	7
8	19.5407	0.0512	0.02427	41.202	0.47427	2.1085	8
9	28.3341	0.0353	0.01646	60.742	0.46646	2.1438	9
10	41.0844	0.0243	0.01123	89.076	C.46123	2.1681	10
11	59.5723	0.0168	0.00768	130.161	0.45768	2.1849	11
12	86.3797	0.0116	0.00527	189.733	0.45527	2.1965	12
13	125.2505	0.0080	0.00362	276.112	0.45362	2.2045	13
14	181.6131	0.0055	0.00249	401.363	0.45249	2.2100	14
15	263.3386	0.0038	0.00172	582.975	C.45171	2.2138	15
16	381.8408	0.0026	0.00118	846.313	0.45118	2.2164	16
17	553.6689	0.0018	0.00081	1228.154	C.45081	2.2182	17
18	802.8193	0.0012	0.00056	1781.822	0.45056	2.2195	18
19	1164.087	0.0009	0.00039	2584.641	0.45039	2.2203	19
20	1687.925	0.0006	0.00027	3748.725	C.45027	2.2209	20
22	3548.857	0.0003	0.00013	7884.133	C.45013	2.2216	22
24	7461.457	0.0001	0.00006	16578.800	C.45006	2.2219	24
25	10819.11	0.0001	0.00004	24040.250	0.45004	2.2220	25
26	15687.70	0.0001	0.00003	34859.350	0.45003	2.2221	26
28	32983.32	0.0000	0.00001	73294.060	0.45001	2.2222	28
30	69347.31	0.0000	0.00001	154103.00	0.45001	2.2222	30
32	145802.5	0.0000	0.00000	324003.60	0.45000	2.2222	32
34	306549.3	0.0000	0.00000	681219.10	C.45000	2.2222	34
35	444496.2	0.0000	0.00000	987768.30	0.45000	2.2222	35

TABLE A-24

50.00% COMPOUND INTEREST FACTORS

	SINGLE PAYMENTS			UNIFORM SERIES PAYMENTS			
N	COMPOUND AMOUNT F/P	PRESENT WORTH P/F	SINKING FUND A/F	COMPOUND AMOUNT F/A	CAPITAL RECOVERY A/P	PRESENT WORTH P/A	N
1	1.5000	0.6667	1.00000	1.000	1.50000	0.6667	1
2	2.2500	0.4444	0.40000	2.500	0.90000	1.1111	2
3	3.3750	0.2963	0.21053	4.750	0.71053	1.4074	3
4	5.0625	0.1975	0.12308	8.125	0.62308	1.6049	4
5	7.5937	0.1317	0.07583	13.187	0.57583	1.7366	5
6	11.3906	0.0878	0.04812	20.781	0.54812	1.8244	6
7	17.0859	0.0585	0.03108	32.172	0.53108	1.8829	7
8	25.6288	0.0390	0.02030	49.258	0.52030	1.9220	8
9	38.4431	0.0260	0.01335	74.886	0.51335	1.9480	9
10	57.6647	0.0173	0.00882	113.329	0.50882	1.9653	10
11	86.4969	0.0116	0.00585	170.994	0.50585	1.9769	11
12	129.7453	0.0077	0.00388	257.491	0.50388	1.9846	12
13	194.6179	0.0051	0.00258	387.236	0.50258	1.9897	13
14	291.9265	0.0034	0.00172	581.854	0.50172	1.9931	14
15	437.8896	0.0023	0.00114	873.780	0.50114	1.9954	15
16	656.8340	0.0015	0.00076	1311.669	0.50076	1.9970	16
17	985.2505	0.0010	0.00051	1968.503	0.50051	1.9980	17
18	1477.875	0.0007	0.00034	2953.753	0.50034	1.9986	18
19	2216.811	0.0005	0.00023	4431.625	0.50023	1.9991	19
20	3325.214	0.0003	0.00015	6648.434	0.50015	1.9994	20
22	7481.723	0.0001	0.00007	14961.450	0.50007	1.9997	22
24	16833.85	0.0001	0.00003	33665.730	0.50003	1.9999	24
25	25250.77	0.0000	0.00002	50499.570	0.50002	1.9999	25
26	37876.13	0.0000	0.00001	75750.310	0.50001	1.9999	26
28	85221.13	0.0000	0.00001	170440.30	0.50001	2.0000	28
30	191747.4	0.0000	0.00000	383493.10	0.50000	2.0000	30
32	431431.1	0.0000	0.00000	862861.50	0.50000	2.0000	32
34	970718.8	0.0000	0.00000	1941437.0	0.50000	2.0000	34

TABLE A-25

PRESENT WORTH GRADIENT FACTORS(P/G)

N	1%	2%	3%	4%	5%	6%	N
2	0.958	0.958	0.941	0.924	0.906	0.890	2
3	2.895	2.841	2.772	2.702	2.634	2.569	3
4	5.773	5.612	5.437	5.267	5.101	4.945	4
5	9.566	9.233	8.887	8.554	8.235	7.934	5
6	14.271	13.672	13.074	12.506	11.966	11.458	6
7	19.860	18.895	17.952	17.066	16.230	15.449	7
8	26.324	24.868	23.478	22.180	20.968	19.840	8
9	33.626	31.559	29.609	27.801	26.124	24.576	9
10	41.764	38.943	36.305	33.881	31.649	29.601	10
11	50.721	46.984	43.530	40.377	37.496	34.869	11
12	60.479	55.657	51.245	47.248	43.621	40.335	12
13	71.018	64.932	59.416	54.454	49.984	45.961	13
14	82.314	74.783	68.010	61.961	56.550	51.711	14
15	94.374	85.183	76.996	69.735	63.284	57.553	15
16	107.154	96.109	86.343	77.744	70.156	63.457	16
17	120.662	107.535	96.023	85.958	77.136	69.399	17
18	134.865	119.436	106.009	94.350	84.200	75.355	18
19	149.754	131.792	116.274	102.893	91.323	81.304	19
20	165.320	144.577	126.794	111.564	98.484	87.228	20
21	181.546	157.772	137.544	120.341	105.663	93.111	21
22	198.407	171.354	148.504	129.202	112.841	98.939	22
23	215.903	185.305	159.651	138.128	120.004	104.699	23
24	234.009	199.604	170.965	147.101	127.135	110.379	24
25	252.717	214.231	182.428	156.103	134.223	115.971	25
26	272.011	229.169	194.020	165.121	141.253	121.466	26
27	291.875	244.401	205.725	174.138	148.217	126.858	27
28	312.309	259.908	217.525	183.142	155.105	132.140	28
29	333.280	275.674	229.407	192.120	161.907	137.307	29
30	354.790	291.684	241.355	201.061	168.617	142.357	30
31	376.822	307.921	253.354	209.955	175.228	147.284	31
32	399.360	324.369	265.392	218.792	181.734	152.088	32
33	422.398	341.016	277.457	227.563	188.130	156.766	33
34	445.919	357.845	289.536	236.260	194.412	161.317	34
35	469.916	374.846	301.619	244.876	200.575	165.741	35
36	494.375	392.003	313.695	253.405	206.618	170.037	36
37	519.279	409.305	325.755	261.839	212.538	174.205	37
38	544.622	426.738	337.788	270.175	218.333	178.247	38
39	570.396	444.291	349.786	278.406	224.000	182.163	39
40	596.579	461.953	361.742	286.530	229.540	185.955	40
42	650.167	497.560	385.495	302.437	240.234	193.171	42
44	705.288	533.474	408.989	317.869	250.412	199.911	44
46	761.870	569.618	432.177	332.810	260.079	206.192	46
48	819.829	605.921	455.017	347.244	269.242	212.033	48
50	879.089	642.316	477.472	361.163	277.910	217.456	50

TABLE A-25

PRESENT WORTH GRADIENT FACTORS(P/G)

N	7%	8%	9%	10%	15%	20%	N
2	0.873	0.857	0.841	0.826	0.756	0.694	2
3	2.506	2.445	2.386	2.329	2.071	1.852	3
4	4.794	4.650	4.511	4.378	3.786	3.299	4
5	7.646	7.372	7.111	6.862	5.775	4.906	5
6	10.978	10.523	10.092	9.684	7.937	6.581	6
7	14.714	14.024	13.374	12.763	10.192	8.255	7
8	18.788	17.806	16.887	16.028	12.481	9.883	8
9	23.140	21.808	20.570	19.421	14.755	11.434	9
10	27.715	25.977	24.372	22.891	16.979	12.887	10
11	32.466	30.266	28.247	26.396	19.129	14.233	11
12	37.350	34.634	32.158	29.901	21.185	15.467	12
13	42.330	39.046	36.072	33.377	23.135	16.588	13
14	47.371	43.472	39.962	36.800	24.972	17.601	14
15	52.445	47.886	43.806	40.152	26.693	18.509	15
16	57.526	52.264	47.584	43.416	28.296	19.321	16
17	62.592	56.588	51.281	46.581	29.783	20.042	17
18	67.621	60.842	54.885	49.639	31.156	20.680	18
19	72.598	65.013	58.386	52.582	32.421	21.244	19
20	77.508	69.090	61.776	55.406	33.582	21.739	20
21	82.339	73.063	65.050	58.109	34.645	22.174	21
22	87.079	76.926	68.204	60.689	35.615	22.555	22
23	91.719	80.672	71.235	63.146	36.499	22.887	23
24	96.254	84.300	74.142	65.481	37.302	23.176	24
25	100.676	87.804	76.926	67.696	38.031	23.428	25
26	104.981	91.184	79.586	69.794	38.692	23.646	26
27	109.165	94.439	82.123	71.777	39.289	23.835	27
28	113.226	97.569	84.541	73.649	39.828	23.999	28
29	117.161	100.574	86.842	75.414	40.315	24.141	29
30	120.971	103.456	89.027	77.076	40.753	24.263	30
31	124.654	106.216	91.102	78.639	41.147	24.368	31
32	128.211	108.857	93.068	80.108	41.501	24.459	32
33	131.643	111.382	94.931	81.485	41.818	24.537	33
34	134.950	113.792	96.693	82.777	42.103	24.604	34
35	138.135	116.092	98.358	83.987	42.359	24.661	35
36	141.198	118.284	99.931	85.119	42.587	24.711	36
37	144.144	120.371	101.416	86.178	42.792	24.753	37
38	146.972	122.358	102.815	87.167	42.974	24.789	38
39	149.688	124.247	104.134	88.091	43.137	24.820	39
40	152.292	126.042	105.376	88.952	43.283	24.847	40
42	157.180	129.365	107.643	90.505	43.529	24.889	42
44	161.660	132.355	109.645	91.851	43.723	24.920	44
46	165.758	135.038	111.410	93.016	43.878	24.942	46
48	169.498	137.443	112.962	94.022	44.000	24.958	48
50	172.905	139.593	114.325	94.889	44.096	24.970	50

TABLE A-25

PRESENT WORTH GRADIENT FACTORS (P/G)

N	25%	30%	35%	40%	45%	50%	N
2	0.640	0.592	0.549	0.510	0.476	0.444	2
3	1.664	1.502	1.362	1.239	1.132	1.037	3
4	2.893	2.552	2.265	2.020	1.810	1.630	4
5	4.204	3.630	3.157	2.764	2.434	2.156	5
6	5.514	4.666	3.983	3.428	2.972	2.595	6
7	6.773	5.622	4.717	3.997	3.418	2.946	7
8	7.947	6.480	5.352	4.471	3.776	3.220	8
9	9.021	7.234	5.889	4.858	4.058	3.428	9
10	9.987	7.887	6.336	5.170	4.277	3.584	10
11	10.846	8.445	6.705	5.417	4.445	3.699	11
12	11.602	8.917	7.005	5.611	4.572	3.784	12
13	12.262	9.314	7.247	5.762	4.668	3.846	13
14	12.833	9.644	7.442	5.879	4.740	3.890	14
15	13.326	9.917	7.597	5.969	4.793	3.922	15
16	13.748	10.143	7.721	6.038	4.832	3.945	16
17	14.108	10.328	7.818	6.090	4.861	3.961	17
18	14.415	10.479	7.895	6.130	4.882	3.973	18
19	14.674	10.602	7.955	6.160	4.898	3.981	19
20	14.893	10.702	8.002	6.183	4.909	3.987	20
21	15.078	10.783	8.038	6.200	4.917	3.991	21
22	15.233	10.848	8.067	6.213	4.923	3.994	22
23	15.362	10.901	8.089	6.222	4.927	3.996	23
24	15.471	10.943	8.106	6.229	4.930	3.997	24
25	15.562	10.977	8.119	6.235	4.933	3.998	25
26	15.637	11.005	8.130	6.239	4.934	3.999	26
27	15.700	11.026	8.137	6.242	4.935	3.999	27
28	15.752	11.044	8.143	6.244	4.936	3.999	28
29	15.796	11.058	8.148	6.245	4.937	4.000	29
30	15.832	11.069	8.152	6.247	4.937	4.000	30
31	15.861	11.078	8.154	6.248	4.938	4.000	31
32	15.886	11.085	8.157	6.248	4.938	4.000	32
33	15.906	11.090	8.158	6.249	4.938	4.000	33
34	15.923	11.094	8.159	6.249	4.938	4.000	34
35	15.937	11.098	8.160	6.249	4.938	4.000	35
36	15.948	11.101	8.161	6.249	4.938	4.000	36
37	15.957	11.103	8.162	6.250	4.938	4.000	37
38	15.965	11.105	8.162	6.250	4.938	4.000	38
39	15.971	11.106	8.162	6.250	4.938	4.000	39
40	15.977	11.107	8.163	6.250	4.938	4.000	40
42	15.984	11.109	8.163	6.250	4.938	4.000	42
44	15.990	11.110	8.163	6.250	4.938	4.000	44
46	15.993	11.110	8.163	6.250	4.938	4.000	46
48	15.995	11.111	8.163	6.250	4.938	4.000	48
50	15.997	11.111	8.163	6.250	4.938	4.000	50

TABLE A-26

ANNUAL COST GRADIENT FACTORS(A/G)

N	1/2%	1%	2%	3%	4%	5%	6%	N
2	0.461	0.486	0.493	0.492	0.490	0.487	0.485	2
3	0.954	0.984	0.985	0.980	0.974	0.967	0.961	3
4	1.453	1.480	1.474	1.463	1.451	1.439	1.427	4
5	1.954	1.971	1.959	1.941	1.922	1.902	1.883	5
6	2.448	2.463	2.441	2.413	2.386	2.358	2.330	6
7	2.942	2.952	2.920	2.881	2.843	2.805	2.767	7
8	3.440	3.440	3.395	3.345	3.294	3.244	3.195	8
9	3.931	3.926	3.867	3.803	3.739	3.675	3.613	9
10	4.425	4.410	4.336	4.256	4.177	4.099	4.022	10
11	4.916	4.893	4.801	4.705	4.609	4.514	4.421	11
12	5.405	5.374	5.263	5.148	5.034	4.922	4.811	12
13	5.894	5.853	5.722	5.587	5.453	5.321	5.192	13
14	6.385	6.331	6.177	6.021	5.866	5.713	5.563	14
15	6.873	6.807	6.630	6.450	6.272	6.097	5.926	15
16	7.360	7.281	7.079	6.874	6.672	6.473	6.279	16
17	7.846	7.754	7.524	7.293	7.066	6.842	6.624	17
18	8.331	8.225	7.967	7.708	7.453	7.203	6.960	18
19	8.816	8.694	8.406	8.118	7.834	7.557	7.287	19
20	9.300	9.162	8.842	8.523	8.209	7.903	7.605	20
22	10.266	10.092	9.704	9.318	8.941	8.573	8.216	22
24	11.228	11.016	10.553	10.095	9.648	9.214	8.795	24
25	11.707	11.476	10.973	10.476	9.992	9.523	9.072	25
26	12.186	11.934	11.390	10.853	10.331	9.826	9.341	26
28	13.141	12.844	12.213	11.593	10.991	10.411	9.857	28
30	14.092	13.748	13.024	12.314	11.627	10.969	10.342	30
32	15.041	14.646	13.822	13.017	12.241	11.500	10.799	32
34	15.986	15.537	14.607	13.702	12.832	12.006	11.227	34
35	16.458	15.980	14.995	14.037	13.120	12.250	11.432	35
36	16.928	16.421	15.380	14.369	13.402	12.487	11.630	36
38	17.867	17.299	16.140	15.018	13.950	12.944	12.006	38
40	18.802	18.170	16.887	15.650	14.476	13.377	12.359	40
45	21.126	20.320	18.702	17.155	15.705	14.364	13.141	45
50	23.429	22.429	20.441	18.557	16.812	15.223	13.796	50
55	25.711	24.498	22.105	19.860	17.807	15.966	14.341	55
60	27.973	26.526	23.695	21.067	18.697	16.606	14.791	60
65	30.214	28.515	25.214	22.184	19.491	17.154	15.160	65
70	32.435	30.463	26.662	23.214	20.196	17.621	15.461	70
75	34.635	32.372	28.042	24.163	20.821	18.017	15.706	75
80	36.814	34.242	29.356	25.035	21.372	18.352	15.903	80
85	38.973	36.073	30.605	25.835	21.857	18.635	16.062	85
90	41.112	37.866	31.792	26.566	22.283	18.871	16.189	90
95	43.230	39.620	32.918	27.235	22.655	19.069	16.290	95
100	45.328	41.336	33.985	27.844	22.980	19.234	16.371	100

TABLE A-26

ANNUAL COST GRADIENT FACTORS(A/G)

N	7%	8%	9%	10%	12%	15%	18%	N
2	0.483	0.481	0.478	0.476	0.472	0.465	0.459	2
3	0.955	0.949	0.943	0.936	0.925	0.907	0.890	3
4	1.415	1.404	1.392	1.381	1.359	1.326	1.295	4
5	1.865	1.846	1.828	1.810	1.775	1.723	1.673	5
6	2.303	2.276	2.250	2.224	2.172	2.097	2.025	6
7	2.730	2.694	2.657	2.622	2.551	2.450	2.353	7
8	3.146	3.099	3.051	3.004	2.913	2.781	2.656	8
9	3.552	3.491	3.431	3.372	3.257	3.092	2.936	9
10	3.946	3.871	3.798	3.725	3.585	3.383	3.194	10
11	4.330	4.239	4.151	4.064	3.895	3.655	3.430	11
12	4.702	4.596	4.491	4.388	4.190	3.908	3.647	12
13	5.065	4.940	4.818	4.699	4.468	4.144	3.845	13
14	5.417	5.273	5.133	4.995	4.732	4.362	4.025	14
15	5.758	5.594	5.435	5.279	4.980	4.565	4.189	15
16	6.090	5.905	5.724	5.549	5.215	4.752	4.337	16
17	6.411	6.204	6.002	5.807	5.435	4.925	4.471	17
18	6.722	6.492	6.269	6.053	5.643	5.084	4.592	18
19	7.024	6.770	6.524	6.286	5.838	5.231	4.700	19
20	7.316	7.037	6.767	6.508	6.020	5.365	4.798	20
22	7.872	7.541	7.223	6.919	6.351	5.601	4.963	22
24	8.392	8.007	7.638	7.288	6.641	5.798	5.095	24
25	8.639	8.225	7.832	7.458	6.771	5.883	5.150	25
26	8.877	8.435	8.016	7.619	6.892	5.961	5.199	26
28	9.329	8.829	8.357	7.914	7.110	6.096	5.281	28
30	9.749	9.190	8.666	8.176	7.297	6.207	5.345	30
32	10.138	9.520	8.944	8.409	7.459	6.297	5.394	32
34	10.499	9.821	9.193	8.615	7.596	6.371	5.433	34
35	10.669	9.961	9.308	8.709	7.658	6.402	5.449	35
36	10.832	10.095	9.417	8.796	7.714	6.430	5.462	36
38	11.140	10.344	9.617	8.956	7.814	6.478	5.485	38
40	11.423	10.570	9.796	9.096	7.899	6.517	5.502	40
45	12.036	11.045	10.160	9.374	8.057	6.583	5.529	45
50	12.529	11.411	10.429	9.570	8.160	6.620	5.543	50
55	12.921	11.690	10.626	9.708	8.225	6.641	5.549	55
60	13.232	11.902	10.768	9.802	8.266	6.653	5.553	60
65	13.476	12.060	10.870	9.867	8.292	6.659	5.554	65
70	13.666	12.178	10.943	9.911	8.308	6.663	5.555	70
75	13.814	12.266	10.994	9.941	8.318	6.665	5.555	75
80	13.927	12.330	11.030	9.961	8.324	6.666	5.555	80
85	14.015	12.377	11.055	9.974	8.328	6.666	5.555	85
90	14.081	12.412	11.073	9.983	8.330	6.666	5.556	90
95	14.132	12.437	11.085	9.989	8.331	6.667	5.556	95
100	14.170	12.455	11.093	9.993	8.332	6.667	5.556	100

TABLE A-26

ANNUAL COST GRADIENT FACTORS(A/G)

N	20%	25%	30%	35%	40%	45%	50%	N
2	0.455	0.444	0.435	0.426	0.417	0.408	0.400	2
3	0.879	0.852	0.827	0.803	0.780	0.758	0.737	3
4	1.274	1.225	1.178	1.134	1.092	1.053	1.015	4
5	1.641	1.563	1.490	1.422	1.358	1.298	1.242	5
6	1.979	1.868	1.765	1.670	1.581	1.499	1.423	6
7	2.290	2.142	2.006	1.881	1.766	1.661	1.565	7
8	2.576	2.387	2.216	2.060	1.919	1.791	1.675	8
9	2.836	2.605	2.396	2.209	2.042	1.893	1.760	9
10	3.074	2.797	2.551	2.334	2.142	1.973	1.824	10
11	3.289	2.966	2.683	2.436	2.221	2.034	1.871	11
12	3.484	3.115	2.795	2.520	2.285	2.082	1.907	12
13	3.660	3.244	2.889	2.589	2.334	2.118	1.933	13
14	3.817	3.356	2.968	2.644	2.373	2.145	1.952	14
15	3.959	3.453	3.034	2.689	2.403	2.165	1.966	15
16	4.085	3.537	3.089	2.725	2.426	2.180	1.976	16
17	4.198	3.608	3.135	2.753	2.444	2.191	1.983	17
18	4.298	3.670	3.172	2.776	2.458	2.200	1.988	18
19	4.386	3.722	3.202	2.793	2.468	2.206	1.991	19
20	4.464	3.767	3.228	2.808	2.476	2.210	1.994	20
22	4.594	3.836	3.265	2.827	2.487	2.216	1.997	22
24	4.694	3.886	3.289	2.839	2.493	2.219	1.999	24
25	4.735	3.905	3.298	2.843	2.494	2.220	1.999	25
26	4.771	3.921	3.305	2.847	2.496	2.221	1.999	26
28	4.829	3.946	3.315	2.851	2.498	2.221	2.000	28
30	4.873	3.963	3.322	2.853	2.499	2.222	2.000	30
32	4.906	3.975	3.326	2.855	2.499	2.222	2.000	32
34	4.931	3.983	3.329	2.856	2.500	2.222	2.000	34
35	4.941	3.986	3.330	2.856	2.500	2.222	2.000	35
36	4.949	3.988	3.330	2.856	2.500	2.222	2.000	36
38	4.963	3.992	3.332	2.857	2.500	2.222	2.000	38
40	4.973	3.995	3.332	2.857	2.500	2.222	2.000	40
45	4.988	3.998	3.333	2.857	2.500	2.222	2.000	45
50	4.995	3.999	3.333	2.857	2.500	2.222	2.000	50
55	4.998	4.000	3.333	2.857	2.500	2.222	2.000	55
60	4.999	4.000	3.333	2.857	2.500	2.222	2.000	60
65	5.000	4.000	3.333	2.857	2.500	2.222	2.000	65
70	5.000	4.000	3.333	2.857	2.500	2.222	2.000	70
75	5.000	4.000	3.333	2.857	2.500	2.222	2.000	75
80	5.000	4.000	3.333	2.857	2.500	2.222	2.000	80
85	5.000	4.000	3.333	2.857	2.500	2.222	2.000	85
90	5.000	4.000	3.333	2.857	2.500	2.222	2.000	90
95	5.000	4.000	3.333	2.857	2.500	2.222	2.000	95
100	5.000	4.000	3.333	2.857	2.500	2.222	2.000	100

APPENDIX B
CONTINUOUS COMPOUNDING

This appendix is designed to assist you in the understanding of continuously compounded interest. Two uses of continuous compounding are presented. First, the payments or receipts are, as in all chapters of the book, assumed to occur discretely at the end of an interest period. Second, we will allow cash to flow uniformly throughout the period.

CRITERIA

To complete the material of this appendix you must be able to do the following:

1. Use the effective interest formula, Eq. (4.3), to compute the effective rate for continuous compounding, given the nominal interest rate.
2. Derive the formulas for the interest factors for continuous compounding and discrete cash flows.
3. Compute the present worth, future worth, or equivalent uniform annual series for continuous compounding, given the nominal rate and a monetary value at a specified time.

4 Derive the factor formulas for funds flow and perform present-worth, future-worth, and funds-flow series computations using these factors, given the nominal rate and monetary values at specified times.

EXPLANATION OF MATERIAL

B-1 Effective Interest for Continuous Compounding

We have often used the effective interest formula, Eq. (4.3), which is

$$i = \left(1 + \frac{r}{t}\right)^t - 1 \qquad \text{(B-1)}$$

where i = effective interest rate
  r = nominal interest rate
  t = number of compounding periods

For continuous compounding we must allow t to approach infinity. This will require that we take the limit of Eq. (B-1). If we let $r/t = 1/h$, we can write

$$i = \left(1 + \frac{r}{t}\right)^t - 1 = \left(1 + \frac{1}{h}\right)^{hr} - 1$$

$$\lim_{t \to \infty} i = \lim_{h \to \infty} \left[\left(1 + \frac{1}{h}\right)^h\right]^r - 1 \qquad \text{(B-2)}$$

But the natural logarithm base e is defined as

$$\lim_{h \to \infty}\left(1 + \frac{1}{h}\right)^h = e = 2.71828 + \qquad \text{(B-3)}$$

Therefore, the limit of Eq. (B-2) is $(e^r - 1)$. If we have $r = 15\%$ per year, then the effective continuous rate is $(e^{0.15} - 1) = 0.16183$. Naturally, for any finite t value, the effective interest rate formula can be used. For convenience, effective rates are tabulated in Table B-1.

Example B-1 Compute the effective interest rates from Eq. (B-1) for a nominal rate of 10% per year for the number of compounding periods used in Table B-1. Compare your results with those shown in the table.

SOLUTION We have $r = 0.10$. Table B-2 presents the results. For continuous compounding (t of infinity), effective $i = (e^r - 1)$. Comparison indicates that the values are the same as those of Table B-1.   ////

B-2 Interest Factors for Continuous Compounding

If the cash flows occur at the end of specified time periods and interest is compounded continuously, the base amount used for interest computation is increased every

Table B-1 EFFECTIVE RATES FOR SPECIFIC NOMINAL RATES*

Nominal rate, $r\%$	Semiannually $(t = 2)$	Quarterly $(t = 4)$	Monthly $(t = 12)$	Weekly $(t = 52)$	Daily $(t = 365)$	Continuously $(t = \infty; e^r - 1)$
0.25	0.250	0.250	0.250	0.250	0.250	0.250
0.50	0.501	0.501	0.501	0.501	0.501	0.501
0.75	0.751	0.752	0.753	0.753	0.753	0.753
1.00	1.003	1.004	1.005	1.005	1.005	1.005
1.50	1.506	1.508	1.510	1.511	1.511	1.511
2	2.010	2.015	2.018	2.020	2.020	2.020
3	3.023	3.034	3.042	3.044	3.045	3.046
4	4.040	4.060	4.074	4.079	4.081	4.081
5	5.063	5.095	5.116	5.124	5.126	5.127
6	6.090	6.136	6.168	6.180	6.180	6.184
7	7.123	7.186	7.229	7.246	7.247	7.251
8	8.160	8.243	8.300	8.324	8.325	8.329
9	9.203	9.308	9.381	9.409	9.413	9.417
10	10.250	10.381	10.471	10.506	10.516	10.517
11	11.303	11.462	11.572	11.614	11.623	11.628
12	12.360	12.551	12.683	12.734	12.745	12.750
13	13.423	13.648	13.803	13.864	13.878	13.883
14	14.490	14.752	14.934	15.006	15.022	15.027
15	15.563	15.865	16.076	16.158	16.177	16.183
16	16.640	16.986	17.227	17.322	17.345	17.351
17	17.723	18.115	18.389	18.497	18.524	18.530
18	18.810	19.252	19.562	19.684	19.714	19.722
19	19.903	20.397	20.745	20.883	20.917	20.925
20	21.000	21.551	21.939	22.093	22.132	22.140
21	22.103	22.712	23.144	23.315	23.358	23.368
22	23.210	23.883	24.359	24.549	24.598	24.608
23	24.323	25.061	25.586	25.796	25.849	25.860
24	25.440	26.248	26.824	27.054	27.113	27.125
25	26.563	27.443	28.073	28.325	28.390	28.403
26	27.690	28.646	29.333	29.609	29.680	29.693
27	28.823	29.859	30.605	30.905	30.982	30.996
28	29.960	31.079	31.888	32.213	32.298	32.313
29	31.103	32.309	33.183	33.535	33.626	33.643
30	32.250	33.547	34.489	34.869	34.968	34.986
31	33.403	34.794	35.807	36.217	36.327	36.343
32	34.560	36.049	37.137	37.578	37.693	37.713
33	35.723	37.313	38.478	38.952	39.076	39.097
34	36.890	38.586	39.832	40.339	40.472	40.495
35	38.063	39.868	41.198	41.740	41.883	41.907
40	44.000	46.410	48.213	48.954	49.150	49.182
45	50.063	53.179	55.545	56.528	56.788	56.831
50	56.250	60.181	63.209	64.479	64.816	64.872

*Formula used: $(1 + r/t)^t - 1$.

moment by the interest earned in the last moment. The continuous-compounding interest factors are obtained by taking the limit (as t increases toward infinity) of each compound factor as written in the effective rate form. We have previously used the equation $F = P(F/P,i\%,n)$ where $(F/P,i\%,n) = (1 + i)^n$. For a nominal rate r compounded t times per period, we can write the F/P factor as

Table B-2 EFFECTIVE RATES FOR $r = 0.10$ PER YEAR

Compounding frequency	t	$[(1 + 0.10/t)^t - 1]100\%$
Annually	1	10.000
Semiannually	2	10.250
Quarterly	4	10.381
Monthly	12	10.471
Weekly	52	10.506
Daily	365	10.516
Continuously	∞	10.517

$$(F/P, i\%, n) = \left(1 + \frac{r}{t}\right)^{tn}$$

Using the same definition of e as in the previous section, Eq. (B-3), we have

$$\lim_{t \to \infty} (F/P, i\%, n) = \lim_{t \to \infty} \left(1 + \frac{r}{t}\right)^{tn} = e^{rn}$$

Now, e^{rn} is the single-payment compound-amount factor for continuous compounding and discrete cash flows.

One further derivation is illustrated here. For the F/A factor, we have used

$$(F/A, i\%, n) = \frac{(1 + i)^n - 1}{i}$$

If we take the limit, we have

$$\lim_{t \to \infty} (F/A, i\%, n) = \lim_{t \to \infty} \left[\frac{(1 + r/t)^{tn} - 1}{(1 + r/t)^t - 1}\right] = \frac{e^{rn} - 1}{e^r - 1}$$

where $i = [1 + r/t]^t - 1$ is the effective interest-rate equation. Table B-3 presents the results of similar manipulations for the F/P, A/F, and A/P factors. We use the nominal interest symbol r to replace i in the factor notations to distinguish continuous compounding. The other three factor formulas can be obtained by reciprocal relations.

Table B-3 FACTOR FORMULAS FOR CONTINUOUS COMPOUNDING, DISCRETE CASH FLOW

Factor	Formula	Equation
$(F/P, r\%, n)$	e^{rn}	$F = P(F/P, r\%, n)$
$(A/F, r\%, n)$	$\dfrac{e^r - 1}{e^{rn} - 1}$	$A = F(A/F, r\%, n)$
$(A/P, r\%, n)$	$\dfrac{e^r - 1}{1 - e^{-rn}}$	$A = P(A/P, r\%, n)$

B-3 Computations Using Continuous Compounding

Tables of interest factors for continuous compounding are not included here for several reasons: the formulas can be used to get the factor value; continuous compounding is not usually used; and the factors are available in other books (see the Bibliography at the end of this Appendix). It is worthwhile, however, for us to compare results of annual and continuous compounding to see the magnitude of the difference between the two. This is done in Examples B-2 and B-3. Also of interest is the nominal rate, when compounded continuously, which yields a stated annual effective rate. See Example B-4 for this discussion.

Example B-2 Mr. Jones and Mr. Blunder both plan to invest $5,000 for ten years at 10% per year. Compute the future worth for both men if Mr. Jones figures interest compounded annually and Mr. Blunder assumes continuous compounding.

SOLUTION

Mr. Jones: For annual compounding, as we have done previously,

$$F = P(F/P,10\%,10) = 5,000(2.5937) = \$12,969$$

Mr. Blunder: For continuous compounding, the nominal rate r is 10%.

$$(F/P,10\%,10) = e^{rn} = e^{0.1(10)} = 2.71828$$
$$F = P(e^{rn}) = \$13,591$$

COMMENT Continuous compounding represents a $622, or 4.8%, increase in earnings. This is not a tremendous difference, as might have been expected. Just for comparison, a savings and loan association might compound daily, which yields an effective rate of 10.516% ($F = \$13,590$), whereas 10% continuous compounding is only a very slight increase of 10.517% (Table B-1).   ////

Example B-3 Compare the present worth of $2,000 a year for ten years at 10% per year (*a*) compounded annually and (*b*) compounded continuously.

SOLUTION

(*a*) For annual compounding,

$$P = 2,000(P/A,10\%,10) = 2,000(6.1445) = \$12,289$$

(*b*) For continuous compounding, $r = 10\%$ and the P/A factor equation (reciprocal of A/P relation in Table B-3) is

$$(P/A,r\%,n) = \frac{1 - e^{-rn}}{e^r - 1}$$

$$P = 2,000\left(\frac{1 - e^{-1}}{e^{0.1} - 1}\right) = 2,000(6.0104) = \$12,021$$

You should have expected continuous-compounding present worth to be less, since momentary accumulation of interest requires a smaller investment *now* to accrue the same amount as annual compounding at a later date.   ////

Example B-4 Use the effective interest-rate equation for continuous compounding or Table B-1 to determine the nominal rate r needed to return effective rates of 2%, 5%, 10%, 15%, 20%, and 50% compounded annually.

SOLUTION Table B-1 can be used to compute a linearly interpolated nominal rate. However, if i' identifies the effective continuous-compounding rate, we have

$$i' = e^r - 1$$
$$r = \ln(1 + i')$$

The results below show that as i' increases the difference between the annual effective rate (i') and the continuously compounded nominal rate (r) grows.

i'	r
2%	1.98%
5	4.88
10	9.53
15	13.98
20	18.23
50	40.55

////

B-4 Uniform Flow of Funds

If we assume that the cash flow is spread uniformly over each period, we are using the *funds-flow* or continuous cash-flow assumption. In funds-flow computations, the interest rate is compounded continuously. Figure B-1 illustrates the continuous flow of an amount A' over one year. The P value can be found using the relation

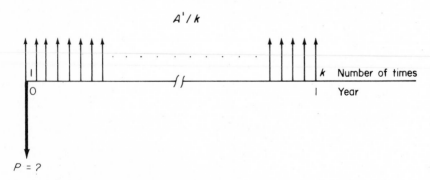

FIG. B-1 Continuous flow of A' through one year.

Table B-4 FUNDS–FLOW FACTOR FORMULAS,
CONTINUOUS COMPOUNDING

Factor	Formula	Equation
$(F/P,r\%,n)$	e^{rn}	$F = P(F/P,r\%,n)$
$(A'/F,r\%,n)$	$\dfrac{r}{e^{rn}-1}$	$A' = F(A'/F,r\%,n)$
$(A'/P,r\%,n)$	$\dfrac{re^{rn}}{e^{rn}-1}$	$A' = P(A'/P,r\%,n)$

$$P = \frac{A'}{k}(P/A, r\%, 1)$$

where k is the number of times that the flow of a fraction of A' occurs. The nominal rate r is present in the P/A factor since interest is compounded continuously. Using the P/A factor formula we have

$$P = \frac{A'}{k}\left[\frac{(1 + r/k)^k - 1}{(r/k)(1 + r/k)^k}\right]$$

If we take the limit as k approaches infinity, we are allowing interest to compound continuously *and* requiring the time between cash flows in Fig. B-1 to approach zero. Therefore, due to the definition of e, Eq. (B-3),

$$\lim_{k\to\infty} P = \lim_{k\to\infty} A'\left(\frac{(1 + r/k)^k - 1}{r(1 + r/k)^k}\right) = A'\left(\frac{e^r - 1}{re^r}\right)$$

Allowing the funds to flow uniformly for n periods results in the expression

$$(P/A',r\%,n) = \frac{e^{rn} - 1}{re^{rn}}$$

where P/A' is used to distinguish this factor from the common P/A notation. P/A' is referred to as the *funds-flow present-worth* factor.

Similar analysis on all factors gives the results of Table B-4. You can compare these formulas with the ones in Table B-3 for continuous compounding, discrete cash flow at the end of each period and see that the mathematical expressions are quite different, except for the F/P factor.

Finally, we should compare results using the funds-flow assumption with those using discrete cash flows. This comparison is presented in the following two examples.

////

Example B-5 Solve (*a*) Example B-2 and (*b*) Example B-3 using the funds-flow assumption.

SOLUTION

(a) Since the $(F/P,r\%,n) = e^{rn}$ factor is used to compute F in both cases (Tables B-3 and B-4), $F = \$13,591$ in both cases.

(b) For funds flow in Example B-3,

$$P = 2,000(P/A',10\%,10)$$

where $(P/A',10\%,10) = \dfrac{e^{0.1(10)} - 1}{0.1e^{0.1(10)}} = 6.3212$

Therefore, $P = \$12,642$, which is $\$621$, or 5.2%, greater than $P = \$12,021$ for discrete cash flow, continuous compounding. ////

Example B-6 Two investment plans have been proposed for the Buy Now–See Later Corporation. Plan D requires an investment of $\$10,000$ with an expected annual positive cash flow of $\$2,500$ for ten years. Plan Z requires $\$50,000$ and promises a cash flow of $\$3,000$ for 30 years. Compute the rate of return for each plan using (a) discrete cash flows, continuous compounding, and (b) funds flow.

SOLUTION

(a) For discrete cash flows we can use the EUAC (A) values for each plan as follows:

$$\text{EUAC}_D = -10,000(A/P,r\%,10) + 2,500 = 0$$

Thus,

$$(A/P,r\%,10) = \frac{e^r - 1}{1 - e^{-10r}} = 0.2500$$

Interpolation gives $r = 19.39\%$. Likewise,

$$\text{EUAC}_Z = -50,000(A/P,r\%,30) + 3,000 = 0$$

which is satisfied if $r = 4.21\%$. Clearly the return is in the favor of Plan D.

(b) For funds flows, we have

$$\text{EUAC}_D = -10,000(A'/P,r\%,10) + 2,500 = 0$$

which yields

$$(A'/P,r\%,10) = \frac{re^{10r}}{e^{10r} - 1} = 0.2500$$

at $r = 22.32\%$. Similarly,

$$\text{EUAC}_Z = -50,000(A'/P,r\%,30) + 3,000 = 0$$

which is correct if $r = 4.39\%$. Again, Plan D is favored.

COMMENT Note that the funds-flow r value is greater than the discrete cash-flow r value. This is apparent when you realize that the momentary accumulation of interest is present under the funds-flow assumption. ////

The funds-flow assumption more closely approximates reality in many situations; however, due to the tabulation availability and ease of the assumption of discrete cash flows, the latter are used more often than funds flow. It should be realized that if discrete cash flows are assumed but funds *actually* flow uniformly, the computed return values will be reduced.

BIBLIOGRAPHY

Barish, pp. 62–66; no tables.
DeGarmo and Canada, pp. 140–146; pp. 546–561 (tables).
Grant and Ireson, pp. 537–545; pp. 624–627 (tables).
Ostwald, pp. 327–331, 471; pp. 477–481 (tables).
Reisman, pp. 10–39; pp. 435–515 (tables).
Smith, pp. 55–60, 76–77; p. 541 (summary table).
Thuesen, Fabrycky, and Thuesen, pp. 67–75; pp. 462–477 (tables).

APPENDIX C
ANSWERS TO PROBLEMS

The final solution to each problem is given in this appendix. The section number given after the answer in the right margin or in place of a word answer is the section in the text to which you should first refer if your answer is incorrect. Since the text material is cumulative from chapter to chapter, material from several sections is usually applicable, but only the most advanced section is referenced.

CHAPTER 1

P1.1	(Sec. 1.1)	
P1.2	(Sec. 1.1)	
P1.3	Period = 6 months	(Sec. 1.2)
P1.4	Interest = $600	(Sec. 1.2)
P1.5	Principal = $1,666.67	(Sec. 1.2)
P1.6	Rate = 10.5%	(Sec. 1.3)
P1.7	(Sec. 1.3)	
P1.8	(Sec. 1.4)	
P1.9	Rate of return = 15%	(Sec. 1.5)
P1.10	(Sec. 1.5)	

P1.11	Invest at 6% simple for extra interest of $18.98	(Sec. 1.6)
P1.12	(*a*) Interest = $27.41 (*b*) 4.57% of P	(Sec. 1.6)
P1.13	(*a*) Interest = $27.00 (*b*) 4.50% of P	(Sec. 1.6)
P1.14	$1,280 is owed	(Sec. 1.6)
P1.15	$620 is owed	(Sec. 1.6)
P1.16	Principal = $756.50	(Sec. 1.6)
P1.17	Rate = 9.54%	(Sec. 1.6)
P1.18	Invest for 4 years	(Sec. 1.6)
P1.19	Principal = $150	(Sec. 1.6)
P1.20	Receive money in 5.36 years	(Sec. 1.6)
P1.21	Rate = 5.67%	(Sec. 1.6)

CHAPTER 2

P2.1 $P = \$1,000$ every 2 years, $F = ?$, $i = 10\%$, $n = 9$ years (Sec. 2.1)

P2.2 $P = \$709.90$, $A_1 = \$100$, $n_1 = 5$, $A_2 = \$200$, $n = 2$,
$i = 6\%$ (Sec. 2.1)

P2.3 $P = \$1,400$, $F = \$4,200$, $i = 6\%$, $n = ?$ (Sec. 2.1)

P2.4 (Sec. 2.2)

P2.5

Year	Payment	Income	Cash flow
0	0	$2,000	$+2,000
1–4	0	0	0
5	$F = ?$	0	$-F = ?$

(Sec. 2.3)

P2.6

Year	Deposit	Withdrawal	Cash flow
May 1, 1975	$500	0	$-500
May 1, 1976–1985	0	$A = ?$	$+A = ?$

(Sec. 2.3)

P2.7

Year	Deposit	Withdrawal	Cash flow
0	$500	$ 0	$-500
1	0	0	0
2	500	0	-500
3	0	0	0
4	500	0	-500
5	0	300	+300
6	500	300	-200
7	0	300	+300
8	500	300	-200
9	0	300	+300
10	500	300	-200
11–13	0	300	+300

(Sec. 2.3)

Note: In P2.8–P2.20, we tabulate or explain the cash flows rather than draw the cash-flow diagram, but the numbers should be the same as yours.

P2.8

Year, k	Disbursement*	Income	Cash flow
0	$2,500	$ 0	$−2,500
1-7	$100 + 25(k − 1)$	750	$+650 − 25(k − 1)$
8	275	900	+625

*Substitute $k = 1, 2, \ldots, 7$ into the relation. For $k = 7$, disbursement is $100 + 25(6) = \$250$.　　　(Sec. 2.4)

P2.9　$P = -$? in year zero, $F = \$+3,000$ in year 5　　(Sec. 2.4)

P2.10

Year	Deposit	Withdrawal	Cash flow
0-4	$700	$ 0	$ −700
5-8	0	0	0
9	0	3,000	3,000
10-12	0	$A_2 = +$?	$A_2 = +$?

(Sec. 2.4)

P2.11

Year	Deposit	Withdrawal	Cash flow
0-1	$0	$ 0	$ 0
2	$P_1 = -$?	0	$P_1 = -$?
3	0	0	0
4	$P_2 = -$?	0	$P_2 = -$?
4-8	0	100	100
9	0	500	500

(Sec. 2.4)

P2.12　Plan 1: $P_1 = \$-351.80$ in year zero, $P_2 = \$-351.80$ in year 3, $F = +$? in year 6; Plan 2: $A = \$-136.32$ in years 1-6, $F = +$? in year 6　　(Sec. 2.4)

P2.13　$P = -$?, $F = \$+580$ in year 8　　(Sec. 2.4)

P2.14

Year	Deposit	Withdrawal	Cash flow
0	$ 0	$0	$ 0
1-5	100	0	−100
6-14	0	0	0
15	0	$F = +$?	$F = +$?

(Sec. 2.4)

P2.15　$P = +$? in year zero, $F_1 = \$-1,200$ in year 5, $F_2 = \$-2,200$ in year 8　　(Sec. 2.4)

P2.16　$P = +$? in year zero, $A = \$-85$ in years 3-8　　(Sec. 2.4)

P2.17　$P = \$-10,000$ in year zero, $F = +$? in year 10, $n = 10$, $i = 12\%$　　(Sec. 2.4)

P2.18　$P = \$-4,100$ in year zero, $F = \$+7,500$ in year 5, $n = 5$, $i = ?\%$　　(Sec. 2.4)

P2.19

Year, k	Deposit	Withdrawal	Cash flow
0	$500	$0	$−500
1-6	$500 + 50(k)$	0	$−500 − 50(k)$
6	0	$F = +$?	$F = +$?

(Sec. 2.4)

P2.20 $P_1 = \$-4,500$ in year zero, $P_2 = \$-3,300$ in year 3,
$P_3 = \$-6,800$ in year 5, $A = +$? for years 1-8, $n = 8$,
$i = 8\%$ (Sec. 2.4)

CHAPTER 3

P3.1   SPPWF: Move F to year $(n-1)$ to get a factor of

$$\frac{1}{(1 + i)^{n-1}}$$

USPWF: Move all A values back one year to get a P/A
factor in year zero of

$$\frac{(1 + i)^n - 1}{i(1 + i)^{n-1}}$$

USCAF: Move all A values back one year to get an F/A
factor in year n of

$$\frac{1 + i}{i} [(1 + i)^n - 1]$$ (Sec. 3.3)

P3.2	(*a*) 0.02718	(*b*) 1.4693	(*c*) 3.8372	(*d*) 0.11017	(Sec. 3.4)
P3.3	(*a*) 7.6954	(*b*) 5,513,383	(*c*) 0.5133	(*d*) 0.00001	(Sec. 3.5)
P3.4	(*a*) 3.1684	(*b*) 0.16493	(*c*) 0.00308	(*d*) 0.00225	(Sec. 3.5)
P3.5	(*a*) 0.7509	(*b*) 10.3126	(*c*) 44.374	(*d*) 0.00488	(Sec. 3.5)

P3.6 $F = \$7,882.70$ (Sec. 3.6)
P3.7 $A = \$808.32$ (Sec. 3.6)
P3.8 $P = \$2,564.14$ (Sec. 3.6)
P3.9 $F = \$25,281.42$ (Sec. 3.6)
P3.10 $A = \$611.42$ (Sec. 3.6)
P3.11 $P = \$6,810$ four years from now, $n = 14$ (Sec. 3.6)
P3.12 $P = \$8,233.77$ (Sec. 3.6)
P3.13 $P = \$3,233.58$ (Sec. 3.6)
P3.14 $P = \$3,331.50$ (Sec. 3.6)
P3.15 $A = \$2,374.20$ (Sec. 3.6)
P3.16 $F = \$27,640.64$ (Sec. 3.6)
P3.17 $F = \$15,016.50$ (Sec. 3.6)
P3.18 $P = \$5,956.63$ (Sec. 3.6)
P3.19 $F = \$588,097.86$ (Sec. 3.6)
P3.20 $P = \$2,019.22$ (Sec. 3.6)
P3.21 $i = 4.36\%$ (Sec. 3.7)
P3.22 $i = 3.61\%$ (Sec. 3.7)

P3.23	$i = 4.003\%$	(Sec. 3.7)
P3.24	$i = 12.05\%$	(Sec. 3.7)
P3.25	$i = 10.78\%$	(Sec. 3.7)
P3.26	$i = 6.87\%$	(Sec. 3.7)
P3.27	$(a)\,i = 0\%$ $(b)\,i = 5.18\%$	(Sec. 3.7)
P3.28	$n = 9.68$ years	(Sec. 3.8)
P3.29	$n = 20.95$ years	(Sec. 3.8)
P3.30	$n = 15$ years	(Sec. 3.8)
P3.31	$n = 7$ years	(Sec. 3.8)
P3.32	$n = 6.82$ years	(Sec. 3.8)

CHAPTER 4

P4.1	$(a)\,i = 13\%$ $(b)\,i = 4\%$	(Sec. 4.1)
P4.2	Year	(Sec. 4.1)
P4.3	% per compounding period	(Sec. 4.2)
P4.4	Nominal $i = r = 18.0\%$ per year, $i = 19.56\%$	(Sec. 4.3)
P4.5	$r = 8.0\%$ per year, $i = 8.16\%$ per year	(Sec. 4.3)
P4.6	$i = 12.36\%$	(Sec. 4.3)
P4.7	$i = 16.99\%$	(Sec. 4.3)
P4.8	$r = 21\%$ per year, $(P/A,21\%,10) = 4.0681$; $i = 23.14\%$, $(P/A,23.14\%,10) = 3.8019$	(Sec. 4.3)
P4.9	$r = 1.47\%$ per quarter	(Sec. 4.3)
P4.10	$i = 4\%$, $n = 16$, $F = \$4,682.50$	(Sec. 4.4)
P4.11	$r = 18.44\%$ per year, $i = 20.08\%$	(Sec. 4.4)
P4.12	$P = \$5,080.62$	(Sec. 4.4)
P4.13	By Eq. (3.9) for F/A value, $F = \$17,252.90$	(Sec. 4.4)
P4.14	$A = \$141.24$ per month	(Sec. 4.4)
P4.15	$P = \$5,906.15$	(Sec. 4.4)
P4.16	$P = \$18,133.30$	(Sec. 4.4)
P4.17	$F = \$80,374.10$	(Sec. 4.4)
P4.18	$r = 43.68\%$, $i = 53.57\%$ using the F/P factor	(Sec. 4.4)
P4.19	$r = 0.656\%$ per month	(Sec. 4.4)
P4.20	$A = \$542.50$ per month	(Sec. 4.4)
P4.21	$P = \$3,570.50$	(Sec. 4.4)
P4.22	$A = \$1,436.76$ per 6 months	(Sec. 4.4)
P4.23	$A = \$177.72$ per month, or $F = \$2,253.76$ per year	(Sec. 4.4)
P4.24	Actual X and Y flows assumed:	

		Actual		Bank	
Month	Quarter	X	Y	X	Y
1		0	0	0	0
2		X_1	0	X_1	0
3	1	0	Y_1	0	$Y_1 + Y_2$

Month	Quarter	Actual X	Actual Y	Bank X	Bank Y
4		X_2	0	X_2	0
5		0	Y_2	0	0
6	2	0	0	0	$Y_3 + Y_4$
7		0	Y_3	0	0
8		0	Y_4	0	0
9	3	X_3	0	X_3	Y_5
10		X_4	0	X_4	0
11		0	Y_5	0	0
12	4	0	0	0	0

(Sec. 4.5)

P4.25 $i = 0.415\%$ per month compounded monthly, $F = \$6,801.51$ (Sec. 4.5)

P4.26   (a) $i = 0.417\%$, $F = \$6,805.61$ (b) $t = 30$, $i = 0.4178\%$, $F = \$6,807.25$ (Sec. 4.5)

P4.27 $i = 3\%$ per 6 months, $F = \$332.74$ (Sec. 4.5)

P4.28 $i = 3\%$ per 6 months, $F = \$338.64$ (Sec. 4.5)

P4.29 $D = \$258.46$ per month, $A = \$1,570.14$ per 6 months (Sec. 4.5)

P4.30 $i = 1.489\%$ per quarter compounded quarterly, $F = \$10,328.38$ (Sec. 4.5)

P4.31   For $A = \$600$ per quarter, $i = 1.489\%$ per quarter and $F = \$5,057.55$; then at $i = 0.494\%$ per month, $A = \$199.05$ per month (Sec. 4.5)

P4.32   Net $A = \$600$ per 6 months, $F = \$2,472.90$ (Sec. 4.5)

P4.33 $A = \$455.63$, $(F/A,3\%,n) = 32.9218$; make 140 deposits (Sec. 4.5)

CHAPTER 5

P5.1   (a) $A = \$-600$ for years 4–10, $P = ?$ in year 3, $F = ?$ in year 10   (b) $A_1 = \$-2,000$ for months 1–6 and $A_2 = \$+1,500$ for months 3–6, $P_1 = ?$ in month 0 and $F_1 = ?$ in month 6, $P_2 = ?$ in month 2 and $F_2 = ?$ in month 6
(c)

Month	Deposit	Withdrawal
December	$P_1 = ?$	$\$\ 0$
January	$A_1 = \$20$	0
February	$A_1 = \ 20$	$P_4 = ?, 0$
March	$A_1 = \ 20$	$A_4 = 10$
April	$F_1 = ?, A_1 = 20, P_2 = ?$	$A_4 = 10$
May	$A_2 = \ 75$	$F_4 = ?, A_4 = 10, P_5 = ?$
June	$A_2 = \ 75$	$A_5 = 25$
July	$A_2 = \ 75$	$A_5 = 25$
August	$F_2 = ?, A_2 = 75, P_3 = ?$	$F_5 = ?, A_5 = 25, P_6 = ?$
September	$A_3 = \ 25$	$A_6 = 10$
October	$A_3 = \ 25$	$A_6 = 10$
November	$F_3 = ?, A_3 = 25$	$F_6 = ?, A_6 = 10$
December	0	250

(Sec. 5.1)

P5.2	$P = \$12,403$	(Sec. 5.2)
P5.3	$A = \$141$	(Sec. 5.2)
P5.4	$A = \$2,667$	(Sec. 5.2)
P5.5	$F = \$15,162$	(Sec. 5.2)
P5.6	$F = \$34,752$	(Sec. 5.2)
P5.7	$P = \$8,041$	(Sec. 5.2)
P5.8	$A = \$1,846$ for years 25–54	(Sec. 5.2)
P5.9	$P = \$253$	(Sec. 5.2)
P5.10	$P = \$1,342$	(Sec. 5.2)
P5.11	$A = \$172$	(Sec. 5.2)
P5.12	$F = \$24,002$ in year 28	(Sec. 5.2)
P5.13	$F = \$15,912$	(Sec. 5.2)
P5.14	$P = \$53,831$ in year 4	(Sec. 5.2)
P5.15	$F = \$13,576$	(Sec. 5.3)
P5.16	$A = \$2,781$	(Sec. 5.3)
P5.17	Lost an equivalent $2,645 in month 16	(Sec. 5.3)
P5.18	$P = \$1,477,332$	(Sec. 5.3)
P5.19	(*a*) $i = 6.09\%, P = \$41,683$ (*b*) $F = \$75,282$	(Sec. 5.3)
P5.20	$A = \$8,831$	(Sec. 5.4)
P5.21	$A = \$5,184$	(Sec. 5.4)
P5.22	$A = \$5,688$	(Sec. 5.4)
P5.23	$A = \$3,339$	(Sec. 5.4)
P5.24	Savings at purchase time $= \$-959$; not justified	(Sec. 5.4)
P5.25	$A = \$-64.59$ per month for May through September	(Sec. 5.4)

CHAPTER 6

P6.1

Year	Cash flow	Year	Cash flow	
1	$\$+5,000$	7	$\$+5,400$	
2	$+5,500$	8	$+4,600$	
3	$+6,000$	9	$+3,800$	
4	$+6,500$	10	$+3,000$	
5	$+7,000$	11	$+2,200$	
6	$+6,200$	12	$+1,400$	(Sec. 6.1)

P6.2

Year	Cash flow	
1	$\$\ -60$	
2	-60	
3	-60	
4	-100	
5	-110	
6	-120	
7	-130	(Sec. 6.1)

P6.3   (Sec. 6.2)

P6.4 $(A/G,20\%,12) = 3.484$ (Sec. 6.3)

P6.5   (a) 16.028   (b) 1.723   (c) 3.945   (d) 13.057   (Sec. 6.3)

P6.6   (a)

Year	Cash flow
1	$200
2	222
3	244
4	266
5	288
6	310
7	332
8	354

(b) $G = \$22$ (c) $P_G = ?$ in year 0   (d) $n = 8$ years   (Sec. 6.4)

P6.7 $P_G = ?$ in year 3, $(A/G,25\%,4) = 1.225$ (Sec. 6.4)

P6.8   Base $= \$100$, $n = 5$; base $= \$150$, $n = 4$ (Sec. 6.4)

P6.9   (a)

Year	Cash flow	Year	Cash flow
1	$500	6	$300
2	500	7	300
3	550	8	300
4	600	9	300
5	650		

(b) $P_G = ?$ in year 1   (c) $n = 4$ (Sec. 6.4)

P6.10

Year	Machine A	Year	Machine B	Year	Machine B
6	$2,500	1	$2,000	9	$2,500
7	3,000	2	2,500	10	3,000
8	3,500	3	3,000	11	3,500
9	4,000	4	3,500	12	4,000
10	4,500	5	4,000		
Base $= \$2,500$		Base $= \$2,000$		Base $= \$2,500$	
$G_1 = \$500$		$G_2 = \$500$		$G_3 = \$500$	
$P_{G1} = ?$ in year 5		$P_{G2} = ?$ in year zero		$P_{G3} = ?$ in year 8	
$n = 5$		$n = 5$		$n = 4$	

(Sec. 6.4)

P6.11

Example	G	n	$(P/G,4\%,n)$	$(A/G,4\%,n)$
6.1	$-7,500	8	22.180	3.294
6.2	G	5	8.554	1.922

(Sec. 6.4)

P6.12   (a) $P = \$629$ (b) $A = \$144$ for years 1–6   (Sec. 6.5)

P6.13 $F = \$10,970$ (Sec. 6.5)

P6.14 $A = \$1,149$ (Sec. 6.5)

P6.15 $P_G = \$137.50$ (Sec. 6.6)

P6.16 $A = \$39.19$ (Sec. 6.6)

P6.17 $P_A = \$27,024$, $P_B = \$27,837$ (Sec. 6.6)

P6.18 $P = \$1,285$ (Sec. 6.6)

P6.19	$P = \$2,633$	(Sec. 6.6)
P6.20	$A = \$598$	(Sec. 6.7)
P6.21	$(P/G,9\frac{1}{2}\%,17) = 48.931$	(Sec. 6.7)
P6.22	$P = \$6,093, A = \$992$	(Sec. 6.7)
P6.23	$P = \$1,556$	(Sec. 6.7)
P6.24	$P = \$31,830, A = \$7,650$	(Sec. 6.7)
P6.25	$P = \$7,279, F = \$11,602$	(Sec. 6.7)

CHAPTER 7

P7.1	$\$1,300$	(Sec. 7.2)
P7.2	Additional depreciation $= \$1,160$, tax credit $= \$406$	(Sec. 7.2)
P7.3	(a) $D_2 = \$1,117$ (b) $BV_9 = \$5,266$	(Sec. 7.3)
P7.4	(a) $P = \$350,000$ (b) $SV = \$27,500$ (c) $D = \$10,750$ (d) $BV_{20} = \$135,000$	(Sec. 7.4)
P7.5	(a) $D = \$1,250$ (b) $BV_m = 12,000 - 1,250(m)$ where $m = 1, 2, \ldots, 8$	(Sec. 7.4)

P7.6

Year, m	D_m	BV_m
1	$\$2,222$	$\$9,778$
2	1,944	7,834
3	1,667	6,167
4	1,389	4,778
5	1,111	3,667
6	833	2,834
7	556	2,278
8	278	2,000

(Sec. 7.5)

P7.7

Year, m	D_m	BV_m
2	$\$6,661$	$\$68,287$
7	4,702	40,860
12	2,743	23,228
18	392	15,000

(Sec. 7.5)

P7.8   (Sec. 7.6)

P7.9

Year, m	D_m	BV_m
1	$\$3,000$	$\$9,000$
2	2,250	6,750
3	1,688	5,063
4	1,266	3,797
5	949	2,848
6	712	2,136
7	136	2,000
8	0	2,000

(Sec. 7.6)

P7.10

	D_m		
Year, m	SL	SYD	DDB
4	$6,857	$12,190	$15,327
9	6,857	10,286	11,420
18	6,857	6,857	6,725
26	6,857	3,810	0*

*Can't depreciate below SV = $80,000 by law; BV_{23} = $82,681, thus D_{24} = $2,681 and $D = 0$, thereafter.   (Sec. 7.6)

P7.11 BV_{13} = $230,857 (SL)
   = $176,381 (SYD)
   = $148,918 (DDB)   (Sec. 7.6)

P7.12   Year 2, D_2 = $4,667 (SYD); D_2 = $4,622 (DDB)

P7.13   Year 9, BV_9 = $12,000 (SYD); BV_9 = $12,414 (DDB)   (Sec. 7.6)

P7.14   SL (D = $1,750$) exceeds (a) SYD in year 7 (D_7 = $1,615$) and (b) DDB in year 6 (D_6 = $1,541$)   (Sec. 7.6)

P7.15

	D_m		BV_m	
Year, m	DDB	DB*	DDB	DB
2	$8,099	$6,719	$64,790	$67,896
7	4,494	4,192	35,954	42,357
12	2,494	2,615	19,952	26,424
18	0†	1,485	15,000	15,000

*d = 0.09005.
†Can't depreciate below SV = $15,000.   (Sec. 7.6)

P7.16   (a) Year 14 where D_{14} = $3,600(SL)$, and D_{14} = $3,323$ (SYD); note that D_{13} is the same for both methods, coincidently
  (b) BV_{13} = $86,600 by SYD method   (Sec. 7.6)

P7.17   (a) Year 16 where D_{16} = $3,600(SL)$ and D_{14} = $3,550(DDB)$
  (b) BV_{15} = $44,376 by DDB method   (Sec. 7.6)

P7.18   (a) SV = $14,033 (b) SV = $118,135   (Sec. 7.6)

P7.19   (a) SV = $−337,357 (not a usable value) (b) SV = $107,430   (Sec. 7.6)

P7.20   Total first cost = $1,278,000, total salvage value = $165,000, average life = 10 years

	Depreciation	
Year	(a) Group	(b) Composite
1–8	$111,300	$134,500
9–10	55,650	14,500
11–12	55,650	4,000

(c) composite life = 8.28 years   (Sec. 7.7)

P7.21 Total first cost = $1,278,000, total salvage value = $82,500, life (actual) = 8 years   (*a*) and (*b*) depreciation is $149,438 for years 1–8 for both methods   (*c*) composite life = 8 years   (Sec. 7.7)

P7.22 *P* = $207 tax saved, SL method can be used only through year 19   (Sec. 7.8)

P7.23 (*a*) No actual dollar difference (note that depreciation for DDB runs out in year 8)   (*b*) *P* = $7,814 in favor of DDB   (Sec. 7.8)

P7.24 (*a*) No actual dollar difference   (*b*) *P* = $5,066 in favor of SYD   (Sec. 7.8)

P7.25 Extra tax for Machine B is $300 per year   (Sec. 7.8)

P7.26 Extra tax for Machine D is a total of $2,747.50 for the 9 years   (Sec. 7.8)

P7.27 (*a*)

Year	Depletion	Method
1	$ 70,000	Factor
2	91,000	Factor
3	104,400	Allowance
4	108,000	Allowance
5	100,800	Allowance

(*b*) 13.55%   (Sec. 7.9)

P7.28 (*a*)

Year	Depletion	Method
1	$ 70,000	Factor
2	91,000	Factor
3	104,400	Allowance
4	129,600	Factor
5	120,960	Factor

(*b*) 14.74%   (Sec. 7.9)

P7.29 Depreciation factor = $0.025 per ton

Year	Depreciation	Book value
1	$4,963	$77,037
2	6,204	70,833
3	4,343	66,490

(Sec. 7.9)

CHAPTER 8

P8.1 PW_A = $30,985; PW_B = $28,865; select B   (Sec. 8.1)

P8.2 PW_{10} − PW_{12} = $14,410; select 12-inch line   (Sec. 8.1)

P8.3 Purchase PW = $13,804; rent PW = $8,335; select rent   (Sec. 8.1)

P8.4 Manual PW = $7,768; automatic PW = $4,757; select automatic   (Sec. 8.1)

P8.5 Purchase PW = $7,145; lease PW = $7,034; select lease   (Sec. 8.1)

P8.6   No drains PW = $7,091; drains PW = $5,830; select
        drains   (Sec. 8.1)
P8.7   Valves PW = $13,805; no valves PW = $19,871; select
        valves   (Sec. 8.1)
P8.8   Build PW = $62,223; lease PW = $49,239; select lease   (Sec. 8.1)
P8.9 PW_G = $190,344; PW_H = $217,928; select Machine G   (Sec. 8.2)
P8.10 PW_G = $204,700; PW_H = $217,928; select Machine G   (Sec. 8.2)
P8.11   Manual PW = $13,937; automatic PW = $7,413; select
        automatic   (Sec. 8.2)
P8.12 PW_A = $39,138; PW_B = $42,842; select Proposal A   (Sec. 8.2)
P8.13 PW_A = $35,497; PW_B = $41,101; select Proposal A   (Sec. 8.2)
P8.14 PW_{Hi} = $18,718; PW_{Lo} = $48,718; select high pressure (Sec. 8.2)
P8.15   Paving PW = $19,489; gravel PW = $23,064; select
        paving   (Sec. 8.2)
P8.16   Recap PW − radial PW = $90; select radial   (Sec. 8.2)
P8.17   Expense capitalized cost = $375,790; income capitalized
        cost = $462,441; total capitalized cost = $−86,652 (net
        profit)   (Sec. 8.3)
P8.18 A = $5,199   (Sec. 8.3)
P8.19 P = $120,213   (Sec. 8.3)
P8.20 P = $644,437   (Sec. 8.3)
P8.21 P = $29,964   (Sec. 8.3)
P8.22 P = $48,244   (Sec. 8.3)
P8.23 P_F = $11,333,997; P_G = $16,207,732; select F   (Sec. 8.4)
P8.24 P_A = $45,973; P_B = $50,322; select A   (Sec. 8.4)

CHAPTER 9

P9.1   EUAC is cost for infinite number of renewals and PW is
        for one life cycle   (Sec. 9.1)
P9.2 A = $2,319   (Sec. 9.2)
P9.3 A = $6,437   (Sec. 9.2)
P9.4   (a) $2,319   (b) $6,437   (Sec. 9.3)
P9.5   (a) $2,319   (b) $6,437   (Sec. 9.4)
P9.6   (a) $EUAC_A$ = $4,589; $EUAC_B$ = $4,482; select B   (b)
        70 years   (Sec. 9.5)
P9.7 $EUAC_{new}$ = $12,537; $EUAC_{used}$ = $13,805; select new
        machine   (Sec. 9.5)
P9.8 $EUAC_1$ = $11,857; $EUAC_2$ = $15,219; select Proposal 1   (Sec. 9.5)
P9.9   Gaseous EUAC = $790; dry EUAC = $1,300; select
        gaseous chlorine   (Sec. 9.5)
P9.10 $EUAC_{R-11}$ = $53; $EUAC_{R-19}$ = $41; select R-19   (Sec. 9.5)
P9.11   Savings = $13.77   (Sec. 9.5)

P9.12 Spray EUAC = \$68,000; immersion EUAC = \$14,099; select immersion   (Sec. 9.5)

P9.13 $EUAC_{10}$ = \$3,051; $EUAC_{12}$ = \$1,795; select 12-inch line   (Sec. 9.5)

P9.14 Purchase EUAC = \$270; rent EUAC = \$163; select rent   (Sec. 9.5)

P9.15 Manual EUAC = \$1,055; automatic EUAC = \$646; select automatic   (Sec. 9.5)

P9.16 (*a*) $EUAC_G$ = \$35,115; $EUAC_H$ = \$40,204; select Machine G   (*b*) $EUAC_G$ = \$37,763; $EUAC_H$ = \$40,204; select Machine G   (Sec. 9.5)

P9.17 (*a*) $EUAC_A$ = \$1,552; $EUAC_B$ = \$1,693; select A   (*b*) difference = \$242   (Sec. 9.5)

P9.18 $EUAC_A$ = \$2,005; $EUAC_B$ = \$5,615; select A   (Sec. 9.5)

P9.19 Scram-um EUAC = \$3,419; Catch-um EUAC = \$5,312; select Scram-um   (Sec. 9.5)

P9.20 Perpetual cost = \$7,044   (Sec. 9.6)

P9.21 Perpetual semiannual cost = \$5,057   (Sec. 9.6)

P9.22 EUAC = \$2,652   (Sec. 9.6)

P9.23 EUAC = \$166,334   (Sec. 9.6)

P9.24 EUAC = \$767,912   (Sec. 9.6)

P9.25 EUAC = \$—5,199 (annual profit)   (Sec. 9.6)

P9.26 $EUAC_F$ = \$669,151; $EUAC_G$ = \$972,464; select F   (Sec. 9.6)

CHAPTER 10

P10.1 Net cash flow (*S* − *R*): year 0, \$—4,000; 1–4, \$—500; 5, \$+6,500; 6–9, \$—500; 10, \$+500   (Sec. 10.1)

P10.2

Year	B	A	A − B
0	\$—11,000	\$—15,000	\$ —4,000
1	+500	+1,700	+1,200
2	—9,000	+1,700	+10,700
3	+500	—11,300	—11,800
4	—9,000	+1,700	+10,700
5	+500	+1,700	+1,200
6	+2,000	+3,700	+1,700

(Sec. 10.1)

P10.3 Net cash flow (*Y* − *X*): year 0, \$—9,000; 1–3, \$+1,800; 4, \$—700; 5–7, \$+1,800; 8, \$—700; 9–11, \$+1,800; 12, \$+3,300   (Sec. 10.1)

P10.4 Net cash flow (desert − grass): year 0, \$—2,340; all years 1, 2, 4, 5, 7, 8, 10, 11, 13, 14, 16, 17, 18, \$+110; years 3, 9, 15, \$+130; years 6, 12, \$—220   (Sec. 10.1)

P10.5 i = 6.3%   (Secs. 10.2, 10.3)

P10.6 i = 5.97%   (Secs. 10.2, 10.3)

P10.7 $i = 1.9\%$ (Secs. 10.2, 10.3)

P10.8 $i = 40.9\%$ (Secs. 10.2, 10.3)

P10.9 $i = 10.6\%$ (Secs. 10.2, 10.3)

P10.10 Assuming maintenance cost occurs in year 12 also, annual nominal $r = 16.5\%$; effective monthly, $i = 1.37\%$ (note: $i = 17.78\%$ is effective annual rate)   (Secs. 10.2, 10.3)

P10.11 $i = 3.6\%$ (Secs. 10.2, 10.3)

P10.12 $i_A = 8.6\%$, $i_B = 17.3\%$, $i_C = 9.6\%$, overall $i = 10.4\%$ (Secs. 10.2, 10.3)

P10.13 Nominal rate, $r = 18.7\%$ (Secs. 10.2, 10.3)

P10.14 $i = 37.2\%$ (Secs. 10.2, 10.3)

P10.15 $i = 25.6\%$ (Secs. 10.2, 10.3)

P10.16 $i < 0\%$ (Sec. 10.4)

P10.17 $i < 0\%$; select R   (Sec. 10.5)

P10.18 $i = 58.2\%$; select A   (Sec. 10.5)

P10.19 $i = 11.86\%$; select X   (Sec. 10.5)

P10.20 $i < 0\%$; select grass   (Sec. 10.5)

P10.21 Incremental $i = 7.1\%$; EUAC without expansion = $25,000, EUAC with expansion = $987; do not expand   (Secs. 10.5, 10.6)

P10.22 $i = 11.85\%$; select X   (Sec. 10.6)

P10.23 $i = 1.27\%$; select grass   (Sec. 10.6)

P10.24 $i = 38.0\%$ on increment, as it makes no difference since $38\% > 10\% > 8\%$; select Machine A   (Sec. 10.6)

P10.25 $i < 0\%$ on increment, but now select Machine M, regardless of MARR   (Sec. 10.6)

P10.26 (a) $i = 0\%$; select Roofer B   (b) $i = 3\%$; select Roofer B   (Sec. 10.6)

CHAPTER 11

P11.1   (Sec. 11.1)

P11.2   (a) disbenefit   (b) cost   (c) benefit   (d) disbenefit (e) cost   (Sec. 11.1)

P11.3   B/C = 80,000/123,012 = 0.65 by EUAC; no   (Sec. 11.2)

P11.4   (a) EUAC = $489,380, B $-$ C = $-139,380; no (b) B/C = 0.72; no   (Sec. 11.2)

P11.5   Disbenefit EUAC = $24,072; from B $-$ C $-$ disbenefit = 0, B = $513,452, and gradient of B is $G = \$18,017$ per year   (Sec. 11.2)

P11.6   EUAC = $155,770, B/C = 1.12; yes   (Sec. 11.2)

P11.7   EUAC = $173,560, B/C = 0.92; no   (Sec. 11.2)

P11.8   (a) Benefit is user cost savings = $576,000 per year, cost is extra EUAC = $2,117,320 for transmountain, B/C = 0.27; select long route
(b) B $-$ C = $-1,541,320; select long route   (Sec. 11.3)

P11.9 $(B/C)_A = 0.77$, $(B/C)_B = 1.28$, $(B/C)_C = 0.58$; build
at site B only   (Sec. 11.3)

P11.10   Resurface EUAC = \$296,850; new EUAC = \$918,603;
annual user costs: resurface, \$720,000, new, \$600,000,
B/C = 0.99; select resurface   (Sec. 11.3)

P11.11   (a) $EUAC_E = \$200,000$; $EUAC_W = \$415,000$, B/C =
1.00; select Location W   (b) $B - C = 0$; select
Location W   (Sec. 11.3)

P11.12   (a) Lined EUAC = \$337,920, not lined EUAC =
\$122,905, B/C = 0.56; do not line   (b) $B - C =$
\$−95,015; do not line   (Sec. 11.3)

P11.13   (Sec. 11.4)

P11.14 $0 = -P + A(P/A,i\%,n')$ (Sec. 11.5)

P11.15   (Sec. 11.6)

P11.16 $n' = 5.7$ years   (Sec. 11.6)

P11.17   (a) $n' = 15.6$ years   (b) no   (Sec. 11.6)

P11.18 $n' = 34.4$ years   (Sec. 11.6)

P11.19 $n' = 3.9$ years   (Sec. 11.6)

P11.20   (a) 2.7 years   (b) yes   (c) $n' = 3.4$ years; yes   (Sec. 11.6)

P11.21 $n' \gg 18$ years; grass is more economic   (Sec. 11.7)

P11.22   (a) $n' = 5.3$ years, Apartment 2 for $n = 5$ (b) $n >$
5.3   (c) PW = \$39,937, $PW_2 = \$41,186$; select 2   (Sec. 11.7)

P11.23   (a) $n' = 14.7$ months = 1.2 years   (b) lease EUAC =
\$275 per month, buy EUAC = \$225 per month; select
buy   (Sec. 11.7)

P11.24 $n' = 2.4$ years   (a) Machine 2   (b) Machine 1   (Sec. 11.7)

CHAPTER 12

P12.1   Don't use it   (Sec. 12.1)

P12.2   (Sec. 12.1)

P12.3   (a) $P = \$18,000$, $n = 8$, SV = \$1,000, AOC = \$150
(b) yes, but it is \$−8,000   (Sec. 12.1)

P12.4   (a) Downtown: $P = \$25,000$, $n = 10$, AOC = \$1,080,
SV = 0; suburban: $P = \$750$, $n = 10$, AOC = \$6,600,
SV = \$750   (b) Sunk cost is negative, since profit is
\$25,000 − \$21,576 = \$3,424   (Sec. 12.1)

P12.5   (a) SC = \$2,925   (b) SC = \$5,300   (Sec. 12.1)

P12.6 $EUAC_{old} = \$79,731$, $EUAC_{new} = \$154,342$; select old   (Sec. 12.2)

P12.7 $EUAC_{old} = \$79,731$, $EUAC_{new} = \$131,968$; select old   (Sec. 12.2)

P12.8 $EUAC_{old} = \$87,117$, $EUAC_{new} = \$196,340$; select old   (Sec. 12.2)

P12.9 $EUAC_{old} = \$4,088$, $EUAC_{new} = \$2,646$; select new   (Sec. 12.2)

P12.10 Downtown EUAC = $5,149, suburban EUAC = $2,606
with sales price of $25,000 now; select suburban (Sec. 12.2)

P12.11 Two mover EUAC = $21,024, conveyor and old mover
EUAC = $24,236, conveyor and new mover EUAC =
$27,757; select two mover plan (Sec. 12.2)

P12.12 In same order as P12.11 answer, EUAC values are
$31,565, $46,314, $54,858; select two mover plan (Sec. 12.2)

P12.13 Plan A vs. Plan B: $EUAC_A$ = $8,047, $EUAC_B$ = $7,265;
Plan A vs. Plan C: $EUAC_A$ = $6,925, $EUAC_C$ = $6,348;
select C (Sec. 12.2)

P12.14 Old range + microwave plan: P = $565, AOC = $15,
SV = $25, EUAC = $148; Combination: P = $650,
AOC = $12, SV = $25, EUAC = $166; select range +
microwave oven (Sec. 12.2)

P12.15 RV = $11,527 (Sec. 12.3)

P12.16 RV = $51,607 (Sec. 12.3)

P12.17 RV = $54,622, a 5.84% increase (Sec. 12.3)

P12.18 RV = $137.70 (Sec. 12.3)

P12.19 RV = $−26,620; thus, if you have to *pay* less then
$26,620 to get rid of it, do so and buy the challenger (Sec. 12.3)

P12.20

Year	Capital recovery plus loan EUAC	Operating EUAC	Total EUAC
8	$880	$1,382	$2,262
9	799	1,446	2,256
10	754	1,512	2,266
11	709	1,581	2,290

Keep the car nine years (Sec. 12.4)

P12.21

Year	Capital recovery plus loan average	Average AOC	Total EUAC
7	$821	$1,355	$2,176
8	719	1,430	2,148
9	639	1,509	2,148
10	575	1,584	2,169

Keep the car eight or nine years (Sec. 12.4)

P12.22 Keep asset full 10 years for EUAC of $10,837; decrease
in EUAC for year 4 is not the minimum cost life (Sec. 12.4)

P12.23 12.6 years (Sec. 12.4)

P12.24 Keep blower total of 14 years for EUAC of $1,881 (Sec. 12.4)

P12.25 X = 2,486 hours per year (Sec. 12.5)

P12.26 P = $16,891 (Sec. 12.5)

P12.27 X = 94.8 days per year (Sec. 12.5)

P12.28 X = 566,935 yards per year (Sec. 12.5)

P12.29 (*a*) n = 14.52 years (*b*) price = $94,450 (Sec. 12.5)

P12.30 (*a*) $X = 333$ samples per year   (*b*) $X = 202$ samples
per year   (*c*) outside testing   (Sec. 12.5)

P12.31 $P = \$288,557$ (Sec. 12.5)

P12.32 $P = \$5,100$ (Sec. 12.5)

P12.33 Extra investment outside city is $851   (Sec. 12.5)

P12.34 Annual savings = $49.33 per year   (Sec. 12.5)

P12.35 (*a*) $X = 25.9$ days per year   (*b*) using 52 weeks per
year, $P = \$34,744$ (Sec. 12.5)

P12.36 $P = \$6,739$ (Sec. 12.5)

CHAPTER 13

P13.1   (Sec. 13.1)

P13.2 $I = \$125$; six months   (Sec. 13.2)

P13.3 $I = \$150$; three months   (Sec. 13.2)

P13.4 $I = \$18.75$; monthly   (Sec. 13.2)

P13.5 $P = \$7,282$ (Sec. 13.3)

P13.6 $P = \$43,567$ (Sec. 13.3)

P13.7 $P = \$14,130$ (Sec. 13.3)

P13.8   Price = $19,200; $P = \$16,524$; no, $i < 8\%$ nominal   (Sec. 13.3)

P13.9 $P = \$7,377$ (Sec. 13.3)

P13.10 Quarterly, $P = \$7,241$; monthly, $P = \$7,214$ (Sec. 13.3)

P13.11 $P = \$12,918$ (Sec. 13.3)

P13.12 $n = 2.61$ years   (Sec. 13.3)

P13.13 $n = 2.67$ years   (Sec. 13.3)

P13.14 Effective $i = 3.15\%$ per semiannual period; nominal $i =$
$r = 6.3\%$ (Sec. 13.4)

P13.15 Nominal $i = 6.4\%$ (Sec. 13.4)

P13.16 Nominal $i = 5.8\%$ (Sec. 13.4)

P13.17 Effective $i = 6.4\%$ (Sec. 13.4)

P13.18 Effective $i = 6.56\%$ (Sec. 13.4)

P13.19 Bond rate = $b = 4.98\%$ (Sec. 13.4)

P13.20 $b = 6.8\%$ (Sec. 13.4)

P13.21 $b = 6.91\%$ (Sec. 13.4)

P13.22 $A = \$15,471$ (Sec. 13.4)

CHAPTER 14

P14.1   (*a*) Current assets = $77,000, fixed assets = $1,055,000,
liabilities = $603,000, equity = $529,000   (*b*)
$1,132,000 = \$603,000 + \$529,000$ (Sec. 14.1)

P14.2   (*a*) Cost of materials = $41,000, prime cost = $91,000, cost of goods sold = $141,000   (*b*) $21,000 decrease   (Sec. 14.2)

P14.3   (*a*) Net profit = $45,000   (*b*) before taxes: $500,000 − $435,000 = $65,000; after taxes: $500,000 − $455,000 = $45,000   (*c*) 9%   (Sec. 14.2)

P14.4   (*a*) Current ratio = 1.79, acid-test ratio = 1.07, equity ratio = 0.47   (*b*) 72%   (Sec. 14.3)

P14.5   (*a*) Operating ratio = 1.15; income ratio = 9%   (*b*) 9%   (Sec. 14.3)

P14.6   Competitor's total expenses = $857,143, net profit = $270,000; horseradish operating ratio = 1.19 and income ratio = 6.25%; the income percent is 2.88 times greater for the competitor   (Sec. 14.3)

P14.7   Horseradish: (*a*) net worth = $285,714, return on invested capital = 17.5%, which is   (*b*) 2.80 times the income ratio; competitor: (*a*) Return = 50.4%, which is (*b*) 2.80 times income ratio   (Sec. 14.3)

P14.8   (*a*) $8.33/hour (M101), $25/hour (M102), $6.25/hour (M103, M104)   (*b*) $7.81/hour   (Sec. 14.4)

P14.9   (*a*) Space (February); DLC (March, April), Space (May), material (June)   (*b*) Rate decreases because of changing basis   (Sec. 14.4)

P14.10   (*a*) $0.41 per square foot   (*b*) $6.73 per hour   (*c*) $1.89 per dollar   (Sec. 14.4)

P14.11   (*a*) $5,831 (M101), $8,750 (M102), $4,063 (M103), $8,125 (M104); variance = $1,769 over   (*b*) $23,430; variance = $1,570 under   (Sec. 14.5)

P14.12   (*a*) $12,500(1), $23,750(2), $13,125(3), $483,000(4), $188,715(5), $29,000(6)   (*b*) $90 over   (Sec. 14.5)

P14.13   Preparation $3,230, subassembly $6,730, final $4,649   (Sec. 14.5)

P14.14   (*a*) Cost to make is $65,910 per month; cheaper than buying   (*b*) preparation overhead rate = $5 per hour, cost of present condition = $65,910 per month, cost of new equipment = $71,669 per month; continue the present condition   (Sec. 14.5)

CHAPTER 15

P15.1   (*a*) Capital gain   (*b*) taxable income deduction   (*c*) operating loss   (*d*) taxable income deduction   (*e*) capital loss   (*f*) taxable income   (Sec. 15.1)

P15.2   Dough: $343,420; broke: $89,500   (Sec. 15.2)

P15.3 Dough: $349,920 (−1.86%); broke: $96,000 (−6.77%) (Sec. 15.2)

P15.4 (*a*) 45.4% (*b*) 48% (*c*) 45.8% (*d*) $20,800 (Sec. 15.2)

P15.5 Income tax = $745,010 (Sec. 15.2)

P15.6 (*a*) 39.9% (*b*) 44.7% (*c*) $35,760 (Sec. 15.2)

P15.7 Taxes would decrease $460 (Sec. 15.2)

P15.8 Effective tax rate = 42.9%, net LTG = $300, taxes = $34,394 (Sec. 15.3)

P15.9 (*a*) LTL = $1,000 (*b*) STL = $200 (*c*) STG = $800 (*d*) LTG = $300 (*e*) STL = $1,750, taxes = $12,460; note that only $1,100 in losses is deductible due to total of capital gains (Sec. 15.3)

P15.10 Increase taxes by $900 (Sec. 15.3)

P15.11

Year	Depreciation (*a*)	Depreciation (*b*)	Taxes (*a*)	Taxes (*b*)
1972	$17,600	$17,600	$ 76,200	$ 76,200
1973	17,600	17,600	86,200	86,200
1974	17,600	17,600	78,700	79,660
1975	13,200	13,200	98,400	101,130
1976	8,800	8,800	60,600	60,600
1977	8,800	0	40,600	44,150
1978	4,400	0	72,800	75,150
			$513,500	$523,090

(*a*) Gains and losses on individual assets not recognized at sale time (*b*) gains and losses accounted for in year of sale (Sec. 15.3)

P15.12 Taxes would decrease by $2,093 (Sec. 15.4)

P15.13 Taxes = $119,025 (Sec. 15.4)

P15.14 Tax credit is (*a*) $700 (*b*) $128 (*c*) 0 (*d*) $222 (*e*) $84 (Sec. 15.4)

P15.15 (*a*) $104,070 (*b*) $105,206 (*c*) $106,240 (*d*) $104,638 (*e*) $105,155 (*f*) $105,723 (*g*) $105,172; recovery in year 1973 only is best (Sec. 15.5)

P15.16 (*a*) SYD has PW_{tax} value 5.6% less than SL (*b*) declining balance has PW_{tax} value 0.37% less than SYD (Sec. 15.6)

P15.17 SL, PW_{tax} = $43,616; declining balance, PW_{tax} = $30,722 (Sec. 15.6)

P15.18 A: total taxes = $147,514, PW_{taxes} = $80,402; B: total taxes = $142,502, PW_{taxes} = $78,970; difference in PW_{tax} = $1,432 less for B (Sec. 15.6)

P15.19 SV = $838 (Sec. 15.6)

P15.20 *n* = 8.48 years (Sec. 15.6)

P15.21 Use DDB for *n* = 7 with PW_{taxes} = $30,959 (Sec. 15.6)

P15.22 Use *n* = 5 for PW_{taxes} = $5,227 (Sec. 15.6)

P15.23 Before-tax rate = 16.5% (Sec. 15.7)

P15.24 (*a*) $i = 12\%$ EUAC of CFAT$_A$ = $\$-48,814$, EUAC$_A$ =
$\$521$, EUAC of CFAT$_B$ = $\$-49,102$, EUAC$_B$ = $\$1,204$;
select A   (*b*) $i = 24\%$, EUAC$_A$ = $\$-6,799$, EUAC$_B$ =
$\$-9,723$; select A   (*c*) same decision would be made   (Sec. 15.7)

P15.25 Tax rate = 57.9%   (Sec. 15.7)

CHAPTER 16

P16.1

Year	CFAT	Remark
0	$\$-350,000$	Purchase
1–8	$+17,460$	Regular
8	$+373,800$	Sale; gain

(Sec. 16.1)

P16.2

Year	CFAT	Remark
0	$\$-309,833$	Tax credit
1–6	$+39,167$	Regular
6	$+167,500$	Sale; gain

(Sec. 16.1)

P16.3 Equal annual amount due = $\$30,767$ with $\$21,667$ on
principal and $\$9,100$ in interest

Year	CFAT	Remark
0	$\$-179,833$	60% equity
1–6	$+13,132$	Regular
6	$+167,500$	Sale; gain

(Sec. 16.1)

P16.4 (*a*)

Year	CFAT	Remark
0	$\$-10,000$	Purchase
1	$+4,136$	SYD
2	$+3,752$	SYD
3	$+3,368$	SYD
4	$+2,984$	SL
5	$+2,600$	No depreciation
6	$+2,600$	No depreciation
6	$+2,000$	Sale

(*b*) In year 3, CFAT = $\$+3,560$ and in year 4, CFAT =
$\$+2,792$ for a net total CFAT difference of zero   (Sec. 16.1)

P16.5 (*a*)

Year	CFAT	Remark
0	$\$-309,833$	Tax credit
1	$+80,333$	DDB
2	$+58,667$	Remaining depreciation
3–6	$+24,000$	No depreciation
6	$+167,500$	Sale; gain
	$\$+75,333$	

(*b*) Use DDB, but CFAT difference = $\$0$ (Sec. 16.1)

P16.6 (*a*)

Year	CFAT	Remark
0	$ −5,000	Purchase
1	+3,058	SYD; loan
2	+2,674	SYD; loan
3	+2,290	SYD; loan
4	+1,906	SL; loan
5	+1,522	No depreciation; loan
6	+2,600	No depreciation; no loan
6	+2,000	Sale
	$+11,626	

(*b*) Difference is zero   (Sec. 16.1)

P16.7

	CFAT	
Year	A	B
0	$−15,000	$−22,000
1–10	−900	+100
10	+3,000	+5,000

Cost PW_A = $19,796, cost PW_B = $18,756; select B. Note that Cost PW means that costs are positive cash flows.   (Sec. 16.2)

P16.8 $EUAC_{new}$ = $6,196, $EUAC_{used}$ = $6,813; select new machine   (Sec. 16.2)

P16.9 $EUAC_A$ = $2,751, $EUAC_B$ = $3,123; select A   (Sec. 16.2)

P16.10 (*a*) Cost PW of Roundee = $1,661, cost PW of Scored = $2,910; select Roundee   (*b*) EUAC of Roundee = $244, EUAC of Scored = $427; select Roundee (don't forget capital gain, tax rate = 30%, in year 12)   (Sec. 16.2)

P16.11 Yes (*a*) Cost PW of Roundee = $3,351, cost PW of Scored = $3,824   (*b*) EUAC of Roundee = $492, EUAC of Scored = $561   (Sec. 16.2)

P16.12 Cost PW_A = $11,550, cost PW_B = $11,725; select A   (Sec. 16.2)

P16.13 Cost PW_A = $10,296, cost PW_B = $9,841; select B, a switch from P16.12   (Sec. 16.2)

P16.14 PW_A = $10,371, PW_B = $37,595; select B   (Sec. 16.2)

P16.15 *i* = 5.69%   (Sec. 16.3)

P16.16 (*a*) *i* = 6.07%   (*b*) *i* = 6.34%   (*c*) due to partial debt financing and tax benefit of interest   (Sec. 16.3)

P16.17 (*a*) *i* = 28.3%   (*b*) *i* = 28.4%, or an increase in *i* of 0.1%   (Sec. 16.3)

P16.18 *i* = 48.6%; debt financing increases *i* by approximately 69%   (Sec. 16.3)

P16.19 (*a*) *i* = 9.76%   (*b–d*) select B   (*e*) select A   (Sec. 16.3)

P16.20 i_A = 4.01%, i_B = 5.94%   (Sec. 16.3)

P16.21 *i* = 0.85%   (Sec. 16.3)

P16.22 (a) $D = \$23,881$ (b) $D = \$48,905$ (Sec. 16.3)

P16.23 (a) $0 = -10,000 + 2,180(P/A,i\%,4) + 5,380(P/F,i\%,5) + 4,780(P/A,i\%,2)\ (P/F,i\%,6) + 10,380(P/F,i\%,6) + 9,980 (P/A,i\%,2)\ (P/F,i\%,8) + 10,580(P/F,i\%, 11) + 14,580 (P/F,i\%,12)$ (b) $i = 35.5\%$ (Sec. 16.3)

P16.24 $i_A = 17.5\%; i_B = 24.53\%$ (Sec. 16.3)

P16.25 (a) $P = \$86,215$ (b) $SV = \$100,000 = P$ (c) $D = \$13,255$ (Sec. 16.3)

P16.26 $i = 1.76\%$ (Sec. 16.3)

P16.27 They need only \$2,244; therefore, none of the \$1,500 is required.   (Sec. 16.3)

P16.28 (a) $EUAC_D = \$1,062$, $EUAC_C = \$3,433$; select defender by \$2,371   (b) $EUAC_D = \$540$, $EUAC_C = \$1,611$; select defender by \$1,071   (Sec. 16.4)

P16.29 Plan 1: actual $P_A = \$12,000$; new depreciation for A = \$7,500 per year; CFAT for plan: year 0, \$$-$12,000; year 1, \$+2,250; year 2, \$+22,750; years 3–13, \$$-$1,000; year 14, \$0; $EUAC_1 = \$-3,689$; Plan 2: CFAT for plan: year 0, \$$-$38,000, years 1–19, \$$-$325; year 20, \$+675; $EUAC_2 = \$-3,087$; select Plan 2 $EUAC_2 = \$-3,087$; select Plan 2   (Sec. 16.4)

P16.30 (a) $i = 9.0\%$ (b) $RV = \$5,641$ including tax credit on capital loss; actual $RV = \$-3,718$, negative trade-in   (Sec. 16.4)

P16.31 (a) $EUAC_D = \$2,721$, $EUAC_C = \$2,388$; select challenger   (b) no   (Sec. 16.4)

P16.32 Same as P7.27   (Sec. 16.5)

P16.33 $i_A = 15.11\%$, $i_B = 18.33\%$ using 0.5TI as depletion   (Sec. 16.5)

CHAPTER 17

P17.1 (Sec. 17.1)

P17.2 (Sec. 17.1)

P17.3

	1	2	3	4	5
(a) EUAC	\$2,061	\$2,512	\$1,994	\$1,992	\$1,836
(b) ΔRR	>30%	>30%	<3%	<12%	<15%

Select Method 2   (Sec. 17.2)

P17.4 Select Method 3   (Sec. 17.2)

P17.5

Proposals compared	Incremental return	Proposals compared	Incremental return
2 to none	> 15%	5 to 6	<0%
3 to 2	> 40	4 to 6	<0
6 to 3	>100	7 to 6	<0
1 to 6	< 0		

Select Proposal 6   (Sec. 17.2)

P17.6

	3	1	2	5	4
(a) PW	$-134,353	$-117,882	$-113,894	$-108,917	$-104,929
Comparison	–	1 to 3	2 to 1	5 to 2	4 to 5
(b) ΔRR	–	>50%	>35%	>25%	>35%

Select Machine 4 (Sec. 17.2)

P17.7

	14 in.	16 in.	20 in.	24 in.	30 in.
(a) PW	$-61,137	$-60,955	$-59,089	$-59,292	$-73,616
(b) ΔRR	–	>9%	>15%	>7%	<0%

Select 20-inch pipe (Sec. 17.2)

P17.8

	8 yd	10 yd	15 yd	20 yd	40 yd
(a) EUAC	$+3,341	$+3,763	$+1,989	$-785	$-4,811
(b) ΔRR	>35%	>20%	<0%	<0%	<0%

Select 10-yard truck (Sec. 17.2)

P17.9 Use a 12-year evaluation period

	1	2	3	4	5
(a) PW	$-67,355	$-67,789	$-62,384	$-60,292	$-69,140
(b) ΔRR	–	<15%	>25%	>25%	<0%

Select Process 4 (Sec. 17.2)

P17.10 (a)

Function	Comparison	ΔRR	Selected
Pressing	2 to 1	>30%	2
Slicing	2 to 1	>35	2
Weighing	2 to 1	>50	2
Wrapping	2 to 1	<15	1

(b) Investment = $45,000, AOC = $37,000 (Sec. 17.2)

P17.11 (a) Comparing (P-S 3) to (P2 and S2), $i > 100\%$; comparing W-W 3 to (W2 and W1), $i > 25\%$; select P-S 3 and W-W 3 machines (b) investment = $55,000, AOC = $27,000 (c) comparing P-S 3 and W-W 3 to Machine 4, $i > 40\%$; select P-S 3 and W-W 3 (Sec. 17.2)

P17.12

Proposals compared	ΔB/C	Proposals compared	ΔB/C
2 to none	1.23	5 to 6	0.65
3 to 2	5.24	4 to 6	<0
6 to 3	∞	7 to 6	<0
1 to 6	<0		

Select Proposal 6 (Sec. 17.3)

P17.13

Comparison	ΔB/C
1 to 3	2.81
2 to 1	1.80
5 to 2	1.38
4 to 5	1.80

Select Machine 4 (Sec. 17.3)

P17.14

| | | ΔB/C | |
Sizes compared, inches		8%(*a*)	15%(*b*)
16 to 14		1.12	0.76
20 to 16		1.57	–
20 to 14		–	0.98
24 to 20		0.93	–
24 to 14		–	0.85
30 to 20		0.19	–
30 to 14		–	0.31

Selected (*a*) 20 inch   (*b*) 14 inch   (Sec. 17.3)

P17.15 MARR = 8%; all comparisons as in P17.14(*a*) except for
30 in. to 14 in., ΔB/C = 0.26; select 14 in.   (Sec. 17.3)

P17.16

Processes compared	Evaluation period, years	ΔB/C
2 to 1	12	0.93
3 to 1	6	2.57
4 to 3	6	1.35
5 to 4	6	0.31

Select Process 4   (Sec. 17.3)

P17.17 (Sec. 17.4)

P17.18 (*a*) 5, 2, 6   (*b*) 5, 2, 6, 1   (Sec. 17.5)

P17.19 (*a*) E, B, D   (*b*) RR = 14.12%   (Sec. 17.5)

P17.20 E, B, C; RR = 14.10%   (Sec. 17.5)

P17.21 Select 5, 3, 4 in ranked order   (Sec. 17.5)

P17.22 (*a*) B, D, E   (*b*) B, D   (*c*) no, because a return of 0%
after the first life cycle is assumed with the PW method   (Sec. 17.6)

P17.23 (*a*) B, D, E   (*b*) B, D   (*c*) selections are the same as
P17.22 for $190,000 and $170,000 budget (different
ranks)   (Sec. 17.6)

P17.24 Select 5, 4, 3 in ranked order (ranking is different than
P17.21)   (Sec. 17.6)

P17.25 Select 5 only; ranking for Alternative 5 only, since all
others make < 15%   (Sec. 17.6)

P17.26 (*a*) C, D   (*b*) n_1 = 12 years, select C, D; ranks and
answers are the same   (Sec. 17.6)

P17.27 Select A, B, C by each method, but with different
rankings; no   (Sec. 17.6)

P17.28 (Sec. 17.6)

P17.29 Ranks are shown below:

	5%	10%	15%	18%	20%	25%
A	2	2	3	3	3	4
B	3	3	2	2	2	1
C	4	4	4	4	4	3
D	1	1	1	1	1	2

Critical values seem to be around 15% and 25% *for this problem only*

(Sec. 17.6)

CHAPTER 18

P18.1   (*a*) debt   (*b*) equity   (*c*) debt   (*d*) equity   (*e*) debt

(Sec. 18.1)

P18.2   0.5%; no, since return is too small to operate a business for long

(Sec. 18.2)

P18.3   (*a*) raise   (*b*) lower

(Sec. 18.3)

P18.4   (*a*) \$2,083,333   (*b*) 4.48% annual

(Sec. 18.4)

P18.5   Loan $i = 4\%$, bonds $i = 2.9\%$; recommend bonds

(Sec. 18.4)

P18.6   Yes

(Sec. 18.4)

P18.7   (*a*) Interest $= \$3,000$, principal $= \$10,000$, $i = 9.4\%$ (*b*) $i = 5.42\%$

(Sec. 18.4)

P18.8   Loan payments $= \$12,676$, CFAT $= \$3,662$, $i = 7.80\%$

(Sec. 18.4)

P18.9   (*a*) BTCC $= 12.4\%$ (*b*) ATCC $= 6.5\%$

(Sec. 18.5)

P18.10 $i = 12.2\% > 10\%$ from investment; profitable

(Sec. 18.5)

P18.11   BTCC $= 20.9\%$

(Sec. 18.5)

P18.12   Y $= \$1.07$, Y $-$ D $= \$-0.18$ (Take funds from retained earnings to pay dividend)

(Sec. 18.5)

P18.13   Earnings $= 3.8\% = \$0.22$ per share

(Sec. 18.5)

P18.14   ATCC $= 21.3\%$; high

(Sec. 18.5)

P18.15   Scheme 2

(Sec. 18.6)

P18.16   Scheme 1, 12%; Scheme 2, 10.7%

(Sec. 18.6)

P18.17 $i = 12.4\%$; no

(Sec. 18.6)

P18.18   100% equity, MARR $= 10.2\%$, return $i = 8.5\%$, no; 60% debt, actual loan rate $= 2.6\%$, MARR $= 5.95\%$, return $i = 15.9\%$, yes

(Sec. 18.6)

P18.19   Plan 1, MARR $= 15.3\%$, $i = 3.6\%$, no; Plan 2, MARR $= 16.2\%$, $i = 19.9\%$, yes; Plan 3, MARR $= 17.4\%$, $i = 96.6\%$, yes; use of Plan 2 is advised

(Sec. 18.6)

P18.20   On a six-month basis, (1) BTCC $= 3.02\%$ (2) BTCC $= 6.15\%$ (3) BTCC $= 4.59\%$; use Plan 1

(Sec. 18.6)

P18.21   50% equity has ATCC $= 12.5\%$, 15% bonds have ATCC $= 3.3\%$, 35% loan has ATCC $= 9\%$ (*a*) average BTCC $= 19.01\%$ (*b*) average ATCC $= 9.9\%$

(Sec. 18.6)

P18.22   (*a*) $(D/E)_1 = 0.11$, $(D/E)_2 = 0.67$, $(D/E)_3 = 4.0$ (*b*) $(D/E)_1 = 0$, $(D/E)_2 = \infty$, $(D/E)_3 = 1.0$ (*c*) $D/E = 1.0$

(Sec. 18.7)

P18.23   (Sec. 18.7)

P18.24

Firm	Debt*	Stock value*	Rate of return
A	$2.5	$2.5	54%
B	2.0	2.0	67
C	3.0	3.0	40

*$ values in millions.   (Sec. 18.7)

P18.25   (a) Company A: debt = $714,286, equity = $1,785,714;
Company B: debt = $1,142,857, equity = $457,143 (b)
$(D/E)_A = 0.51, (D/E)_B = 5.26$ (Sec. 18.7)

P18.26   Return for A = 25.2%, return for B = 86.9%   (Sec. 18.7)

CHAPTER 19

P19.1   Ten tons, PW = $-134,922; 15 tons, PW = $-137,589;
20 tons, PW = $-140,257; 25 tons, PW = $-142,924;
30 tons, PW = $-166,078 (overtime)   (Sec. 19.2)

P19.2   Income = $14,000, PW_A = $-5,977; income =
$15,000, PW_B = $-1,355; income = $18,000, PW_C =
$+12,514; engineers will decide to buy only for income
of $18,000 per year   (Sec. 19.2)

P19.3   Income = $14,000, PW_A = $-3,317; income =
$15,000, PW_B = $-881; income = $18,000, PW_C =
$6,428; same conclusion as P19.2   (Sec. 19.2)

P19.4   (a) Build PW = $69,223; lease PW = $37,302; same
decision as P8.8 (lease), since costs go in wrong direction
to change decision   (b) Construction cost = $5, Build
PW = $42,223; select build; for construction costs of $6
to $9, select lease (cost of $5.25 per square foot is the
break point)   (Sec. 19.2)

P19.5   Yes, since results are as follows:

G	i	G	i
$300	42.0%	$600	40.4%
400	41.5	700	39.8
500	40.9		

(Sec. 19.2)

P19.6

i	$EUAC_1$	$EUAC_2$	Difference (2 − 1)
8%	$2,173	$2,673	$500
10	2,307	2,907	600
12	2,445	3,152	707
15	2,659	3,536	877

(Sec. 19.2)

P19.7

n	EUAC$_1$	EUAC$_2$
4	\$3,832	–
6	2,811	\$4,610
8	2,307	3,746
10	–	3,239
12	–	2,907

EUAC$_1$ is slightly more sensitive to varying n (Sec. 19.2)

P19.8

Time (days/year)	Cabin EUAC	Trailer EUAC
6	\$2,271	\$2,385
8	2,175	2,375
10	2,079	2,365
12	1,983	2,355
14	1,887	2,345

Purchase cabin for all time values; breakeven point: 3.35 days per year   (Sec. 19.2)

P19.9 (*a*)

Time (days/year)	Cabin EUAC	Trailer EUAC
6	\$2,531	\$2,385
8	2,435	2,375
10	2,339	2,365
12	2,243	2,355
14	2,147	2,345

Select cabin for all time values
(*b*) Breakeven point: 9.4 days per year is sensitive   (Sec. 19.2)

P19.10

Life, years	Rate Per quarter	Rate Nominal annual
10	1.71%	6.84%
12	1.66	6.64
15	1.60	6.40
18	1.56	6.24
20	1.54	6.16

       (Sec. 19.2)

P19.11 $n = 6$, $P = \$1,935$; $n = 10$, $P = \$-956$; yes, decision is sensitive to n (Sec. 19.2)

P19.12 (*a*) $n = 15$, $P = \$264,801$; from P12.31, $n = 20$, $P = \$288,557$ (*b*) $n = 25$, $P = \$301,584$ (Sec. 19.2)

P19.13

Gradient	$n*$
\$ 60	16.3
80	14.1
100	12.7
120	11.5
140	10.7

       (Sec. 19.2)

P19.14

Gradient	n^*
$ 60	19
80	16
100	14
120	13
140	12

(Sec. 19.2)

P19.15

EUAC

Machine	O	R	P
New	$12,503	$12,539	$12,692
Used	13,313	13,541	13,805

New: pessimistic = 4 years, used: pessimistic = 2 years, decision will not change from P9.7

(Sec. 19.3)

P19.16 $EUAC_P = \$102,000$, $EUAC_R = \$68,000$, $EUAC_O = \$51,000$; it won't

(Sec. 19.3)

P19.17

EUAC

AOC or lease cost ($/year)	50 days per year		100 days per year	
	Plan 1	Plan 2	Plan 1	Plan 2
1,800	–	$3,800	–	$5,800
2,000	$3,664	–	–	–
2,500	4,164	4,500	–	6,500
3,200	–	5,200	–	7,200
3,750	5,414	–	–	–
4,000	–	–	$5,664	–
5,000	–	–	6,664	–
7,500	–	–	9,164	–

(*a*) Plan 1   (*b*) Plan 2

(Sec. 19.3)

P19.18

Present worth

Interest rate or Life	Plan M	Plan Q
(*a*) Pessimistic (15%)	$ −6,111	$ +8,927
Reasonable (10%)	+27,704	+51,758
Optimistic (8%)	+47,272	+76,544
(*b*) Pessimistic (22 years)	+31,573	+56,659
Reasonable (20 years)	+27,704	+51,758
Optimistic (16 years)	+17,356	+38,650

(Sec. 19.3)

P19.19

Interest rate or Life	Plan A	Plan B
(*a*) Pessimistic (15%)	$−13,317	$−37,293
Reasonable (10%)	−14,867	−38,600
Optimistic (8%)	−15,916	−39,472
(*b*) Pessimistic (44,22)	−14,908	−38,519
Reasonable (40,20)	−14,867	−38,600
Optimistic (32,10)	−14,716	−38,762

(Sec. 19.3)

P19.20 (Sec. 19.3)

P19.21 $E(X) = 9.84$ (Sec. 19.4)

P19.22 $AOC_{low} = \$1,900$ (Sec. 19.4)

P19.23 $E(PW) = \$-2,463$ (Sec. 19.4)

P19.24 $E(EUAC) = \$-1,169$ (Sec. 19.4)

P19.25 $E(EUAC) = \$+79,696$ (net profit); yes (Sec. 19.5)

P19.26 $E(EUAC) = \$+950$; yes, 8% can be expected (Sec. 19.5)

P19.27 Build for 2.5 inches with EUAC $= \$2,183$ (Sec. 19.5)

P19.28 Build for 2.5 inches with EUAC $= \$1,403$ (Sec. 19.5)

P19.29 Savings $i = 6.35\%$, stocks $i = 10.7\%$, property $i = 8.54\%$; invest in stocks (Sec. 19.5)

BIBLIOGRAPHY

Barish, N. N.: "Economic Analysis for Engineering and Managerial Decision-making," McGraw-Hill Book Company, New York, 1962.

DeGarmo, E. P., and J. R. Canada: "Engineering Economy," 5th ed., Macmillan Publishing Company, Inc., New York, 1973.

Emerson, C. R., and W. R. Taylor: "An Introduction to Engineering Economy," Cardinal Publishers, Bozeman, Mont., 1973.

Fabrycky, W. J., and G. J. Thuesen: "Economic Decision Analysis," Prentice-Hall, Inc., Englewood Cliffs, N.J., 1974.

Grant, E. L., and W. G. Ireson: "Principles of Engineering Economy," 5th ed., The Ronald Press Company, New York, 1970.

Ostwald, P. F.: "Cost Estimating for Engineering and Management," Prentice-Hall, Inc., Englewood Cliffs, N.J., 1974.

Park, W. R.: "Cost Engineering Analysis," John Wiley & Sons, Inc., New York, 1973.

Reisman, A.: "Managerial and Engineering Economics," Allyn & Bacon, Inc., Boston, 1971.

Riggs, J. L.: "Economic Decision Models for Engineers and Managers," McGraw-Hill Book Company, New York, 1968.

Smith, G. W.: "Engineering Economy: Analysis of Capital Expenditures," 2d ed., Iowa State University Press, Ames, 1973.

Taylor, G. A.: "Managerial and Engineering Economy: Economic Decision-making," 2d ed., Litton Educational Publishing, Inc., New York, 1975.

Thuesen, H. G., W. J. Fabrycky, and G. J. Thuesen: "Engineering Economy," 4th ed., Prentice-Hall, Inc., Englewood Cliffs, N.J., 1971.

A/F factor, 30
A/G factor, 87
A/P factor, 30
Accounting:
    ratios (factors), 242–245
    statements, 240–242
Acid-test ratio, 243
Additional depreciation, 110
After-tax analysis:
    general procedure, 274–277
    rate of return, 277
    replacement analysis and, 279
Alternatives:
    comparison: by benefit/cost ratio, 188,
        292, 297
      by capitalized-cost method, 135–
        138
      by equivalent-uniform-annual-cost
        method, 151–154

Alternatives, comparison (*continued*):
    by payout-period method, 191, 196
    by present-worth method, 132–134
    by rate-of-return method, 170–174,
        293–297
    by service-life method, 192–194
  definition of, 7
  independent (non–mutually exclusive),
    190, 298
  multiple, 190, 292
  mutually exclusive, 190, 292
Amortization (*see* Depreciation)
Annual cost, equivalent uniform:
  advantages of, 148
  after-tax analysis using, 276, 279
  alternative evaluation by, 151–154,
    292, 296
  capital-recovery-plus-interest
    method, 150

Annual cost, equivalent uniform
(*continued*):
computation of, 38, 148–151
definition of, 148
salvage present-worth method, 149
salvage sinking-fund method, 148
Assets:
book value, 109, 111, 113, 115
depreciation of, 109–118
replacement studies and, 203–208
sunk cost for, 203
Average cost of capital, 321

Balance sheet:
basic equation for, 241
categories on, 240
description of, 240
example of, 240
ratios from, 243
Benefit cost difference, 187
Benefit/cost ratio:
alternative evaluation by, 188–190,
292, 297
calculation of, 187
classifications for, 186
definition of, 186
for three or more alternatives, 190,
292, 297
Benefits, 186
Bonds:
classification and types of, 226
face value of, 227
interest computation for, 227
present worth of, 228-231
rate of return on, 231
Standard and Poor's rating of, 226
treatment of, in debt financing, 317
Book value:
by declining-balance method, 115
by double declining-balance method,
115
by straight-line method, 111
by sum-of-year-digits method, 113
Borrowed money (*see* Debt capital;
Equity capital)
Breakeven analysis:
charts for, 173, 193, 210, 212, 279
computations for, 172, 209–213, 278

Breakeven analysis (*continued*):
general description, 209, 333
for three or more alternatives, 212
variable cost and, 209–212

Capital:
cost of, 316–322
debt, 315, 317–319
financing with, 283–286, 315,
317–319
debt/equity ratio, 323
equity, 315, 319–321
financing with, 283–286, 315,
319–321
limited, financing with, 316
Capital budgeting:
differences in solutions, 301
general problem of, 298
present-worth method, 301–304
rate-of-return method, 299
solution techniques, 299–304
Capital financing:
cost of capital (*see* Cost of capital)
debt, 315, 317–319
equity, 315, 319–321
Capital gains and losses:
computation of net, 258
short-term and long-term, 256
taxes for, 258, 276
Capital management (*see* Capital
budgeting)
Capital-recovery factor (*A/P*), 29
Capital-recovery-plus-interest method, 150
Capitalized cost:
alternative evaluation by, 135–138
definition of, 135
use in mutually exclusive alternatives,
304–306
Carry-back and carry-forward income
tax provisions, 260, 261
Cash flow:
after taxes, 274
before taxes, 18, 274
discounted, 132
net, definition of, 163
for rate-of-return analysis, 163
tabulation of, 18, 19, 22, 163,
274–276

Cash flow, tabulation of (*continued*):
    for equal-lived alternatives, 163
    for unequal-lived alternatives, 164
    (*See also* Cash-flow diagrams)
Cash flow after taxes, tabulation, 274–276
Cash-flow diagrams:
    construction of, 19–21
    definition of, 19
    importance of, 19
Challenger:
    in multiple-alternative evaluation,
        293, 298
    in replacement analysis, 203
    selecting a service life for, 204–207
Common multiple of lives, 133, 296
Common stock (*see* Stocks)
Composite depreciation, 117
Composite life, 117
Compound amount factors:
    single payment (F/P), 28
    summary of, 32
    uniform series (F/A), 31
Compound interest, 10, 11
Compounding:
    annual, 10, 16
    continuous with continuous cash
        flows, 51, 385–393
    continuous with discrete cash flows,
        51, 385–393
Compounding frequency:
    less frequent than payments, 55–59, 61
    more frequent than payments, 52–55,
        60, 61
    and payment frequency coincide, 51,
        54, 60
Compounding period, 49
Consultant's viewpoint, 203
Continuous compounding, 51, 385–393
    continuous cash flow: compared to
        discrete cash flow, 389
    effective interest rate for, 386
    funds flow, 390–393
Corporation income taxes (*see* Income
    taxes)
Cost allocation (*see* Factory expense)
Cost of capital:
    average, 321
    debt financing, 317–319

Cost of capital (*continued*):
    definition of, 316
    effect of debt/equity ratio on, 323
    effect of income taxes on, 317–319
    equity financing, 319–321
Cost-of-goods-sold statement, 242
Costs:
    capitalized (*see* Capitalized cost)
    depreciation, 109
    EUAC (*see* Annual cost, equivalent
        uniform)
    factory overhead (*see* Factory expense)
    as function of life, 208
    incremental, 170
    sunk, 203
        in replacement studies, 203
    variable, 209–212
Current assets, 240, 243
Current ratio, 243

Debt capital, 283–286, 315, 317–319
Debt/equity ratio, 323
Declining-balance depreciation, 114–116,
    121
Defender:
    meaning of, in multiple-alternative
        evaluation, 293, 298
    in replacement analysis, 203
Depletion:
    compared with depreciation, 119, 281
    definition of, 119
    effect on income taxes, 120, 281
    effect on rate of return, 281
Depletion allowance, 120
Depletion factor, 119, 124
Depreciable property, gains and losses
    from disposal of, 116, 256
Depreciation:
    compared with depletion, 119, 281
    composite method of, 117
    declining-balance, 114–116, 121
    definition of, 109
    double declining-balance, 114–116
    effect of time on, 109
    group method of, 116, 123
    income taxes and, 118, 261, 282
    multiple-asset accounting, 116–118
    rate of, 112, 114, 115

Depreciation (*continued*):
  sinking-fund, 110
  straight-line, 111
  sum-of-year-digits, 112–114
  unit-of-production method of, 120
Diagrams:
  breakeven, 173, 193, 210, 212
  cash-flow, 19–23
Disbenefits, 186, 189
Discounted-cash-flow method, 132, 301
Dividend method, 319
Do-nothing alternative, 188, 295
Double declining-balance method,
    114–116

Earnings/price-ratio method, 319
Effective income tax rate, 258
Effective interest rate:
  computation of, 49–52, 59–61
  definition of, 49
  equation for, 50
Equal-service alternatives, 132
Equity capital, 283–286, 315, 319–321
Equity ratio, 243
Equivalence:
  definition of, 6
  involving several factors, 67–73
  involving single factors, 35–38
Equivalent uniform annual series (*see*
    Annual cost, equivalent uniform)
Estimated rate of return, calculation of,
    167
Estimates of:
  costs, 333, 339
  optimistic, 299, 337–339
  pessimistic, 299, 337–339
  realistic, 299, 337–339
EUAC (*see* Annual cost, equivalent
    uniform)
Expected value:
  computation of, 339
  in economy studies, 340
Expenses (*see* Costs)
Extra investment rate of return, 170–174,
    293–297

F/A factor, 31
F/P factor, 27

Face value, of bonds, 227
Factor depletion, 119
Factors:
  continuous compound interest,
    386–388, 391
  derivation of, 27–31
  discrete compound interest, 353
  standard notation for, 31, 32
Factory expense:
  allocation of, 245–248
  definition of, 242–245
  rate, 246
  variance, 247
Financing:
  debt, 315
  equity, 315
First cost, 111
Fixed assets, 240
Funds flow (*see* Continuous
    compounding)
Future worth:
  computation of, 35
  definition of, 16

Gains and losses (*see* Capital gains and
    losses)
Gordon-Shapiro method, 320
Gradient:
  base amount, 83
  conventional, 84, 93
  decreasing, 97–99
  definition of, 83
  derivation of factors for, 85–88
  equivalent uniform annual series for,
    87, 94–97
  n value, 89–93
  present worth of, 89–97
  shifted, 94–97
  summary of, factors, 88
  uniform, 83
  use of tables, 88
Gross income, 256
Group depreciation, 116, 123

Income ratio, 244
Income statement:
  basic equation for, 242
  categories of, 241

Income statement (*continued*):
  description of, 241
  example of, 241
  ratios from, 244
Income taxes:
  capital gains and losses and, 258, 276
  cash flow after, 274–276
  corporate, 256–264
  definitions, 118, 256
  depletion and, 120, 281
  depreciation and, 109, 118, 261–264
  effect of, in economy studies, 118,
    255, 274–282
  effective rates of, 258
  investment tax credit and, 110, 257,
    260, 276
  present worth of, 262–264
  rate of return considering, 277–282
  rate used for corporation, 118, 258
  replacement studies and, 279
  state and local, 118, 258
Incremental B/C analysis, for multiple
    alternatives, 297
Incremental costs:
  B/C analysis and, 190, 297
  definition of, 170
  present-worth analysis and, 171–173
  rate-of-return analysis and, 170–174
Incremental effective income tax rate,
    258
Incremental rate of return:
  for multiple alternatives, 293–297
  for two alternatives, 170–174
Independent alternatives (*see* Non-
    mutually exclusive [independent]
    alternatives)
Inflated before-tax rate of return, 264
Intangible factors, 7
Interest:
  compound, 10
  computation of, for bonds, 227
  continuous compounding of,
    385–393
  definition of, 4
  factors and symbols: continuous
    compounding, 386–388, 391
    discrete compounding, 16–18,
    353

Interest (*continued*):
  rate of, definition of, 5
  simple, 8–10
Interest factors:
  derivation of, 27–31
  format for use, 31–33
  symbols, 16–18
  tables, use of, 31–35
Interest period, 5, 49, 227
Interest rates:
  definition of, 5
  determination of unknown, 38–41,
    166–170, 231, 277, 293–297
  effective, 49–52, 59–61
  estimation for trial-and-error solutions,
    167
  nominal, 49
Interest tables:
  continuous compounding, 389
  discrete compounding, 355–384
  interpolation in, 33–35, 42
  use of, 31–33
Internal Revenue Service:
  depreciation rules and, 114, 261–264
  publications of, 255
  tax laws of, 257–261
Investment tax credit, 110, 257, 260, 276

Least-cost life, 208
Leverage, 323, 326
Life:
  economic, 208
  remaining (*see* Replacement studies)
Life cycle, 133, 148, 151
Lives:
  common multiple of, 133, 296
  comparing alternatives with equal,
    132
  comparing alternatives with perpetual,
    135–138, 152–154, 297
  comparing alternatives with unequal,
    133, 151, 296

Market value, 109, 203, 320
Minimum attractive rate of return
    (MARR):
  definition of, 7
  establishing the, 8, 314–323

Minimum attractive rate of return
   (MARR) (*continued*):
   use of: for capital budgeting, 299–301
      for mutually exclusive alternatives,
         292–297
   variations in, 316
Minimum cost life of asset, 208
Multiple alternatives:
   comparison of, using incremental
      benefit/cost ratio, 190, 292, 297
   comparison of, using incremental rate
      of return, 293–297
   definition of, 190, 292
   independent, 190, 298
   mutually exclusive, 190, 292–298
Multiple-asset accounting, 116–118
Multiple factors, use of, 65–78
Multiple rate-of-return computation,
   174–176
Mutually exclusive alternatives, 190,
   292–298

Net cash flow (*see* Cash flow)
Net worth, 241
Nominal interest rate, 49
Nonmonetary values, 7, 187
Non–mutually exclusive (independent)
      alternatives:
   capital budgeting and, 298–303
   problems in evaluation of, 301
   selection (*see* Capital budgeting)

Operating loss, 257, 261
Operating ratio, 244
Opportunity cost method, 321
Optimistic estimate, 299, 337–339
Overhead (*see* Factory expense)

P/A factor, 29
P/F factor, 28
P/G factor, 85–88
Payback period:
   example of calculation of, 195, 196
   limitations of, 195–197
Payment period, 49
Payout period (*see* Payback period)
Perpetual investment (*see* Capitalized
   cost)

Pessimistic estimates, 299, 337–339
Placement of future worth, 65–67
Placement of present worth, 65–67, 89
Planning horizon:
   capital budgeting and, 301–303
   replacement studies and, 204–207
Portfolio selection (*see* Capital budgeting)
Present worth:
   after-tax analysis using, 276–279
   alternative evaluation by, 132–135,
      292–296
   bond, 228–231
   computation of, for equal lives, 132
   computation of, for unequal lives,
      133, 296
   of income taxes, 261–264
   use of, in capital budgeting, 301–303
Present-worth factors (*P/F, P/A, P/G*):
   gradient, 85–88
   single-payment, 27–29
   uniform-series, 29–31
   use of tables for, 31–35
Probability:
   expected value and, 339
   use of, in economy studies, 340
Profit-and-loss statement (*see* Income
   statement)
Property, gains and losses from disposal
   of, 258–260
Public projects, economy studies of,
   186–188

Rate of depreciation:
   declining-balance, 114
   double declining-balance, 115
   straight-line, 112
Rate of return:
   after tax, 277, 281
   on bond investments, 231
   computation of, by EUAC, 169
   computation of, by present worth,
      166–169
   on debt capital, 317–319, 324
   definition of, 7
   on equity capital, 323
   on extra investment, 170–173
   guessing for trial-and-error solutions,
      167

Rate of return (*continued*):
    inflated, before taxes, 264
    method of evaluation, 170–174
Reasonable estimate, 299, 337–339
Remaining life, 203–207
Replacement:
    reasons for, 202
    versus augmentation, 207
Replacement studies:
    after-tax, 279
    and augmentation, 207, 213
    before-tax, 202–209
    equal-lived assets, 204, 205
    gains and losses in, 280
    and replacement value, 207
    sunk costs in, 203
    unequal-lived assets, 205–207
    use of market value in, 203–207
    use of study periods in, 204–207
Replacement value, 203–207
Retained earnings, 315
Retirement of assets, 208
Return, minimum attractive rate of, 7
Risk, 316, 339–341

Salvage:
    present-worth method of, 149
    sinking-fund method of, 148
Salvage value, 111, 133
Section 38 property, 260
Sensitivity analysis:
    approach, 333
    of one factor, 334–337
    of several factors, 337–339
    using three estimates, 337–339
Service life:
    comparison of alternatives by, 192–194
    computation of, 191
    definition of, 190
    model, 191
    relation to payback period, 191,
        195–197

Shifted gradients, 94–97
Simple interest, 8–10
Single-payment factors (P/F, F/P):
    derivation of, 27–29
    summary of, 32
Sinking-fund depreciation, 110
Sinking-fund factor (A/F), 30
Standard factor notation, 31, 32
Stocks:
    common, 315
    market value of, 320
    treatment of, in equity financing,
        319–321
Straight-line depreciation, 111
Study period (*see* Planning horizon)
Sum-of-year-digits depreciation, 112–114
Sunk cost, 203
Symbols, in engineering economy, 16–18,
    21

Tables, use of (*see* Interest tables)
Tax credit, 110, 260
Taxable income, 256, 274–276
Taxes (*see* Income taxes)
Time placement of dollars, 17
Time value of money, 4

Uncertainty in estimates, 333, 337–339
Uniform gradient (*see* Gradient)
Uniform-series factor, 29–31
Unit-of-production depreciation, 120
Unknown interest rate, 38–41, 166–170
Unknown years (life), 41, 44

Value, of money with time, 4
Variable cost, in breakeven analysis,
    209–212
Variance, in overhead allocation, 247
Variations in economic decisions,
    333–339
Viewpoint, in replacement analysis,
    203